MARIA CSUROS

Environmental Sampling and Analysis
Lab Manual

LEWIS PUBLISHERS

Boca Raton   New York

Acquiring Editor:	Ken McCombs
Project Editor:	Albert W. Starkweather, Jr.
Direct Marketing Manager:	Arline Massey
Cover design:	Denise Craig
Typesetter:	Pamela Morrell
Manufacturing:	Sheri Schwartz

Library of Congress Cataloging-in-Publication Data

Csuros, Maria.
    Environmental sampling and analysis lab manual / Maria Csuros.
       p. cm.
    Includes bibliographical references and index.
    ISBN 1-56670-178-3 (alk. paper)
    1. Pollution—Measurement. 2. Pollution—Measurement—Laboratory
manuals. I. Title.
TD193.C782 1997
628.5′02′87—dc20

96-37154
CIP

This book contains information obtained from authentic and highly regarded sources. Reprinted material is quoted with permission, and sources are indicated. A wide variety of references are listed. Reasonable efforts have been made to publish reliable data and information, but the author and the publisher cannot assume responsibility for the validity of all materials or for the consequences of their use.

Neither this book nor any part may be reproduced or transmitted in any form or by any means, electronic or mechanical, including photocopying, microfilming, and recording, or by any information storage or retrieval system, without prior permission in writing from the publisher.

The consent of CRC Press does not extend to copying for general distribution, for promotion, for creating new works, or for resale. Specific permission must be obtained in writing from CRC Press for such copying.

Direct all inquiries to CRC Press LLC, 2000 Corporate Blvd., N.W., Boca Raton, Florida 33431.

No claim to original U.S. Government works
International Standard Book Number 1-56670-178-3
Library of Congress Card Number 96-37154
Printed in the United States of America   2 3 4 5 6 7 8 9 0
Printed on acid-free paper

DEDICATION

To My Father

Biography

Maria Csuros is an environmental chemist with many years of varied experience. She received her Ph.D. in Environmental Chemistry from Janus Pannonius University in Pecs, Hungary, and her Master of Science Degree in Chemistry from Jozsef Attila University in Szeged, Hungary. Her postgraduate work involved research in environmental chemistry under a grant from the Environmental and Public Health Insitution, Budapest, Hungary.

Most of her professional life has revolved around environmental testing laboratories and teaching.

Her first encounter with environmental science occurred at the Environmental and Public Health Laboratory, Pecs, Hungary, where she worked as a supervisor of the Water Department. Her main area of interest at that time was the prevention and elimination of methemoglobinemia caused by high nitrate content of private well waters. She spent six years in Benghasi, Lybia, as part of an international team, studying the health effects of brackish drinking water.

After moving to the United States, she continued to dedicate her knowledge and time to environmental analytical work and education. She has designed and developed an Environmental Science program for Pensacola Junior College focused on environmental sampling and analysis. She presently is the coordinator of the this program and teaches chemistry and environmental science courses.

She also conducts various continuing education programs for local, state, and federal agencies as well as regional companies.

Her first book, "Environmental Sampling and Analysis," published by CRC/Lewis Publishers in 1994, and has been on the "Best Sellers" list since.

She lives in Pensacola, Florida, with her husband, who is professor of Microbiology and Anatomy and Physiology. She has two sons, and five grandchildren.

PREFACE

Satisfying the requirements of the environmental laws and regulations is possible only with defendable and accurate laboratory results, based on approved methodologies, updated instrumentation and technologies, well-trained and qualified laboratory personnel, and sufficient QC data.

Environmental analytical laboratories must offer a wide-scale analytical repertoire, from the simple determination of physical properties to the most modern analytical performances of trace organic pollutants. The concern of the microbiological quality of the environment is also increasing and demands extended testing.

Chemists working in the environmental analytical field should be familiar not only with the general laboratory techniques, and with the chemical background of different analytical procedures and instrumentation, but they must also be knowledgeable about approved methodologies, QC requirements, calculations, evaluations of analytical data, and generation of correct, defendable analytical results, along with their regulatory levels.

The large diversity of the literature in this field and the diffuse sources of the necessary information make this wide-scale training difficult. The goal of this text is to give comprehensive and easy-to-read information for any individual who is working in the environmental analytical chemistry arena, and to provide essential information to consultants and regulators about analytical and quality control procedures helpful in their evaluation and decision-making procedures. The text also gives useful information for college and university students taking chemistry and environmental courses, for graduate students, and for chemists who need a ready source of information that meaningfully relates to actual laboratory practice.

The book is designed as the first part of a comprehensive laboratory manual specially designed for environmental professionals. It contains general laboratory considerations, laboratory safety procedures, and laboratory techniques emphasized on the determination of inorganic nonmetallic constituents in environmental samples along with their chemistry, occurrence, source, fate, and their control by regulations and standards. Sample collection, preparation for analysis, step-by-step easy to follow methodology with detailed quality control procedures, as well as calculations for correct reporting units concerning different matrices are discussed for each parameter.

All in all, this book will provide a valuable advantage to environmental education and special training programs; as a practical handbook it will also assist environmental chemists in their everyday chores; and it will give a helping hand to the reader to understand analytical reports.

Acknowledgments

I owe a great debt of gratitude to all the people who participated in making this book possible.

My most sincere appreciation goes to James Tucci, manager of Law Engineering and Environmental Services, Inc., for his valuable suggestions, comments, and technical information. His professional input and respected expertise considerably contributed to the correctness of the information contained in this publication. I can't imagine completing this task without his help and support.

My very special thanks go to Bernie Fuson, project manager at Law Engineering and Environmental Services, for his support, patience, and accuracy in proofreading and correction of the text.

I also am pleased to express my thanks to my student and friend, Ken Connor, for the preparation of the figures and tables.

I gratefully acknowledge the support of the outstanding editorial and production staff of CRC/Lewis Publishers. My personal thanks to Ken McCombs, Acquiring Editor, and Susan Alfieri, for their friendship, understanding and encouragement. I am grateful to the project editor, Albert Starkweather, who guided the project with his careful work and patience through the complexities of production. Thank you also for the superb design work.

It is with particular pleasure that I thank my husband for his patience, support, excellent ideas, and his cheerful spirit. Warm words of thanks also to my sons, Geza and Zoltan, for their love, encouragement and technical assistance.

TABLE OF CONTENTS

LIST OF TABLES AND FIGURES

TABLES

FIGURES

Chapter **1**

INTRODUCTION
TO CHEMICAL ANALYSIS

1.1 THE NATURE OF ANALYTICAL CHEMISTRY

Chemical analysis of a sample of matter is the determination of the chemical composition of that sample. Chemical analysis consists of a set of chemistry laboratory operations designed to reveal the chemical makeup of a material. Analytical chemistry is a branch of chemistry dealing with the separation and analysis of chemical substances.

Analytical chemistry can be divided into two major categories: *qualitative* and *quantitative* analytical chemistry. Qualitative analysis is concerned with *what* is present, quantitative analysis with *how much.* Analysis procedures can be additionally classified into *wet chemistry procedures* and *instrumentation procedures* (see Figure 1.1). Wet techniques are those that employ chemical reactions and classical reaction stoichiometry for results.

Wet chemistry procedures give excellent accuracy, but their major disadvantages are the time requirement and subsequent tedium. Instrumental or "modern" techniques give speed and offer a much greater scope and practicality to the analysis. In addition, the much lower detection limits of the instrumentation analytical methods make possible the quantitation of the minor constituents in a substance. Although a number of purely instrumental methods of analysis have been developed, chemical methods are still vital and widely used, for several reasons. Many instruments require extensive calibration for each type of sample to be analyzed, whereas chemical methods can be quickly adapted to analyzing new types of samples. Few instrumental methods can match the accuracy and precision of chemical methods. For example, instrumental methods are excellent for the analysis of trace constituents, but their accuracy with respect to major constituents is greatly deficient to that achieved by wet chemistry methods. By proper selection, both instrumentation and wet chemistry methods are used equally in laboratories, and they supplement each other. On the other hand, knowledge of chemical methods provides the background to understanding instrumental analysis and analytical problems.

1.2 GENERAL DIRECTION OF CHEMICAL ANALYSIS

The results of a chemical analysis could affect important decisions, such as suitability of a material for an intended purpose, the quality of the environment, the health of

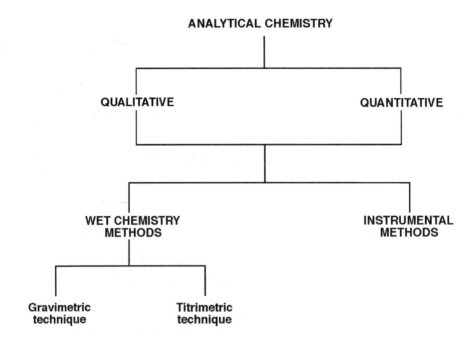

Figure 1.1
Analytical chemistry can be classified as either qualitative or quantitative. All analytical procedures can be additionally classfied into those that are wet procedures and those that are instrumental procedures. Wet chemical methods are divided into gravimetric technique (based on weight) and titrimetric technique (based on volume).

individuals, the freedom of a prisoner on trial, etc. Stringent requirements in a chemical laboratory are the basis of reliable and comparable data from a chemical analysis.

1.2.1   The Responsibility and Psychology of the Analyst

"Chemical analysts and chemical technicians are professionals in every sense of the word. They should look, think, and act as professionals are expected to do. They are entitled to all the privileges of professionals and at the same time are expected to shoulder the responsibilities of professionals." A true analyst, or analytical chemist, has several characteristics. He or she has a knowledge of the methods and instruments used for analysis, and understands the principles of the analysis. Laboratory analysts should have a chemistry background adequate to understand and correctly apply all of the laboratory rules and to evaluate and interpret the results of their analysis. They have to know how to plan and organize laboratory work so the time is used efficiently. Thus, a laboratory technician should be a skilled, well-trained chemist — in sharp contrast with the so-called "determinators" who simply twist the dials of an instrument or follow "cookbook" analytical procedures.

The analyst should be familiar with the test methods described in the laboratory standard operation procedures (SOP) as well as the required quality assurance-quality control (QA/QC) applications as stated in the laboratory comprehensive QA/QC manual. The analyst should run the test by following the approved method and incorporate all QC checks according to the laboratory comprehensive QA/QC program. He or she should be able to recognize problems, initiate and conduct corrective actions, and keep all documentation related to the analysis clean and in order to be ready for inspection at any time. The analyst should be able to protect and defend all of the raw data as well as the reported results.

Laboratory personnel must understand the potential hazards in the laboratory and then become familiar with the precautions and safety rules to be followed in laboratory work. Accidents can be minimized by recognizing their causes and by being alert. They should know basic emergency first-aid treatment for minor injuries and also the location and use of emergency equipment. Laboratory safety procedures are discussed in Chapter 2.

1.2.2   Cleanliness in the Laboratory

Good analysts should be scrupulously clean and neat. An orderly laboratory bench is not prone to mix up samples, use wrong reagents, spill solutions, and break glassware. Much time can be wasted in searching for a small item in a jumble of glassware or in finding a certain reagent bottle that has been misplaced on the shelf. Cleanliness also includes laboratory equipment, ovens, hot plates, hoods, balances, sinks, floors, shelves, cabinets and laboratory clothes.

The result of a chemical analysis depends on clean glassware. Glassware that looks clean may or may not be clean for chemical analysis. Surfaces on which no visible dirt appears are often contaminated by minute quantities of a substance that interfere with the analysis and give false results. The analyst must carefully read the method and consider any special recommendation for glassware washing procedures.

1.2.3   Recording Analytical Data

Data obtained in the analytical laboratory should be recorded. The record should be complete, understandable, and easy to find.

Laboratory notebook — The key to successful work in the analytical laboratory is an orderly notebook with well-organized data. All documentation related to the raw data and the reported results should be in a bound form, easy to identify, and ready for inspection at any time. A bound notebook is preferred to a loose-leaf one. The size of the notebook should be adequate and comfortable to use. Preferably, it should fit easily around the balance and working table, leaving space for glassware, reagents, etc. necessary for the analysis. Larger notebooks, especially when opened, tend to get in the way and can cause spills and other problems. The first two pages should be reserved for an index, which can list the page numbers and the subject. If the pages of the notebook are not numbered, the analyst must number each page in the upper corner.

The title heading should be used to identify and date data on each page. All raw data and any observations should be recorded directly in the notebook at the time the work is performed. Especially forbidden is the recording of data on loose paper with the idea of copying it into the notebook later. All data should be written in ink to avoid smearing and erasures. Nothing should be erased; anything not needed should be crossed out with a single line, followed by an explanatory note, initialed, and dated.

Working sheets — Analytical raw data with related QC work and calculated final results may be documented in specified worksheets. The worksheets are designed for each type of analytical performance, and contain the information relating the sample, such as ID number (identification number), sample type, source, analytical method or method number with reference, method detection limit (MDL), specification of instrument used, ana-

Method No.: 150.1.   Date   : _____
Reference : EPA 600/4-79-020   Analyst: _____
Model of pH meter   : _____

Sample I.D. No.	Sample description	Measured pH	Sample temperature °C	Remarks
	Buffer pH 7.00			
	Buffer pH 4.00			
	Buffer pH 10.00			
	Q.C. sample			
	Duplicate			

QC sample true value: _____

Precision (RPD)   : _____   Accepted limit _____

Accuracy (% R)   : _____   Accepted limit _____

RPD (Relative Percent Deviation) = (A-B)/(A+B) x 200
                        A and B = duplicates

% R (% Recovery) = (Measured value x 100) / True value

Figure 1.2
Working paper for pH measurement.

lytical and QC data, calculated values and proper units. Shortly, the worksheet contains all information necessary for validation of the analytical process, and the reported values. Figures 1.2, 1.3, and 1.4 are examples of laboratory worksheets. The designs of these documentation forms are just examples; each laboratory has its own patterns as approved in its QA/QC programs. As with all documentation, these sheets should be stored in bound form per analytical groups and parameters with document title, date start and end, identification number start and end. Strip charts, documented AA calibration curves, and raw data should be stored in file boxes and identified as mentioned above. Additional examples for working paper formats can be found in Appendix C.

1.2.4 Planning Laboratory Work

It is essential to plan laboratory work ahead of time, so as to use the laboratory time efficiently and to avoid delays in using equipment that must be shared with other analysts. The following questions should be asked before the actual preparation for analysis: which kind of samples will be analyzed, how many samples analyzed, what is the source of the

Parameter: _____ Method No. _____ Reference : _____
Method Detection Limit (MDL) : _____
Titrant and Normality (standardized) _____
Analyzed by _____ Date _____

ID No.	Sample Identification	Sample type	Sample ml	Titrant ml	Result mg/L	Remark
	Blank					
	Q.C. sample					
	Duplicate					
	Spiked sample					

INFORMATIONS RELATED TO QC CHECKS

True value of QC sample _____ ID of duplicate samples _____
ID NO. of spiked sample _____ Sample ml spiked _____
Concentration of spike stock solution _____
ml or ul spike stock added _____ Added spike value _____

Precision as Relative Percent Deviation (RPD) _____
Accuracy as % Recovery (% R) _____
% Recovery of Spike (% R_{spike}) _____

Approval of Supervisor _____ Date _____

Figure 1.3
Working paper for titrimetric analysis.

samples, which tests should be performed, which method should be used, which preservative was used, which pretreatment is necessary, etc.

The following list contains the duties of the analyst from the choosing of proper samples for the particular analysis through the actual analytical performance, QC checks, documentation, calculations, recognition and correction of problems, and giving accurate, defendable results.

- The analyst should be familiar with the method and the method should be followed as described in the laboratory standard operation procedure (SOP).
- If the sample needs pretreatment prior the actual analysis, such as drying, filtration, distillation, digestion, extraction, etc., start as soon as possible, because these are time-consuming procedures. If sample preparation is the duty of another laboratory depart-

Parameter _____ Method No. _____ Reference_____
Method Detection Limit (MDL) _____
Model of spectrophotometer _____ Wavelength _____
Correlation Coefficient of the calibration curve _____
Analyzed by _____ Date _____

ID No	Sample Identifica tion	Sample type	Sample ml	Absor bance	Dilut ion	Result mg/L	Remarks
	Blank						
	Standard 1						
	2						
	3						
	4						
	CCS						
	CVS						
	Duplicate						
	Spiked s.						
	Blank						

INFORMATIONS RELATED TO QC CHECKS

% Recovery of Continuing Calibration Standard (CCS) _____
True Value of Calibration Verification Standard (CVS) _____
% Recovery of CVS _____ ID of Duplicate Sample _____
ID No. of Spiked Sample _____ Sample ml spiked _____
Conc. of Spike Stock solution _____ Added spike value _____
Precision as Relative Percent Deviation (RPD) _____
Accuracy as % Recovery (%R) _____
% Recovery of spike ($\%R_{spike}$) _____

Approval of supervisor _____ Date _____

Figure 1.4
Working paper for spectrophotometric analysis.

ment, collect all of the information needed for calculation of the final result (original volume, weight, final volume after treatment, etc.).

- When samples are stored in a refrigerator, remove them in time to allow the samples to warm up to room temperature prior to starting the analysis.
- Carefully check sample label for identification and select samples according to the appropriate bottle type and preservation for the analysis.
- Collect all standards and reagents needed and check their preparation date. When the expiration date arrives, discard the solution and prepare a fresh one. When reagents are stored in the refrigerator, leave time for warming up to room temperature before using them. If necessary, standardize solutions and document on designated log form.
- Select the worksheet according to the analytical procedure and sample type, and fill out initial information.

- Collect the necessary glassware, check its cleanliness, use special treatment (for example 1+1 HNO_3 rinse for metal analysis) if applicable. Number the glassware with sample ID number.

- Switch instrument to "on" position (if instrumental analysis) to give time to "warm up."

- Start the analytical procedure according to the approved method and incorporate all QC checks required by the laboratory QA/QC program. (Detailed QA/QC requirements and documentation during analytical work are discussed in Chapter 3.) All raw data and information should be documented in the laboratory notebook or worksheet.

- At the end of the analysis, switch off instrument, collect analytical wastes in designated containers, collect dirty glassware and transfer to the washing area, and transfer standards, reagents and samples to their storage area or into the proper refrigerator if needed. Clean the work area.

- Calculate the results and collect all documentation to be ready for further questions or checking. If all the QC data agree with the analytical values, transfer all of the calculated results to the analytical report sheet or computer as specified by the laboratory rules.

1.2.5 Documents to be Saved

The numerous documentation used during analytical work should be saved and stored in the same way as the above-mentioned worksheets, in bound forms or file boxes as described previously in Section 1.2.3.

A list of the stored documentation follows.

- All notebooks, data forms, and log forms that belong to laboratory operation. All calculation related to the sample result.
- Chromatograms, charts, and other instrument response readouts.
- Calibration curves, preparation of calibration standards.
- Target limits for each analytical group and parameters.
- Method detection limits (MDLs).
- Summary log form for accuracy and precision data per parameters.
- QC charts.
- Records concerning receipt of solutions from commercial sources.
- Laboratory custody reports (holding times, sample transmittal forms, sample storage logs, sample disposal log).
- Sample preparation logs.
- Instrument maintenance and instrument performance check logs.
- Copies of final reports.
- Field records (field notebooks, field tests, field custody reports).

Records must be retained for a period of a least 3 years. Drinking water reports require a retention time of up to 10 years.

Chapter 2

SAFETY AND LABORATORY WORK

Before entering into laboratory work, it is important to understand the potential hazards in the laboratory, to be familiar with the precautions and rules, and to recognize and avoid the causes of those hazards. According to the Occupational Safety and Health Act (OSHA)," … each employer has the general duty to furnish each of the employees a workplace free from recognized hazards causing or likely to cause death or serious harm." **No one can start to work in any laboratory without comprehensive safety training!**

Laboratory supervisors and managers, and laboratory instructors in schools are responsible for ensuring adherence to the rules. Safety is the number one concern for everybody working in the laboratory. Accidents can be minimized by recognizing their causes and by being alert at all times. Everybody working in the laboratory should know basic emergency first aid treatment for minor injuries (see Appendix A) and also the location and use of emergency equipment such as safety showers, fire extinguisher, and spill kits. All hazards should be identified whether they are visible or hidden, and appropriate safety signs should be posted wherever applicable in conspicuous locations. Selected safety signs are shown in Figure 2.1.

2.1 LABORATORY HAZARDS

Laboratory hazards can be divided into three groups: (1) chemical hazards, (2) fire hazards, and (3) careless habits.

2.1.1 Chemical Hazards

Virtually all chemicals are toxic to some extent, and care should be taken in handling them. Chemical hazards may be minimized by taking the following precautions.

Cleanliness

- Wash hands periodically and immediately after contact with chemicals and just before leaving the laboratory.
- Never drink from laboratory glassware.
- Clean the working area before and after work.

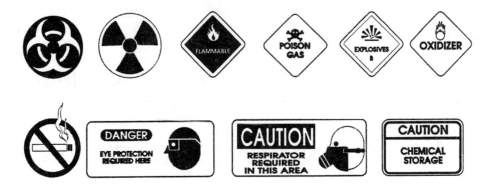

Figure 2.1
Safety signs can inform of possible risks and needed precautions within a facility.

- Laboratory coats and aprons are designed to protect the body from chemical spills. Dirty clothes can be a health hazard and a source of contamination.

Protection of the eyes — The eyes are specially susceptible to injury from chemicals. The most common types of accidents that threaten the eyes are breaking of glass containers of acid, bases, etc. and reactions that go out of control. Both accidents can cause chemicals to be sprayed into the eyes. For this reason, safety glasses, goggles, or face shields should be worn during laboratory work. Eye hazards are constant in a chemical laboratory; therefore eye protective devices should be worn at all times except when perfect vision is required for measurements. Never wear cracked, pitted, or damaged eye protective devices! (An eye protective device is shown in Figure 2.2.)

If any chemical is sprayed into the eyes, immediately flood the eyes with water; use the specially designed eye wash fountain (Figure 2.3). Quick flushing with water from the nearest tap also will minimize the damage. See a physician as soon as possible.

Injuries from contact of certain chemicals with the skin

- Skin contact with certain chemicals, such as those from strong acids or bases, can cause chemical burns.
- Certain chemicals can absorb through the skin. Because many chemicals absorb rapidly through the skin, prompt cleanup is important. Remove contaminated clothing immediately (forget modesty, if necessary!) and flush affected areas with large volumes of water. If more than a minor amount of chemical is involved, medical attention is necessary.

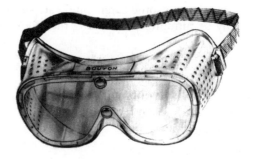

Figure 2.2
Eye protection device.

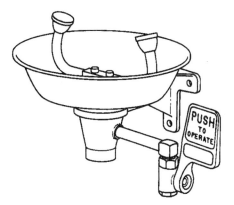

Figure 2.3
Eye wash fountain.

Protection for the body

- Laboratory workers should wear appropriate safety clothing, such as laboratory coats or aprons. The material of the laboratory coats gives protection against acids and bases. A laboratory apron is not affected by ordinary corrosive fluids or other chemicals.

- Never wear open-toed shoes and sandals, because they offer little or no protection against chemical spills or broken glass.

- Keep ties or scarves secured with fasteners.

- Tie long hair up and out of the way.

- When handling corrosive chemicals, use protective gloves. Gloves are selected according to the need; asbestos gloves protect against heat, but they are not advisable for handling corrosive chemicals (acids or bases), because asbestos absorbs the substance and increases the contact time and area. When working with hot objects or organic solvents, rubber or plastic gloves should not be used, as they may soften and dissolve.

Hazards from swallowing toxic chemicals — There should be no food or drinks in the laboratory. It is too easy to contaminate food, especially with traces of chemicals on the hand. To avoid any possibility of swallowing chemical solutions while using a pipet, all pipeting should be done using a pipeting bulb and not the mouth.

Inhalation of volatile liquids and gases — The presence of these substances in the air (even in low concentration) is hazardous. Acute exposure to extremely high concentration of vapors (above the maximum allowable concentration) can cause unconsciousness and even death, if the person is not removed from the area and is not given medical attention. Exposure to solvent and chemical vapors may be avoided by working with such chemicals under chemical hoods, and wearing protective respiratory device (Figure 2.4). Also, good ventilation is essential to a safe laboratory.

Toxicity of metallic elements — Metals with a specific gravity of greater than 5 are called heavy metals. In the metallic state they are harmless, but in vapor state are toxic, and so are their soluble compounds. The most common heavy metals are antimony (Sb), arsenic (As), cadmium (Cd), chromium (Cr), lead (Pb), mercury (Hg), nickel (Ni), silver (Ag), and thallium (Tl).

Figure 2.4
Protective respiratory device.

Chemical dust — Finely powdered chemicals can be inhaled as dust; therefore, such chemicals should also be handled under a laboratory hood.

Chemical spills — *Solid, dry substances* — These can be swept together, brushed into a dustpan or cardboard, and then deposited in the proper waste container.

Acid spills — These should be cleaned by using the proper spill kit and following the enclosed instructions. The materials neutralize and absorb the acid for easy cleanup. Afterwards, the area should be washed with water. Alternatively, soda ash (Na_2CO_3) or sodium bicarbonate ($NaHCO_3$) solution can be used for neutralization, followed by water flushing of the area. *CAUTION:* When water is poured on spills of concentrated sulfuric acid (H_2SO_4), tremendous heat is formed (exothermic reaction) and the acid spatters. Deluge with water to dilute the acid to minimize heat generation and splattering.

Alkaline spills — Alkaline spills are treated similarly to acid spills, using an alkaline spill kit and following the instructions. Alternatively, a weak acid solution, such as diluted acetic acid, can be used for neutralization. The floor should be flushed with water to a floor drain; a mop and bucket also can be used. Flush mop and bucket, replacing water frequently. *CAUTION:* Alkali solutions make the floor slippery! Clean sand can also be used; throw sand over the spill and sweep up. The wet sand is then discarded.

Volatile solvents spills — These evaporate very rapidly because of the extremely large surface area. This kind of spill can create a fire hazard if the solvent is flammable, and it will invariably cause high, dangerous concentrations of fumes in the laboratory. These fumes can have serious physiological effects when inhaled. They can also form explosive mixtures with air. Cleanup procedures are as follows: if minor quantities of solvents are spilled, wipe up the liquid with rags or towels and discard them in the proper waste receptacle. If a large amount of solvent is involved in the spill, use a mop and pail. Squeeze out the mop in the pail and continue as needed.

Spill of oily substances — Oily substances should be cleaned up with a selected nonflammable volatile solvent. Pour some on an absorbent rag, and wipe up the spilled substance. The rag will need to be rinsed in a pail of the solvent to remove all the spilled material, because oily floors are slippery and dangerous. Finally, a thorough detergent-water scrub will clean up oily remains.

Mercury spills — These are one of the most common sources of mercury vapor in the laboratory air. As a result of the spill, mercury may be distributed over a wide area, exposing a large surface area of the metal. In any mercury spill, unseen droplets are trapped in crevices. Unless the laboratory has adequate and reasonable ventilation, the combined mercury vapor concentration may exceed the recommended limit.

CAUTION: Surfaces which are apparently free of mercury will harbor microscopic droplets. Vibrations increase vaporization.

Clean up mercury spills as follows: (1) push droplets together to form pools. To pick up the mercury, use a suction device. (2) If there are many crevices or cracks in the floor which trap small droplets of mercury that cannot be picked up, seal over the cracks with a thick covering of floor wax or an aerosol hair spray. The covering will dramatically reduce vaporization. (3) Sulfur powder can also be used to fix mercury. (4) Mercury spill kits are also available for proper mercury cleanup.

2.1.2 Fire Hazard

A fire in a chemical laboratory can be dangerous and devastating. In case of fire, keep calm and think! Sources of fires can be electrical equipment, friction, mechanical sparks, flames, hot surfaces, and flammable organic compounds. Accidental ignition of volatile organic solvents is perhaps the most common source of laboratory fires. To avoid accidental spills and reduce fire hazards, *keep volatile solvents in small containers and never work with a volatile solvent around an open flame.* The sooner you attack the fire, the easier it is to control. All fires are different. The use of the wrong type of firefighting equipment may increase the intensity of the fire.

Fires are classified according to the type of material being consumed. Class "A" fires are caused by burning of paper, wood, textiles, etc. Almost any type of extinguisher is satisfactory. Class "B" fires are caused by burning of oil, grease, organic solvents, paint, etc. Use dry-chemical, liquid, CO_2, or foam extinguisher. Class "C" fires are electrical fires in equipment. Do not use a water or foam extinguisher! You may become a part of the electrical circuit and be electrocuted! Use CO_2 or dry chemical fire extinguisher only! Class "D" fires are caused by sodium, potassium, magnesium, lithium, and all metal hydrides. Use dry soda ash fire extinguisher, sodium chloride, or dry sand.

Fire in clothing — Smother in fire blanket or heavy toweling. Use an emergency shower.

General fires — Select the proper fire extinguisher according to the type of fire. First cool the area around the fire with extinguisher to prevent the fire from spreading. Then use extinguisher at base of the fire, and finally extinguish any scattered remnants of the fire.

Electrical fires — First disconnect the apparatus by pulling the safety switch to avoid the possibility of being electrocuted. Then use Class C (CO_2 or dry chemical) extinguisher.

Poisonous gas fires — Use proper respirator. Select the proper fire extinguisher. If the fire gets beyond the control of the available fire extinguisher, get out of the room immediately.

Figure 2.5
Safety stepladder.

Close the door to prevent drafts and gas spread. Always be certain that no one is left behind.

As always, in case of fire be certain that the local fire department is immediately notified!

2.1.3 Careless Habits

Most accidents in the laboratory are caused by doing something impulsive that later seems thoughtless. As a general rule, think about what you are doing before doing it!

Hazards from falling objects — Falling objects can cause serious injuries. Do not place heavy objects on high shelves! If a heavy object must be put on a shelf, secure it with a belt or chain. Be careful in moving heavy instruments or heavy objects; use a laboratory cart where possible.

Hazards from falling — Never climb on drums, cartons, or boxes to reach objects located on high shelves. You may be severely injured, and the injury can be compounded by breakage of glassware or chemical splash. Always use a safety stepladder (Figure 2.5), which provides high stability due to its design and does not slip or move because special locking devices ensure that its rubber-tipped legs engage the floor firmly.

Transporting large bottles — Moving large bottles and carboys is a dangerous operation because of bottle breakage and liquid spillage. Always use safety carts and safety bottle carriers when transporting large bottles of chemicals. The safety bottle carriers secure the bottle from shock and breakage.

2.2 SAFE HANDLING OF COMPRESSED GASES

Compressed gas cylinders can be dangerous because gases are contained under very high pressures. Always handle cylinders of compressed gas with safety precautions.

2.2.1   General Precautions When Working with Compressed Gases

General precautions

- Close off main cylinder valve when not in use.
- Close needle valve or auxiliary cut-off valve in the line as well as that located at the cylinder. Do not rely solely on the cylinder valve.
- Replace cylinders in reasonable time. Replacement for corrosive gases should be 3 months or less.
- Always use gases in areas where adequate ventilation is provided.
- Keep cylinders in outside storage, or have manifolds piping low-pressure gas into buildings.
- Use the smallest cylinder practical for the purpose.

Safety Rules of Using Compressed Gases

- Cylinder contents must be properly identified. Do not accept cylinders that do not identify contents by name. Do not rely on color codes for identification. Do not destroy identification tags or labels.
- Cylinder valves must be protected. Accept only cylinders provided with cylinder valve protective caps. Leave caps in place until ready to use the gas.
- Store properly. Provide specifically assigned location, preferably in a fire-resistant, dry, and well-ventilated area, away from sources of ignition or heat. Outdoor storage area should have proper drainage, and should be protected from direct sunlight. Secure cylinders by chaining or other means to keep them from falling accidentally.
- Transport cylinders by means of a suitable hand truck. Do not roll!
- Never drop cylinders or permit them to strike each other.
- Return in condition received. Close valve, replace cylinder valve protective cap, and dust cap. Mark or label cylinder "EMPTY" or "MT."
- Prevent confusing empties with full cylinders. Store empty cylinders apart from full cylinders. Connecting empty cylinder by mistake to pressurized system could cause contamination or violent reaction in cylinder.

2.2.2   Hazardous Properties of Compressed Gases

It is very important that the properties of a compressed gas be well known before the gas is put to use. Hazards presented by compressed gases are flammability, toxicity, and corrosivity. Hazardous properties of compressed gases are listed in Table 2.1.

2.3   STOCKROOM SAFETY RULES

The laboratory stockroom should be adequate and efficiently planned for safe operation.

2.3.1   Checklist for Safety Storage Room

- The safety storage room should be well planned with wide aisles and adequate lighting, and no blind alleys. It should be orderly and clean.

TABLE 2.1

Hazardous Properties of Compressed Gases

Gas	Toxic	Flammable	Corrosive
Acetylene		X	
Air			
Ammonia	X	X	X
Argon			
Arsine	X	X	
Boron trichloride	X		X
Boron trifluoride	X		X
Butane		X	
Butenes		X	
Carbon dioxide			
Carbon monoxide	X	X	
Carbonyl sulfide	X	X	
Chlorine	X		X
Cyanogen	X	X	
Cyclopropane	X	X	
Deuterium		X	
Diborane	X	X	
Dimethylamine	X	X	X
Dimethylether		X	
Ethane		X	
Ethyl acetylene	X	X	
Ethyl chloride	X	X	
Ethylene		X	
Ethylene oxide	X	X	
Fluorine	X		
Helium			X
Hexafluoropropane	X		
Hydrogen		X	
Hydrogen bromide	X		X
Hydrogen chloride	X		X
Hydrogen fluoride	X		X
Hydrogen sulfide	X	X	
Isobutane		X	
Isobutylene		X	
Krypton			
Methane	X	X	
Methylacetylene		X	
Methyl bromide	X	X	
Methyl chloride	X	X	
Methylmercaptan	X	X	
Neon			
Nitric oxide	X		X
Nitrogen			
Nitrous oxide			
Oxygen			
Ozone	X		
Phosgene	X		X
Phosphine	X	X	
Propane		X	
Propylene		X	
Silane	X	X	
Sulfur dioxide	X		X
Sulfur hexafluoride			
Sulfur tetrafluoride	X		X
Trimethylamine	X	X	
Xenon			

- It should be well ventilated, with an emergency exhaust system.
- Exits should be well marked.
- Emergency exits should be available.
- It should be equipped with adequate fire-protection and firefighting equipment.

Storage area

- Heavy items should be stored near the floor.
- Proper storage should be available for glass apparatus and tubing (never projecting beyond shelf limits).
- Fragile and bulky equipment should be secured to shelving.
- Shelving should be fitted with ledges to prevent items from falling off.
- Liquids and hazardous chemicals should be grouped and separated.
- No waste accumulation of any kind should be permitted.

Other safety precautions in the storage room

- Safety ladders should be used. Safety ladders prevent accidents and save time and effort.
- No excessive heat should be permitted, because of fire hazard.
- Good housekeeping should be maintained, with vigilant and constant maintenance for safe storage.
- Commonsense behavior is necessary of all personnel using storage facilities.

2.3.2 Storage of Chemicals

Chemicals are manufactured in varying degrees of purity. Carefully select the grade of the chemical that meets the need of the work to be done. Always recheck the label of the chemical that you are using! The use of a wrong chemical can cause an explosion, or ruin the analytical work. Check the information carefully on the chemical container: name, formula, formula weight, percentage of impurities, analytical grade, health hazards, and safety codes.

Storage of acids — Acids should be stored in original containers in an "ACIDS"-labeled cabinet, grouped by safety color codes. Bottles having impact-resistant plastic coating are preferred.

Storage of flammable solvents — Store in original containers that are specially designed and put in a "FLAMMABLE"-labeled cabinet. Large quantities should be stored in metal safety cans and outside of the laboratory in marked "FLAMMABLE STORAGE AREA."

Storage of solvents — Store solvents in separate solvent cabinet, in the original container, and in a well-ventilated area.

Storage of chemicals used for volatile organic (VOC) analysis — Store in a separate solvent cabinet, in original containers, and in a well-ventilated area. No other chemicals should be stored in this area.

Storage of chemicals — Chemicals should be stored in alphabetical order in the storage room, with record of "date of arrival" and "date of opening" on each container. Phenol

TABLE 2.2

LabGuard Safety Label System

Color Code	Hazard	Storage
White	Contact hazard	Corrosion-proof area
Yellow	Reactivity hazard	Store separately from flammables and combustibles
Blue	Health hazard	Store in a secure "poison"-marked area
Red	Flammable hazard	Store in an area segregated for flammables
Gray	Minimum or no hazard	Store in a general storage area

Note: The storage code color band runs the entire width of the label of the chemical container, so it is perfectly visible regardless of the container's orientation on the shelf. Store chemicals with like colors together according to recommendation. Incompatible substances that should not be stored alongside chemicals with the same color labels should carry a blue color band at the bottom of the label. Recommended protective equipment, warning statements, and first aid instructions are carried within the lablel's safety information portion.

and hydrogen peroxide should be stored in "chemical storage"-marked refrigerator. The LabGuard Safety Label System (see Tables 2.2 and 2.3) on chemical bottles assists the proper storage of chemicals. Each chemical used in the laboratory should be accompanied with the *material safety data sheet* (MSDS). The MSDSs for every substance give the identity of ingredients, the physical and chemical characteristics of the substance, the physical hazards, the reactivity and health hazards involved, and safe handling and safety precautions. In addition, the control measures to reduce harmful exposures are also listed in every MSDS. An example of an MSDS is shown in Figure 2.6.

2.4 LABORATORY SAFETY RULES

1. Safety glasses/corrective glasses shall be worn at all times in the laboratory. Visitors to the laboratory must be appropriately warned and safety glasses made available to them.

2. Participation in practical jokes or "horseplay" in the laboratory is not permitted.

3. Each laboratory worker is expected to cooperate in keeping his/her working area in a neat and orderly condition and to cooperate with others in making this possible throughout the laboratory. A CLEAN LABORATORY IS A SAFE LABORATORY.

TABLE 2.3

LabGuard Safety Label System: Acids and Bases

Color Code	Chemical
Yellow	Sulfuric acid
Red	Nitric acid
Blue	Hydrochloric acid
Brown	Acetic acid
Black	Phosphoric acid
Green	Ammonium hydroxide

Note: Besides the colored lable, the containers also have color-coded polystyrene screw caps that simplify identification and prevent contamination. Acid bottles have dripless pouring rings and finger grips for easy handling, as well as impact-resistant plastic coating. Although the bottle can break, the coating keeps glass fragments and the reagent trapped within.

Material Safety Data Sheet
LaMotte Chemical Products Company
PO Box 329 • Chestertown • Maryland • 21620
Telephone # for information 301-778-3100
In an emergency: Local Poison Control Center

I • Product Identification

Name: Nitric Acid	Code #: 3933

II • Hazardous Ingredients

Name	CAS #	%	PEL	TLV
Nitric Acid	7697-37-2	0.6	1 mg/m^3	1 mg/m^3

III • Non-Hazardous Ingredients except water (7732-18-5)

Name	CAS#	%

IV • Physical Data

Appearance: Clear, colorless liquid	Odor: None
Solubility in water: Soluble	pH: 1
Boiling point: unknown	Melting point: N/A
Vapor pressure (mmHg):Unknown	Vapor Density (Air=1): Unknown

V • Fire & Explosion Data

Flash point (method used):

Flammable limit: LEL:	UEL:
Extinguishing Media:	Not a fire hazard

Special Fire Fighting Procedures:

Unusual Fire & Explosion Hazard:

VI • Reactivity Data

Stability: Stable [X] Unstable [ ]

Conditions to Avoid: N/A

Incompatability (Materials to Avoid): N/A

Hazardous Decomposition Products: N/A

VII • Health Hazard Data

Toxicity: Unknown

Primary Route of Entry: Inhalation [ ] Skin [X] Ingestion [ ]

Target Organ: N/A

Signs & Symptoms of Exposure: Irritating to eyes, nose and skin.

Medical Condition Aggravated by Exposure: N/A

Carcinogenicity: [X] N/A [ ] NTP [ ] IARC

VIII • Emergency First Aid Procedures

Eye Contact: Flush thoroughly with water for 15 minutes. Consult a physician.

Skin Contact: Flush with water, remove affected clothing and wash skin with soap and water.

Ingestion: Rinse mouth, drink glass of water and consult physician.

Inhalation: N/A

IX • Spill & Disposal Procedures

Spill & Leak: Cover spill with sodium bicarbonate. Scoop up slurry and wash down drain with excess water.

Disposal: Add slowly to a solution of soda ash and slaked lime. Pour this neutralized solution down drain with excess water.

X • Precautionary Measures

In Handling: Gloves [X] Eye Protection [X]Other: Lab Coat

Ventilation: Normal [X] Mechanical [ ] Respiratory Protection [ ]

Work/Hygienic Practices: Avoid contact with skin and clothing

XI • Special Precautions: N/A

Date: 8/6/90

The above information is believed to be correct but does not claim to be all inclusive and should be used only as a guide.

† This is a toxic chemical subject to reporting requirements of section 313 of EPCRA and 40CFR372.

Figure 2.6
Material safety data sheet.

4. The proper techniques shall be utilized when lifting, pushing, pulling or carrying materials to prevent injuries.

5. The location of fire extinguishers, safety showers, eye wash stations, and spill kits must be known to all workers.

6. All laboratory workers must know how and when to use the items listed in #5 above.

7. Eating, drinking, and smoking in the laboratory are never allowed. Never use laboratory containers (beakers or flasks) for drinking.

8. No food or edible substance for human consumption is to be stored in refrigerators in the laboratory.

9. All chemicals used in the laboratory must have a material safety data sheet (MSDS).

10. All chemicals should be clearly labeled. Do not use material from unlabeled containers. Clearly identify chemicals before using them.

11. In the case of chemicals splashed in the eyes, use the eye wash station and report the incident to the laboratory supervisor.

12. Respirators must be used when working with hot acids or solvents not under a hood.

13. Pouring of volatile liquids should be done only in a well-ventilated hood remote from sources of ignition.

14. Only the minimum amount of flammable liquids necessary for making the test shall be kept on work benches.

15. Heavy reagent containers, such as 5 gallon containers, must not be carried or placed on a shelf by one person alone.

16. Face shields, rubber gloves, and protective rubber aprons should be used when preparing, transporting, or pouring corrosive chemicals such as concentrated acids and bases.

17. If it is necessary to dilute acid with water, always add the acid to the water, stirring constantly. Never add water to the acid, as this produces a violent reaction.

18. When drawing liquid into a pipette, always use a suction bulb. Mouth-pipetting is never allowed.

19. Pouring mercury into a sink or drain is strictly prohibited. Mercury will remain in the trap and continue to vaporize and contaminate the air.

20. In the event of acid spills on a person, flush thoroughly with water immediately. Be aware that acid–water mixtures will produce heat. Removing clothes from the affected area while flushing may be important so as not to trap hot acid–water mixtures against the skin. Acids or acid–water mixtures can cause very serious burns if left in contact with skin only for a very short period of time.

21. Weak acids should be used to neutralize base spills, and weak bases should be used to neutralize acid spills. Such solutions should be available in the laboratory in case of emergency. Acid and base spill kits are also available.

22. There must be no unsupervised or unauthorized work going on in the laboratory.

23. Never wear open-toed shoes or sandals, because they offer little or no protection against chemical spills and broken glassware.

24. Keep ties and scarfs secured with fasteners. Do not wear medallions or other hanging objects.

25. Tie long hair up and out of the way.

26. Asbestos gloves should be worn when handling or working with hot materials.

27. Gloves will be worn when opening jars, bottles, or other containers when necessary to exert pressure.

28. A face shield will be worn whenever handling a container with more than 1 L of acid, alkali, or corrosive liquid.

29. Chemicals should never be transported, transferred, poured, or otherwise handled at the height above one's head.

30. Any injury sustained, regardless of how superficial, should be reported to the laboratory supervisor or instructor in schools, and appropriate first aid action should be taken.

31. A leak check shall be made on all gas lines and connections whenever a line is broken and reconnected.

32. Report immediately to the laboratory supervisor any failure of exhaust fans to evacuate vapors completely, defective electrical equipments, faulty or empty fire extinguisher, worn or defective rubber gas burner hose, or other gas hazards.

33. If it is necessary to reach a high shelf, use a step ladder provided for this purpose.

34. Operations involving explosives or flammable mixtures shall not be left unattended.

35. When transporting a large quantity of bottles, do so with a basket or receptacle designed for this purpose.

36. Damaged glassware shall not be used.

37. Glassware shall not be placed close to the edge of the laboratory bench; a passerby may knock it off.

38. Goggles or face shields shall be worn when working with glass apparatus which is under pressure or vacuum.

39. When making a vacuum distillation, use a shield to guard against explosion and fire hazard.

40. All broken glass shall be cleaned up immediately and placed in a container provided for broken glass. Never dispose of broken glassware in a regular garbage container!

Chapter **3**

QUALITY ASSURANCE OF CHEMICAL MEASUREMENTS

3.1 INTRODUCTION

A laboratory is expected to be able to specify the quality of its data in quantitative terms. This requires the existence of some degree of quality assurance.

3.1.1 Quality Assurance (QA)

Quality assurance is a definite plan for laboratory operation that specifies standard procedures that help to produce data with defensible quality and reported results with a high level of confidence. It is a necessary part of data production, and it serves as a guide for the operation of the laboratory for production data quality. The basic requirements of a quality assurance program is to recognize possible errors, understand the measurement system used, and develop techniques and plans to minimize errors. It also includes evaluating what was done, and reporting evaluated data which are technically sound and legally defensible. Quality assurance is the statistical control of quality.

The elements of the quality assurance are *quality control* and *quality assessment*. The activities related to each are summarized below.

Quality Assurance

Quality Assessment	Quality Control
Reference material	GLP, GMP, SOP
Replicates	Calibration
Splits	Standardization
Spikes	Instrument maintenance
Surrogates	Facilities maintenance
Collaborative tests	Education and training
Statistical analysis	Inspection and validation

3.1.2 Quality Control (QC)

Quality control is a mechanism established to control errors. It is a set of measures within a sample analysis methodology to assure that the process is in control. It helps to provide quality that is satisfactory, adequate, dependable, and economical. It consists of the use of a series of procedures that must be rigorously followed. Good quality control systems should include provisions for inspection, both periodical and unannounced, to ascertain how well the procedures are functioning. Large laboratories have a quality control officer or group, independent of the laboratory management, that oversees the operation of the system.

The elements of quality control are

- Competent personnel with adequate educational background with specific training and experience.
- Suitable and properly maintained laboratory equipment and facilities. Modern equipment increases the need for QA/QC.
- Properly selected methodology.
- Good laboratory practices (GLP) — see below.
- Good measurement practices (GMP) — see below.
- Standard operation procedures (SOP) — see below.
- Documentation.
- Inspection and validation.

Good laboratory practices (GLP)

- Laboratory safety
- Cleaning, housekeeping, temperature, humidity control, glassware cleaning
- Storage, handling, labeling, shelf-life, and disposal of chemicals
- Sample custody (documentation, routing, storage, preparation, retention)
- General laboratory operations
- Use, maintenance, and calibration of equipment
- Statistical procedures
- Data reporting, format, documentation

Good measurement practices (GMP) — Guidelines should be written for each measurement technique, addressing subjects such as maintenance and records for equipment, and specified calibration procedures. General instruction manuals should be available with the requirements and precautions for each technique.

Standard operation procedures (SOP) — SOPs are written for basic operations to be done in the laboratory. SOPs should include sampling, measurement, calibration, and data processing in a standard format.

The difference between GLPs, GMPs, and SOPs — GLPs and GMPs (practice) provide guidance, allowing some judgment of the user. They are written in indicative mode. SOPs (procedure) provide direction. Every departure from the laboratory SOP must be justified. They are written in imperative mode.

3.1.3 Quality Assessment

Quality assessment is the mechanism to verify that the system is operating within acceptable limits. It is the overall system of activities whose purpose is to provide assurance that the overall quality control job is being done effectively. Quality assessment is a process to determine the quality of the laboratory measurements through internal and external quality control evaluations, and includes performance evaluation samples, laboratory comparison samples, and performance audits.

3.2 LABORATORY QUALITY ASSURANCE PROGRAM

In *Quality Assurance in Chemical Measurements,* John K. Taylor states that "the objective of quality assurance programs for analytical measurements is to reduce errors to tolerable limits and provide a means of ensuring that the measurements generated have a high probability of being of acceptable quality" (Taylor, J.K., 1988). Acceptable quality is variable and depends on the analytical parameter and measurement process. Each laboratory involved in environmental analytical processes has to develop its own quality assurance program, which should be delineated in a QA/QC manual, and should be comprehensive enough to apply to most of the operation. This program should be continually reviewed and updated as needed. The laboratory should also have a separate quality assurance "project plan" for each of the projects. Understanding the quality assurance program helps the analysts to be responsible participants in the laboratory system, who produce defendable, precise, and accurate analytical data.

3.2.1 Sampling

The sample is one of the critical elements of the analytical process. The quality of any data produced by any analytical system primarily depends on the sample analyzed. A sample must be representative of the whole so that the final result of the chemical analysis represents the entire system that it is intended to represent. Samples collected at a particular time and place are called *grab* or *individual samples*. This type of sample represents conditions at the time it was collected. Therefore, a grab sample should not be used as a basis for a decision about pollution abatement. However, some sources are quite stable in composition, and may be represented well by single grab samples. If results for an entire system are to be reported, a series of smaller samples are collected in a single container and blended for analysis. The mixing process averages the variations in sample composition and minimizes analytical effort and expense. These types of samples are called *composite samples*. When a time factor is being taken into consideration, grab samples are collected in suitable intervals, chosen according to the expected changes. Composite samples reflect the average characteristics during the sampling period, and in most cases a 24-hour period is standard. The volume of the subsamples should be constant, at least 200 mL.

Depending on the analytical parameter or parameter group, the required material and type of sample bottles used and the preservation of the samples are different. Recommended containers, preservatives, and holding times for environmental analytical parameters are shown in Table 3.1. Sample preservation may be accomplished by ready, commercially available prepreserved bottles or by addition of preservative to the sample in the field after sample collection. The pH of the preserved sample should be checked by narrow-range pH paper on an aliquot of preserved sample poured into a small disposable container. More preservative should be added as needed until the pH of the sample is satisfactory. In this case, a corresponding blank sample should be taken with the same amount of preservative. The pH result and the additional quantity of the preservative should be documented.

Documentation during sample collection and transportation includes the *chain of custody form* (Figure 3.1) and *field notebook* containing all information about field activities.

TABLE 3.1

Recommended Volumes, Containers, Preservatives, and Holding Times for Collecting Samples

Parameters	Volume (mL)	Container	Preservative	Holding Time
Physical properties				
Color	50	P, G	Cool, 4°C	48 hours
Conductance	100	P, G	Cool, 4°C	28 days
Hardness	100	P, G	HNO_3 to pH <2	6 months
Odor	200	G only	Cool, 4°C	24 hours
pH	25	P, G	None required	Analyze immediately
Residue filtrable	100	P, G	Cool, 4°C	48 hours
Residue nonfiltrable	100	P, G	Cool, 4°C	7 days
Residue total	100	P, G	Cool, 4°C	7 days
Residue volatile	100	P, G	Cool, 4°C	7 days
Settleable matter	1000	P, G	Cool, 4°C	48 hours
Temperature	1000	P, G	None required	Analyze immediately
Turbidity	100	P, G	Cool, 4°C	48 hours
Metals				
Dissolved	200	P, G	Filter on site HNO_3 to pH <2	6 months
Suspended	200	P, G	Filter on site HNO_3 to pH <2	6 months
Total	100	P, G	HNO_3 to pH <2	6 months
Chromium +6	200	P, G	Cool, 4°C	24 hours
Mercury dissolved			HNO_3 to pH <2	28 days
Total	100	P, G	HNO_3 to pH <2	28 days
Inorganic nonmetallic				
Acidity	100	P, G	Cool, 4°C	14 days
Alkalinity	100	P, G	Cool, 4°C	14 days
Bromide	100	P, G	None required	28 days
Chloride	100	P, G	None required	28 days
Chlorine	1000	P, G	None required	Analyze immediately
Cyanides	500	P, G	Cool, 4°C NaOH to pH >12	14 days
Fluoride	300	P, G	None required	28 days
Iodide	100	P, G	Cool, 4°C	24 hours
Nitrogen ammonia	400	P, G	Cool, 4°C H_2SO_4 to pH <2	28 days
Kjeldahl	500	P, G	Cool, 4°C H_2SO_4 to pH <2	28 days
Nitrate plus nitrite	100	P, G	Cool, 4°C H_2SO_4 to pH <2	28 days
Nitrate	100	P, G	Cool, 4°C	48 hours
Nitrite	50	P, G	Cool, 4°C	48 hours
Dissolved oxygen probe	300	G bottle and top	None required	Analyze immediately
Winkler	300	G bottle and top	Fix on site, store in dark	8 hours
Phosphorus ortho-P dissolved	50	P, G	Filter on site Cool, 4°C	48 hours
Hydrolyzable	50	P, G	Cool, 4°C $HsSO_4$ to pH <2	28 days
Total	50	P, G	Cool, 4°C H_2SO_4 to pH <2	28 days
Total dissolved	50	P, G	Filter on site Cool, 4°C H_2SO_4 to pH <2	24 hours

TABLE 3.1 *(continued)*

Recommended Volumes, Containers, Preservatives, and Holding Times for Collecting Samples

Parameters	Volume (mL)	Container	Preservative	Holding Time
Silica	50	P only	Cool, 4°C	28 days
Sulfide	500	P, G	Cool, 4°C	7 days
			2 mL zinc acetate + 2 N NaOH to pH >9	
Sulfite	100	P, G	None required	Analyze immediately
Sulfate	100	P, G	Cool, 4°C	28 days
Organics				
BOD	1000	P, G	Cool, 4°C	48 hours
COD	50	P, G	Cool, 4°C	28 days
Oil and grease	1000	G only	Cool, 4°C H_2SO_4 to pH <2	28 days
Organic carbon	50	P, G, G brown	Cool, 4°C H_2SO_4 to pH <2	28 days
Phenolics	500	G only	Cool, 4°C H_2SO_4 to pH <2	28 days
Surfactants	500	P, G	Cool, 4°C	48 hours
Purgeable halocarbons	40	G, teflon-lined septum	Cool, 4°C 0.008% $Na_2S_2O_3$*	14 days
Purgeable aromatics	40	G, teflon-lined septum	Cool, 4°C 0.008% $Na_2S_2O_3$* HCl to pH <2	14 days
Acrolein and acrylonitrile	40	G, teflon-lined septum	Cool, 4°C 0.008% $Na_2S_2O_3$* pH 4 to 5	14 days
Phenols	1000	G, teflon-lined cups	Cool, 4°C 0.008% $Na_2S_2O_3$*	7 days until extraction, 40 days after extraction
Phthalate esters	1000	G, teflon-lined cups	Cool, 4°C 0.008% $Na_2S_2O_3$*	7 days until extraction, 40 days after extraction
Nitrosamines	1000	G, teflon-lined cups	Cool, 4°C 0.008% $Na_2S_2O_3$* store in dark	7 days until extraction, 40 days after extraction
PCBs	1000	G, teflon-lined cups	Cool, 4°C	7 days until extraction, 40 days after extraction
Nitroaromatics and isophorone	1000	G, teflon-lined cups	Cool, 4°C store in dark	7 days until extraction, 40 days after extraction
Polynuclear aromatic hydrocarbons	1000	G, teflon-lined cups	Cool, 4°C store in dark	7 days until extraction, 40 days after extraction
TCDD (Dioxin)	1000	G, teflon-lined cups	Cool, 4°C 0.008% $Na_2S_2O_3$*	7 days until extraction, 40 days after extraction
Chorinated hydrocarbons	1000	G, teflon-lined cups	Cool, 4°C	7 days until extraction, 40 days after extraction
Pesticides	1000	G, teflon-lined cups	Cool, 4°C pH 5 to 9	7 days until extraction, 40 days after extraction
Soil, sediment, sludge				
Organic extractable	8 oz.	Widemouth G teflon-lined cup	Cool, 4°C	ASAP
Organic volatile	8 oz.	Widemouth G teflon-lined cup	Cool, 4°C	ASAP
Metal	1 pint	P	Cool, 4°C	6 months
Fish Samples				
		Wrap in Al foil**	Freeze	ASAP

TABLE 3.1 *(continued)*

Recommended Volumes, Containers, Preservatives, and Holding Times for Collecting Samples

Parameters	Volume (mL)	Container	Preservative	Holding Time
Chemical wastes				
	8 oz.	Widemouth G,** teflon-lined cap	None	ASAP
Bacteriology				
Total and fecal coli Fecal streptococcus	100	P, G sterile	Cool, 4°C 0.008% $Na_2S_2O_3$	6 hours
Radiological				
Alpha, Beta, Radium	1000	P, G	HNO_3 pH <2	6 months

Note: ASAP = as soon as possible; P = polyethylene container; G = glass container. Sample preservation should be performed immediately upon sample collection. For composite samples, each aliquot should be preserved at the time of collection. When use of an automated sampler makes it impossible to preserve each aliquot, then chemical samples may be preserved by maintaining 4°C until compositing and sample splitting is completed. Samples should be analyzed as soon as possible after collection. The holding times listed should be the maximum time that samples may be held before analysis and still be considered valid. For dissolved parameters, samples should be filtered immediately on site before preservation. Reference 40 CFR Part 136.

* Required if residual chlorine is present.
**Plastic containers may be used if only metals are required.

Field I.D. _____   Site Name: _____

Date Sample Received _____   Address _____

Sampler(s) _____   Laboratory: _____

Sample Container Description

Sample Identity	Date Sampled								Total	Remarks

Total Number of Containers [ ]

Relinquished By: _____ Organization: _____ Received By: _____ Organization: _____

Date: _____ Time: _____ Date: _____ Time: _____

Relinquished By: _____ Organization: _____ Received By: _____ Organization: _____

Date: _____ Time: _____ Date: _____ Time: _____

Delivery Method: _____   (attach shipping bills, if any)

Use extra sheets if necessary

Figure 3.1
Chain of custody form.

Field Sequence No. _____

Field Sample No. _____ Date _____ Time _____

Sample Location _____

Sample Source _____

Preservative used _____

Analyses required _____

_____

Collected by _____

Remarks _____

_____

_____

Final pH checked _____

Additional preservative used (if applicable) _____

Figure 3.2
A sample label should be affixed to all sample containers. This is an important part of sample identification. It should be waterproof, and all of the information should be written in waterproof ink.

Each sample should be labeled with the exact sample identification (including preservation and preservation checks). A typical sample label form is shown in Figure 3.2.

It should be noted that the effectiveness of sample collection and related field activities is supported by field quality control checks.

Equipment blanks — Equipment blanks are used to detect any contamination from sampling equipment. They are prepared in the field before sampling begins, by rinsing the equipment with analyte-free water, filling the appropriate sample bottle with analyte-free water, and preserving with appropriate preservative.

Field blanks — Field blanks are collected at the end of the sampling event by filling the appropriate sample bottle with analyte-free water and preserving the same manner as the samples.

Trip blanks — These are used to verify contaminations that may occur during sample collection and transportation. Trip blanks are blanks of analyte-free water that are prepared by the laboratory; the blank is transported to the field and remains unopened during the sampling event and is transported back to the laboratory with the samples.

Duplicate samples — These samples are collected for checking the preciseness of the sampling process. Duplicate samples are collected at the same time and from the same source as the study samples.

Split samples — These samples are taken to check analytical performance. The sample is taken in one container, mixed thoroughly, and split into another container. Both halves are now samples that represent the same sampling point.

3.2.2 Sample Custody

Sample custody is the process of protecting the samples collected and analyzed. If a sample decomposes or is contaminated prior to the actual analysis, the results are unreliable. Proper sample handling is essential. All the paperwork involved in the sample custody process defends and secures the quality of the reported data. It shows how samples are collected, preserved, stored, transported to the laboratory, treated, numbered, and tracked during the analytical process. All records have to be maintained so that they are easy to find and ready at all times for immediate inspection. All documentation must be signed or initialed by the responsible person and recorded with waterproof ink, without any erasures or marking. All corrections must be made with one line marked through the error, accompanied with a signature or initial, date, and the corrected form.

An important rule to remember is that "it did not happen if it is not documented." Since a large number of the work done in today's laboratories potentially could go to litigation, all aspects of the sample must be documented. This starts with the purchase of the bottles used to collect the samples and ends in the laboratory with the record of the individual who mailed the data package.

Documentation must be related to

- Sample collection, field activities
- Sample receipt, sample distribution, and holding time
- Sample preparation prior analysis
- Analytical methods
- Reagents and standards preparation
- Calibration procedures and frequency
- Analytical data and calculations
- Detection limits
- Quality control requirements and quality control routine checks related to the analysis
- Data validation and reduction
- Data reporting

3.2.3 Sample Preparation for Analysis

Once sampling and sample preservation schemes have been properly executed, the sample is ready for analysis. Most of the time, however, it is still not in a state in which the chosen analytical technique can be properly applied. Sample preparation techniques, such as distillation, digestion, extraction, etc. will be discussed in Chapter 5 and in specific analytical methodologies. A sample holding time log form is shown in Figure 3.3, and a sample preparation log form is shown in Figure 3.4.

3.2.4 Analytical Methodology

The above requirements are closely related to the measurement methodology to be used. The method used should be adequate for the intended purpose and it must be properly utilized. The necessary characteristics of a suitable method include adequate sensitivity, selectivity, accuracy, and precision. The method selection should also entail a large dynamic measurement range, easy operation, required operational skill, low cost, detection limit, biases, interferences, and calibration requirements. A method, once adopted, must be used in a reliable and consistent manner in order to provide reproducible data. This is best accomplished by following detailed, written standard operation procedures (SOPs). In environmental testing, all analyses must be performed using methods approved by the U.S. Environmental Protection Agency (EPA) or state Department of Environmental Protection (DEP), specified according to the sample matrices. Before apply-

Sample ID	Matrix	Analysis Required	Holding Time (days)			Date of Preparation				Storage Sample Prepared	Sign
			prep	anal	dispo	rec.	prep	anal	dispo		

Sample ID = Sample Identification Number
Analysis Required = Analysis required
prep. = Prepared
anal. = Analysis
dispo. = Disposal

rec. = Received
Sign. = Signature of Logger

Holding Time Explanation:

prep. = number of days between the date sample received and the date sample prepared
anal. = number of days between the date sample prepared and the date of actual analysis
dispo. = number of days between the date sample received and the date sample disposed

Storage Designations:

R.T. = Room Temperature in designated area
Ref. O. = Refrigerator, designated for organic samples
Ref. I. = Refrigerator, designated for inorganic samples
Fr. = Freezer, designated for special samples

Figure 3.3
Sample holding time log.

Sample I.D. No.	Matrix	Test for	Method No. Analysis	Method No. Prep.	Date Sample Rec'd	Date Sample Prep.	Sample Size ml.	g	Final Volume ml.	Signature

Method References:

Figure 3.4
Sample preparation log (per analyte group).

ing any method that is different from the approved method, it should be reviewed and accepted by the EPA or DEP. When any modification of a method has been suggested, the modification also needs the approval of the EPA or DEP prior to use. Each method should be described by method number and reference. Approved analytical methods and references are listed in Chapter 14.

3.2.5 Analytical System Calibration

Calibration and standardization of analytical systems are necessary to ensure that the produced data are acceptable or, in other words, successful calibration demonstrates that the data produced during the analytical performance is correct. Obviously the analyst must thoroughly understand each of the calibrations required for a particular measurement. This requires a knowledge of the standards needed and their relation to the measured process, the frequency of the calibration, the effect on a measurement system due to lack of the calibration, and even the shock of the system resulting from recalibration. Calibrations are performed at the beginning of the analysis to make sure that the instrument is working properly. Calibrations must be performed according to the nature of the analytical methodology.

Initial calibration — Initial calibration is based on the response of the instrument on different concentrations of calibration standards against calibration blanks. A calibration blank is a volume of analyte-free water used to zero the instrument. It is also used at the end of the analysis to check if any contamination or drifting occurs. The number and the optimum concentration range of the calibration standards used for each particular method is given by the approved methodology and described by the laboratory SOP. When the number of the standards are not described, a three-point calibration is satisfactory. The response of the instrument should be linear with the concentration of the introduced standards. The concentration of the standards and the response of the instrument should be plotted on a *calibration curve*, or the instrument software automatically prepare it. The approval of the calibration curve depends on the value of the correlation coefficient calculated by the linear regression. The optimum value for the correlation coefficient is >0.9950.

Calibration of the instruments varies with the type and model of the equipment. Detailed operation and calibration procedures for each instrument are available in the laboratory SOP and the manufacturer's instructions.

Initial calibration verification standard (ICVS) — An initial calibration verification standard (ICVS) is analyzed immediately after the calibration curve to check the calibration. It is a mid-range calibration standard. Recovery of this standard must be within ±10% of the value. If the recovery is not within the limits, the instrument must be recalibrated.

Continuing calibration — The initial calibration has to be proven during the analytical process by continuing calibration.

Continuing calibration verification standard (CCVS) — A continuing calibration verification standard (CCVS) is analyzed once every ten samples. It represents the value of the midpoint initial calibration. The deviation from the original value should be within ±15%. When continuing calibration fails to meet acceptance of the criteria, initial calibration should be repeated.

Calibration, initial and continuing calibration verification standards are prepared by appropriate dilution from the calibration stock solutions. Calibration stock solutions are

TABLE 3.2

Storage and Preparation Frequency of Selected Stock and Standard Solutions

Test	Calibration Stocks and Standards	Storage	Preparation Frequency
pH	pH 4.00, 7.00, 10.00 buffers	Room temp.	Expiration date indicates
Conductance	0.01 M KCl	Room temp., glass stop bottle	6 months
Turbidity	400 NTU stock	Refrigerate	1 month
	Dil. standards	Refrigerate	1 week
Bromide Br	500 ppm stock	Room temp.	3 months
Cyanide CN	1000 ppm stock	Refrigerate	1 month, check weekly
Fluoride F	100 ppm stock	Room temp.	3 months
Ammonia–Nitrogen NH$_3$-N	1000 ppm stock	Refrigerate	3 months
Nitrate–Nitrogen NO$_3$-N	1000 ppm stock	Refrigerate, preserve with chloroform	6 months
Nitrite–Nitrogen NO$_2$-N	250 ppm	Refrigerate, preserve with chloroform	3 months
Phosphorus PO$_4$-P	50 ppm	Refrigerate	3 months
Silica SiO$_2$	10 ppm	Refrigerate in tightly stoppered plastic bottle	1 month
Sulfate SO$_4^2$	100 ppm	Room temp.	6 months
Metals	1000 ppm	Room temp.	Expiration date indicates
	10 to 100 ppm standards	Room temp., preserve with 0.5% HNO$_3$	1 month
COD	500 ppm	N/A	Prepare fresh for each
TOC	100 ppm and 10 to 100 ppm standards	Refrigerate, brown glass bottles	3 months
Oil and grease	"Reference oil," calibrate each time used	Freezed in sealed container	3 months
Total phenols	1000 ppm	Refrigerate in glass bottle	3 months
Trace organics	Concentrate depends on methods and analytes	Freeze in individually packaged vials	Expiration date indicates
Tannin and lignin	1000 ppm stock and 10 ppm standard	Room temp.	6 months

Note: Standards must be stored according to the recommendation of the method or according to the statement of the supplier. If there is no guidance for the storage time, change the standards when the reading of the values are decreased.

either commercially available or "in-house prepared" by the analyst. Table 3.2. contains the recommended storage and preparation frequency for selected calibration stock and standard solutions used in environmental testing.

Reference (independent) standard — A reference standard is analyzed once every ten samples. Since the reference standard is prepared from a source different from the calibration and CCV standards, it is used to verify that the CCV and the calibration standards are actually at the concentrations claimed by the vendor and/or the analyst. It should be a certified or independently prepared standard from a source other than that used for the preparation of calibration standards. The value is accepted within ±10% deviation from the 100% recovery.

3.2.6   Detection Limits

Detection limits are the smallest concentration of an analyte of interest that can be measured with a stated probability of significance.

TABLE 3.3

Student's t Variate Table

df	80% $t_{.90}$	90% $t_{.95}$	95% $t_{.975}$	98% $t_{.99}$	99% $t_{.996}$	99.73% $t_{.9985}$
1	3.078	6.314	12.706	31.821	63.657	235.80
2	1.886	2.920	4.303	6.965	9.925	19.207
3	1.638	2.353	3.182	4.541	5.841	9.219
4	1.533	2.132	2.776	3.747	4.604	6.620
5	1.476	2.015	5.571	3.365	4.032	5.507
6	1.440	1.943	2.447	3.143	3.707	4.904
7	1.415	1.895	2.365	2.998	3.499	4.530
8	1.397	1.860	2.306	2.896	3.355	4.277
9	1.383	1.833	2.262	2.821	3.250	4.094
10	1.372	1.812	2.228	2.764	3.169	3.975
11	1.363	1.796	2.201	2.718	3.106	3.850
12	1.356	1.782	2.179	2.681	3.055	3.764
13	1.350	1.771	2.160	2.650	3.012	3.694
14	1.345	1.761	2.145	2.624	2.977	3.636
15	1.341	1.753	2.131	2.602	2.947	3.586
16	1.337	1.746	2.120	2.583	2.921	3.544
17	1.333	1.740	2.110	2.567	2.898	3.507
18	1.330	1.734	2.101	2.552	2.878	3.475
19	1.328	1.729	2.093	2.539	2.861	3.447
20	1.325	1.725	2.086	2.528	2.845	3.422
25	1.316	1.708	2.060	2.485	2.787	3.330
30	1.310	1.697	2.042	2.457	2.750	3.270
40	1.303	1.684	2.021	2.423	2.704	3.199
60	1.296	1.671	2.000	2.390	2.660	3.130
	1.282	1.645	1.960	2.326	2.576	3.000

Note: Columns to be used in calculating corresponding two-sided confidence interval; df = degree of freedom $(n - 1)$, n = number of measurements.

After John Keenan Taylor, *Quality Assurance in Chemical Measurements,* CRC Press/Lewis Publishers, Boca Raton, FL, 1988, 267.

Method detection limit (MDL) — MDL is the smallest concentration of the analyte of interest that can be measured and reported with 98% confidence that the concentration is greater than zero. To determine the MDL for a certain analyte, first estimate the MDL by using the minimum detection concentration described by the method, or based on previous practice. Prepare a standard in this concentration and analyze seven portions of this solution. From the received values, calculate the standard deviation(s) and select from the Student's t variate statistical table (Table 3.3) the value of t related to the seven analytical results at 98% confidence limit (df = degree of freedom, which is the number minus one, so 7 − 1 = 6 at the level of 98%). So, the calculation will be

$$\text{MDL} = s \times 3.14 \tag{3.1}$$

In this formula, s is the standard deviation for the seven analytical results, and 3.14 is the value of t. For example, the estimated detection limit for a titrimetric procedure is 2.0 mg/L. The prepared 2 mg/L concentration standard is analyzed seven times with the following results: 1.8, 2.0, 2.5, 2.0, 2.4, 1.5, and 3.0 mg/L. The calculated standard deviation for these data is 0.499. Therefore, the MDL will be 0.499 × 3.14 = 1.57 mg/L, or using the rounded form, 1.6 mg/L. So, the estimated 2.0 mg/L MDL is acceptable.

Practical quantitation limit (PQL) — PQL is the smallest concentration of the analyte of interest that can be reported with a specific degree of confidence. It is the lowest level achievable among laboratories within specified limits during routine laboratory operation. The value of PQL is about five times that of the MDL. The calculation is

$$PQL = s \times 10 \tag{3.2}$$

The variable s is the standard deviation determined for calculation of the MDL above. The 10 standard deviations corresponds to an uncertainty of ±30% in the measured value at the 98% confidence limit.

Instrument detection limit (IDL) — IDL is the concentration of the analyte that produces a signal greater than five times the signal-to-noise ratio of the instrument. To determine IDL, analyze analyte-free water seven times. Calculate the standard deviation (s) for these responses and multiply by three.

$$IDL = s \times 3 \tag{3.3}$$

An operating analytical instrument usually produces a signal greater than three standard deviations of the mean noise.

3.2.7   Statistics in Chemical Analysis

The satisfactory quality of analytical data may be characterized by accuracy, precision, and bias. Whenever chemical measurements are made, both precision and accuracy are considered. However, it is possible that a measurement could show good precision, but poor accuracy.

Accuracy — Accuracy is the degree of agreement of a measured value with the true or expected value of the quantity of concern. Accuracy is measured and expressed as percentage of recovery (%R).

$$\%R = (\text{analytical value} \times 100)/(\text{true value}) \tag{3.4}$$

Precision — Precision is the degree of mutual agreement among individual measurements as the result of repeated application under the same condition. Precision measures the variation among measurements and may be expressed in different terms.

Standard deviation (s) — Deviation (d) shows how much each measurement differs from the mean. Mathematically the deviation is calculated as follows:

$$d = m - e \tag{3.5}$$

where d = deviation, m = mean, and e = individual measurement. The variables m and e are given by absolute values, which means that the value of d is calculated without regard to sign, so it is always a positive value.

$$s = (d_1 + d_2 + d_3 + ...)/(n - 1) \tag{3.6}$$

where s = standard deviation, n = number of measurements, and d_1, d_2, d_3 = deviations of individual measurements.

Relative standard deviation (RSD) —

$$\%\mathrm{RSD} = (s \times 100)/m \tag{3.7}$$

where s = standard deviation and m = mean.

Relative percent difference (RPD) — RPD is the difference between the values of duplicate analyses divided by the mean of the duplicate values and multiplied by 100. The formula used for calculation is as follows:

$$\mathrm{RPD} = [(A - B)]/[(A + B)/2] \times 100 \text{ or } [(A - B)/(A + B)] \times 200 \tag{3.8}$$

where A = result on analysis one and B = result on duplicate analysis.

Confidence limit (CL) — A more formal confidence limit determination method is a calculation taking into account the mean, the standard deviation, the number of measurements, and a "probability factor" (t). The values of t depend on the number of measurements (n) or more correctly on the number of degrees of freedom ($n - 1$), and based on the analyst's own probability or reliability requirements. That is, the values of t differ according to whether the analyst wants a 90% probability of being accurate, a 95% probability, a 99% probability, etc. Table 3.3 summarizes the values of t for different probability levels and for a number of different degrees of freedom.

$$\mathrm{CI} = (m \pm ts)/n \tag{3.9}$$

where m = mean, s = standard deviation, and n = number of measurements. The value for t depends on the number of degrees of freedom ($n - 1$) for a 98% level of confidence, and taken from the Student's t table.

Confidence limit or target limit — These limits are established by individual laboratories for the precision and accuracy of each analytical method. The results of the daily analytical performances are accepted only when the values of these quality control checks are inside of these limits. If these checks fail criteria, all samples analyzed in the batch must be repeated after the correction of the problem. The outline of the conducted correction action has to be documented.

Confidence limit of a measurement — Reporting results with confidence limits helps to estimate the reliability of the result. Standard deviation is calculated to three significant figures and rounded to two when reported as data. Calculate the limits according to formula (3.9) with the 95% confidence.
    For example,

Available data 0.50, 0.55, 0.50, 0.58, 0.55, 0.50
Average (x) 0.53
Standard deviation (s) 0.0316
Number of measurement..6
Degree of freedom ($n - 1$)..6 − 1 = 5
t value from Student t table with 95% limit..............5.571

Confidence limit (ts/n)..(5.571 × 0.0316)/2.449 = 0.072
Reported result ($x \pm 0.07$) ....0.53 ± 0.07

3.2.8 Quality Control Charts

The performance of a measurement system can be demonstrated by the measurement of accuracy and precision. The data so generated may be plotted as a control chart in a manner that indicates whether the measurement system is in a state of statistical control. Control charts are statistical tools for monitoring the performance of a particular task on a continuing basis. Control limits, i.e., the values believed to be credible, are computed from the standard deviation. For example, the two standard deviation ($2s$) limit represents those within which values are expected to lie 95% of the time. The three standard deviation ($3s$) limit represents the 99% confidence level. Departures from the former are warnings of possible trouble, while exceeding the latter usually means corrective action is needed.

The control chart is prepared for each test parameter after 20 determinations have been performed. The mean is plotted with the warning limit being ±$2s$ and the control limits being ±$3s$.

Representation of control limits: accuracy control chart — An accuracy control chart monitors the percentage of recovery (%R) with standard deviation being the limiting factor. An accuracy control chart is shown below.

- Accuracy is expressed as %R. Collect 20 %R data for the parameter of interest.
- Calculate the mean (m) and the standard deviation (s).
- Calculate the upper and lower warning limits (UWL, LWL).
- Calculate the upper and lower control limits (UCL, LCL).

$$UCL \rightarrow m + 3s$$
$$UWL \rightarrow m + 2s$$
$$\Rightarrow m$$
$$LWL \rightarrow m - 2s$$
$$LCL \rightarrow m - 3s$$

%R data are outside of the warning limits (LWL, UWL), indicating that the system is approaching an out-of-control situation and may require a corrective action. Any data falling outside of control limits (LCL, UCL) signifies an out-of-control system. In such a case, analysis must be stopped and corrective action should be taken before further analysis is done. Samples analyzed with the failed QC check sample should be analyzed again.

Representation of control limits: precision control chart — The precision control chart monitors the repeatability of a measurement system, disregarding accuracy. A precision control chart is shown below.

- Precision may be expressed in different terms, as standard deviation (s), relative standard deviation (RSD), or relative percent different (RPD) as discussed previously. It is based on the term that was selected to express precision values. Collect 20 data points.
- Calculate the mean (m) and the standard deviation.

- Calculate the warning and the control limit (WL, CL).

$$CL \rightarrow m + 3s$$
$$WL \rightarrow m + 2s$$
$$\Rightarrow m$$
$$\rightarrow 0$$

Any value falling above the warning limit should act as a signal that the system is approaching an out-of-control situation and may indicate the need for corrective action. Data falling out of the control limit indicate that the system is out of control.

Interpretation of QC charts is shown in Figure 3.5.

3.2.9   Items for the Essential Control of Chemical Measurements

Quality control checks include all practices and activities that provide the accuracy and the precision of the analysis.

PRECISION CONTROL CHARTS

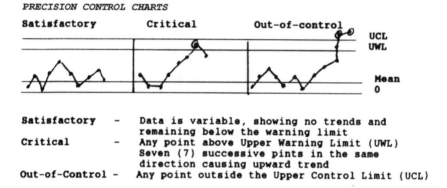

Satisfactory	–	Data is variable, showing no trends and remaining below the warning limit
Critical	–	Any point above Upper Warning Limit (UWL) Seven (7) successive pints in the same direction causing upward trend
Out-of-Control	–	Any point outside the Upper Control Limit (UCL)

ACCURACY CONTROL CHARTS

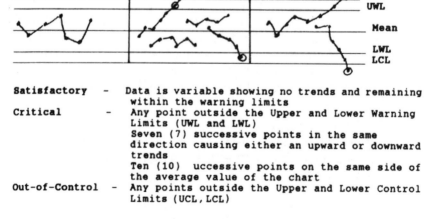

Satisfactory	–	Data is variable showing no trends and remaining within the warning limits
Critical	–	Any point outside the Upper and Lower Warning Limits (UWL and LWL) Seven (7) successive points in the same direction causing either an upward or downward trends Ten (10) uccessive points on the same side of the average value of the chart
Out-of-Control	–	Any points outside the Upper and Lower Control Limits (UCL, LCL)

Figure 3.5
Interpretation of QC charts.

- Reagent blank: Analyte-free water analyzed with the samples.
- Preparation blank or method blank: Analyte-free water processed and analyzed the same way as the samples.
- Reference (independent) standard or QC check standard: A standard solution originated from a source other than the calibration standards. It is analyzed after the calibration of the analytical system to verify the accuracy of the calibration.
- Laboratory control standard (LCS): A reference standard treated exactly as the samples and going through the same pretreatment and preparation procedures as the samples. It verifies any contamination from the pretreatment or an out-of-control event.
- Duplicate samples: *Field duplicates* are samples collected at the same time from the same source, but submitted and analyzed as separate samples. *Laboratory duplicates* are aliquots of the same sample that are prepared and analyzed at the same time. They are used to determine the precision or reproducibility of the analysis. This assures that the analyses are providing consistent results. *Matrix spike duplicates* are aliquots of the same sample spiked with the same concentration of the analyte. They are used to check the precision of the analytical system instead of duplicate samples.
- Spiked sample or matrix spike (MS): Samples that have a specific concentration of the analyte of interest added and analyzed as a sample.
- Reagent water spike: Analyte-free water spiked with the analyte of interest, treated and analyzed as the samples.
- Surrogate spike: Surrogates are organic compounds that are similar to the analytes of interest in chemical composition, extraction, and chromatography, but that are not normally found in environmental samples. Surrogate spikes are used in trace organic determination by gas chromatography (GC) and gas chromatography/mass spectrophotometric (GC/M) methods. Surrogates are spiked into all blanks, standards, samples, and matrix spikes prior to purging or extraction according to the appropriate methods.
- Matrix spike/matrix spike duplicates (MS/MSD): Spiked samples have a specific concentration of the analyte of interest added and analyzed as a sample. They are used to measure the performance of the complete analytical system including any chemical interferences from the sample matrix. Matrix sample duplicate (MSD) are duplicates of the matrix spike sample. The value of the added spike is calculated by using the following formula:

$$c_1 v_1 = c_2 v_2 \qquad \textbf{(3.10)}$$

where c_1 is the value of the stock spike solution in mg/L, v_2 is the volume of the added spike stock solution, c_2 is the concentration of the added spike in mg/L, and v_2 is the volume of the sample spiked. For example, if a 100-mL sample is spiked with 0.5 mL (500 µL) of spike solution, the added spike value (c_2) will be

$$11,000 \times 0.5 = c_2 \times 100$$

$$c_2 = 55 \text{ mg/L}$$

The spiked sample value will be the sample analytical value (mg/L) plus the spike added (mg/L). The percentage of recovery (%R) on the spiked sample is calculated according to the formula:

$$\%R_{sp} = [(\text{Spiked sample value} - \text{Sample value})/(\text{spike value})] \times 100 \qquad \textbf{(3.11)}$$

The precision of the matrix spike duplicates (MSDs) is calculated and expressed as relative percent difference (RPD) from formula (3.8).

- Blind QC standard: Obtained from an independent source with known analytical value. The analyst is not told the true concentration value of the sample. It measures and validates the analytical system and the performance of the analyst.

- Performance evaluation samples: Reference materials obtained from an independent source for which the level(s) of analytes have been validated. Performance testing is done to evaluate the ability of the laboratory to produce a specified quality of data. It measures the capability of the laboratory for certification and also serves as a tool to evaluate the qualification of the laboratory for contractual purposes. Usually, the laboratory participates in performance evaluation programs because it wants to offer its services for evaluation. Along with the analytical data, laboratories should have available documented evidences of their accuracy and precision, a complete standard operation procedure (SOP), and an operational quality assurance (QA/QC) program.

Chapter 4

LABORATORY APPARATUS AND GLASSWARE

4.1 LABWARE MATERIALS

4.1.1 Soft Laboratory Glassware vs. Heat-Resistant Laboratory Glassware

Soft soda-lime glassware is not heat-resistant because it has a low melting point (600–800°C) and a high coefficient of expansion; when it is suddenly subjected to extreme temperature changes, it will break or crack. However, it can be used satisfactorily for such equipment as volumetric flasks, stirring rods, and other containers for normal temperature mixing and reaction operations. It cannot be used over an open flame or with an electric heater. Hard, high-temperature-melting (750–1100°C), heat-resistant, borosilicate laboratory glassware (e.g., Pyrex brand glass) should be used whenever sudden changes in temperature may occur, for example for beakers, reaction flasks, distilling columns, and condensers. It should be used whenever the glassware is subjected to an open flame or to electric heating elements. When choosing glassware, select the type that meets the needs. For general laboratory use, the most suitable material is resistant borosilicate glassware (Pyrex, Kimax, Kimble). Special glassware is also available, such as "resistant to alkali," "low boron content," "exclusion of light," and "high-silica" glass (Vycor).

When selecting stoppers or caps for glassware, the glassware's purpose should be considered. For example, metal caps are never used for corrosives, glass stoppers are not for alkaline solutions, rubber stoppers are excellent for alkaline solutions but unacceptable for organic solvents and metal solutions. Bottles for organic solutions should have Teflon-lined caps.

4.1.2 Plastic Labware

Plastic labware offers practically all the advantages of laboratory glassware, without the problem of breakage that always is present when glass is used. Plastic labware is available in a number of different plastic substances, each type offering specific advantages of clarity, flexibility, temperature stability, resistance to action or effect of chemical substances, resistance to heat, no adhesive surfaces, resistance to the high gravities encountered during centrifugation, and ease of marking. The selection of the most suitable plastic labware depends upon an analysis of all the conditions to which the labware may be exposed.

4.1.3 Porcelain Labware

Porcelain may be heated to higher temperatures than glass (1300°C). It is very resistant against acids and bases, except phosphoric acid. It easily tolerates temperature changes. It is usually used for ignition crucibles and evaporation dishes. Porcelain ware used for ignition at higher temperature than 1000°C is not enamelled.

4.1.4 Platinum Crucible

Platinum is useful in crucibles for specialized purposes. The chemically valuable properties of this soft, dense metal include resistance to attack by most mineral acids including hydrofluoric acid, its inertness with respect to many molten salts, its resistance to oxidation even at elevated temperatures, and its very high melting point. Besides these advantages, platinum has many limitations to its use. Platinum is readily dissolved on contact with aqua regia and with the mixture of oxidizing agents and chlorides. At elevated temperature it is also dissolved by fused alkali oxides, peroxides, and some hydroxides. Also, it should be remembered that contact between heated platinum and other metals or their oxides must be avoided, because platinum readily forms alloys with some metals. Contact with ammonia, chlorine, volatile chlorides, sulfur dioxide, and gases with high carbon content result in surface changes. At red heat, platinum is readily attacked by arsenic, antimony, selenium, tellurium, phosphorus, sulfur, and carbon. And finally, when heated in air for prolonged periods at temperatures greater than 1500°C, there is a significant loss in weight due to the volatilization of the metal.

4.1.5 Nickel Crucible

A nickel crucible is excellent for alkali fusing, but it is very sensitive for even diluted acids. Therefore, never dissolve the fused material with acid solution in this crucible.

4.2 VOLUMETRIC GLASSWARE

The measurement of volumes is very important in quantitative chemical analysis. Containers used for volume measurements must have a calibration line corresponding to the volume of the solution being measured. There are four different glassware used for this purpose: beaker, Erlenmeyer flask, graduated cylinder and volumetric flask. Beakers and Erlenmeyer flasks are never used for accurate solution preparation. The calibration line reflects such poor accuracy that they can only be taken for approximations. Even the graduated cylinder is of insufficient accuracy most of the time.

Volumetric glassware is calibrated either *to contain* (TC) or *to deliver* (TD). There is an imprint on specific pieces of glassware which indicates the particular item is calibrated to contain (TC) or to delivery (TD). Glassware designed TD will do so with accuracy only when the inner surface is so scrupulously clean that water wets it immediately and forms a uniform film upon emptying. The calibration must obviously take this thin film into account in the sense that it will not be a part of the delivered volume. If the glassware is designed TC, it is calibrated that the thin film will be a part of the contained volume. When a liquid is confined in a narrow tube, the surface is found to exhibit a marked curvature, called the *meniscus*. It is common practice to use the bottom of the meniscus in calibrating and using volumetric glassware. Special care must be used to read the meniscus. A useful technique for reading the meniscus is shown in Figure 4.1. Location of the

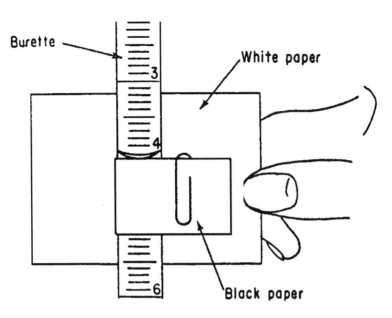

Figure 4.1
Useful technique for reading a meniscus. Render the bottom of the meniscus, which is transparent, more distinct by positioning a black-striped white card behind the glass.

eye in reading any graduated tube is important. The eye must be level with the meniscus of the liquid to eliminate parallax errors, as shown in Figure 4.2.

4.2.1 Volumetric Flask

Volumetric flasks are calibrated to contain a specified volume when filled to the line etched on the neck. They are used in laboratory work in the range of 5 to 5000 mL in capacity. They are designed to contain (TC) the exact volume of liquid when the bottom of the meniscus just touches the etched line across the neck. The accuracy of the delivery by volumetric flasks must be considered poor. On the flask, "TC" is imprinted on the base.

Additional accuracy is achieved with the use of "class A" glassware. Class A volumetric flasks always have a tapered ground glass or plastic stopper, while others may utilize screw caps or plastic snap caps. Always be aware of the type and size of cap used for the volumetric flask, because improperly sealed caps may lead to leakage and inaccurate concentration of the prepared solution. A typical volumetric flask is shown in Figure 4.3.

When a solution is made up in a volumetric flask, it is important that it be well mixed. The accepted practice is to mix the solution thoroughly before the final volume has been adjusted, and mix again after the flask has been filled to the mark; the dissolved solute possibly increases the volume and it is easier to agitate the solution vigorously when the narrow upper portion of the flask has not been filled. Use a dropper or Pasteur pipet to match the bottom of the meniscus to the mark, firmly stopper the flask, and invert repeatedly to assure uniform mixing.

Note: Alkaline solutions cause ground glass stoppers to "freeze" and thus should never be mixed in flasks equipped with such stopper. As a general rule, the heating of calibrated glass equipment should be avoided! Too rapid cooling can permanently distort the glass and cause a change in volume.

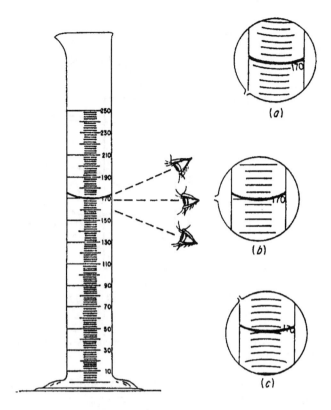

Figure 4.2
Avoiding parallax error in reading the meniscus. (a) Eye level too high, volume too high. (b) Eye level correct, volume correct. (c) Eye level too low, volume too low. The eye must be level with the meniscus of the liquid to eliminate parallax errors.

Cleaning and storage of volumetric glassware — Before use, volumetric flasks should be scrupulously clean. Cleaning with warm detergent solution, and following with several tapwater rinses and three rinses with deionized (DI) water is the general procedure. Volumetric flasks may be used either wet or dry. Do not dry flasks by heating, but only by air. In the case of using wet volumetric flasks, dry the neck of the flask above the mark with a tissue paper when a solid is to be added.

After washing, position the flask in inverted position for quick drainage before storing. Storing upright, cover the top with a beaker, but do not stopper!

Transferring solution into a volumetric flask

Transferring solution with a pipet — When a pipet is filled with the transferring solution, wipe the pipet tip with a clean tissue. Touch the tip of the pipet to the inside of the flask below the mark and allow the pipet to drain. Remove the pipet from the side of the container with a rotating motion to completely remove any drops on the tip. Never blow out the small quantity of liquid inside the tip; the pipet has been calibrated to account for this. Then fill the flask as directed below.

Transferring solution from a beaker — If the transferring solution is in a beaker, pour the solution from the beaker using a stirring rod (as shown in Figure 8.1, Chapter 8) through a funnel placed in the neck of the flask. When the transfer is apparently complete, rinse

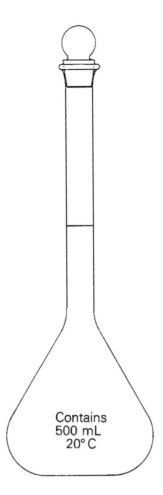

Contains
500 mL
20° C

Figure 4.3
Volumetric flask. The flask contains the stated volume when filled. Volumetric flasks are never used for measuring out solutions into other containers. They are used to make a solution an accurately known volume. Most volumetric flasks have ground-glass or polyethylene stoppers, screw caps, or plastic snap caps.

the sides of the beaker and the tip of the stirring rod with laboratory pure water, and fill up the flask as described below.

Filling up volumetric flasks — Dilute the solution close to the mark, swirl the flask briefly, and add the solvent dropwise with a dropper or Pasteur pipet so the bottom of the meniscus is assigned with the mark, as discussed previously. Stopper the flask well, invert, and mix thoroughly.

Transferring solids into a volumetric flask

When heating is not required — Insert a powder funnel into the flask and slowly pour the solid from the weighing container or paper as much as possible down the neck of the funnel. Rinse the funnel four times with distilled water before removing it, then rinse the neck of the flask to wash any solid into the flask. Shake well and proceed as described in the section above on transferring solution into a volumetric flask.

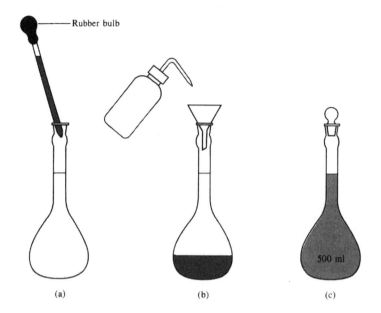

Rubber bulb

(a)   (b)   (c)

Figure 4.4
Correct use of volumetric flask. Transfer solution with a pipet, or transfer the weighed solid into the volumetric flask. Water is added and the solute is brought completely into solution before water is allowed to enter the narrow neck of the flask. Swirl the flask briefly and dilute close to the mark. Add solvent dropwise with a dropper to bring the level of the solution to the bottom of the meniscus. Stopper the flask well, invert several times, and mix the content thoroughly.

When heating is needed — Transfer the solid quantitatively to a beaker and heat to dissolve. Cool to almost room temperature and transfer to the flask dissolve in DI water, and fill up with DI water to the mark as explained previously in the section above on transferring solution into a volumetric flask (see Figure 4.4).

   Note: Do not leave solution in a volumetric flask. After preparation, transfer solution immediately into a completely dry reagent bottle. If the bottle is not dry, rinse four times with the prepared solution before transferring. Label properly. (See Figure 5.7.)

4.2.2   Pipets and Their Use

   Pipets are designed for the transfer of known volumes of liquid from one container to another. Pipets which deliver a fixed volume are called *volumetric* or *transfer pipets*. Other pipets, known as *measuring pipets*, are calibrated in convenient units so that any volume up to maximum capacity can be delivered. The most common style of pipet has the letters "TD" (to deliver) imprinted on them.

Volumetric pipet — Volumetric pipets, like volumetric flasks, have a single calibration line. They deliver even volumes, as 5, 10, 20, ..., 100 mL, etc. as labeled. Volumes from 0.5 to 100 mL may be handled with these pipets. Volumetric pipets labeled as class A have a certain time in seconds, imprinted near the top, which is the length of time necessary to wait before terminating the delivery of the liquid by the pipet. The correct use for a volumetric pipet is outlined in Figure 4.5. *Remember:* Always use a rubber pipet bulb to fill the pipet; do not use the mouth for suction!
   In any use of the volumetric pipet, the following points are important.

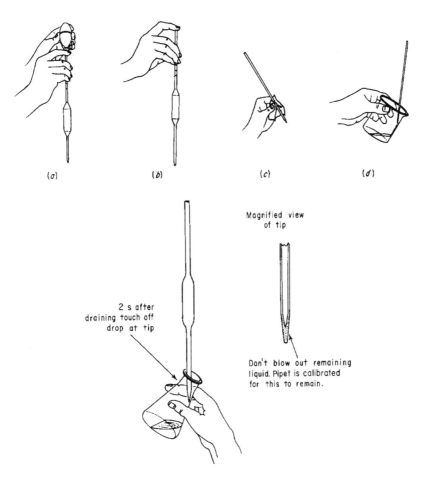

Figure 4.5
Technique for using a volumetric pipet. (a) Draw the liquid up in the pipet using a pipet bulb until past the graduation mark. (b) Use forefinger to maintain liquid level above the graduation mark. Release pressure on the index finger to allow the meniscus to approach the calibration mark. At the mark, apply pressure to stop the liquid flow. (c) Tilt pipet slightly and wipe away any drops on the outside surface. (d) Allow pipet to drain freely. Remove the last drop by touching the wall of the container. Do not blow out pipet.

- Rinse the pipet with distilled water before using. Next, rinse the pipet with the solution to be pipetted, so that the water adhering to the inside of the pipet will not dilute the solution.

- The solution to be transferred is drawn up into the pipet using a pipet bulb. When the level of the solution has risen above the calibration line, the bulb is quickly removed from the pipet and replaced with the index finger to stop the outflow.

- Using the pressure of the index finger, adjust the bottom of the meniscus to coincide with the calibration line. There should be no air bubbles anywhere in the pipet!

- The tip of the pipet is removed from the solution and wiped with a **CLEAN** laboratory tissue (care must be exercised, as the tissue can be a good source of contamination).

- The tip of the pipet is placed into the receiving vessel, and the fingers released, and the solution drained into the container.

- When the solution has completely drained from the pipet, touch the tip against the container's wall so that the last drop of the solution will drain out. Class A volumetric pipets have a certain time in seconds imprinted near the top which is the time that should be allowed to elapse from the time the finger is released until the pipet removed. After this specified time, remove the pipet from the side of the container with a rotating motion to completely remove any drops on the tip.

- A small drop of liquid will remain in the pipet at this point. This remaining liquid is never "blown out"; the pipet has been calibrated to account for this.

Measuring pipets — Pipets that have graduation lines, much like a buret, are called measuring pipets. They are two types of measuring pipets — the *Mohr pipet* and the *serological pipet*. The difference is whether or not the calibration lines stop short of the tip (Mohr pipet) or go all the way to the tip (serological pipet). The serological pipet is better in the sense that the meniscus need be read only once, since the solution can be allowed to drain completely out. In this case, the last drop of the solution is blown out with the pipet bulb. With the Mohr pipet the meniscus must be read twice — once before the delivery begins and again after the delivery is complete. A double or single frosted ring circumscribing the top of the pipet (above the top graduation line) indicates the pipet is calibrated for blow-out. *Disposable pipets* are most often serological "blow-out" type. Disposable pipets are less expensive, because the calibration lines are not permanently affixed to the outside wall of the pipet.

Pipets calibrated "TC" are called *lambda pipets*. This type of pipet is designed to transfer unusually viscous solutions, such as syrups, blood, etc. By transferring these solutions, a thin film remaining inside would represent a significant nontransferred volume. The calibration line is affixed at the factory so that every trace of solution contained within is transferred by flushing the solution out with a suitable solvent. Thus, the pipetted volume is contained within and then quantitatively flushed out.

For extremely small volumes, the so-called *pipettors* are used with various design. These typically handle volumes from 1 µL to 500 µL. Different types of pipets are shown in Figure 4.6.

4.2.3 Burets and Their Use

Burets are calibrated to deliver variable volumes of liquid. The buret is a sort of specialized graduated cylinder. The difference is that the buret has a stopcock at the bottom, which is a valve that can be opened to release solution into a receiving flask. The zero line is at the top of the cylinder, and as the solution is dispensed via stopcock, the volume is easily determined by observing the position of the meniscus. The essential parts of the buret are shown in Figure 4.7. Most burets are equipped with a glass stopcock lubricated with hydrocarbon grease or with a Teflon plastic stopcock, which requires no lubrication. There are a number of greases on the market for lubricating the stopcock of the buret. The amount of grease used should be carefully limited so that excess grease does not pass through the stopcock and plug the tip of the buret. Any material stuck in a buret tip can be usually cleaned with a fine wire inserted from the bottom when the stopcock is open and the buret full of solution. Teflon stopcocks are free from this problem, but Teflon can become deformed and this can cause leakage. Burets of the conventional type must be manually filled. *Automatic filling burets* are filled by pumping a rubber bulb; such burets are used where many titrations with the same solution are to be made. For more accurate work, *Schellbach burets* are employed. These have a white background with a blue stripe and can be read at the point of least magnification. Different types of burets are shown in Figure 4.8.

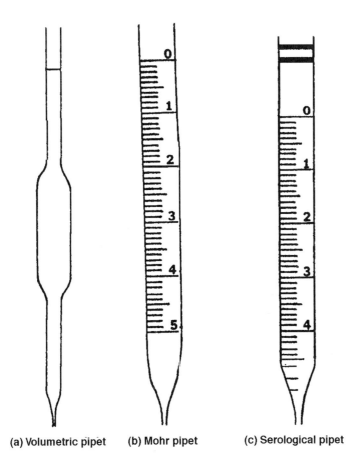

(a) Volumetric pipet   (b) Mohr pipet   (c) Serological pipet

Figure 4.6
Different types of pipets. Volumetric pipet (a) delivers a fixed volume. Like a volumetric flask, it has a single calibration line. Volumetric pipets are not calibrated to "blow out." Measuring pipets (b and c) have graduation lines much like a buret and are used for measuring volumes of solutions more accurately than could be done with graduated cylinders. There are two types of measuring pipets — the Mohr pipet (b) and the serological pipet (c). The difference is whether or not the calibration lines stop short of the tip (Mohr pipet) or go all the way to the tip (serological pipet). The serological pipet is better in the sense that the meniscus need be read only once, since the solution can be allowed to drain completely out. A double or single frosted ring circumscribing the top of the pipet indicates the pipet is calibrated to "blow out." Disposable pipets are most often of the serological "blow out" type.

Before using a buret, it must be rinsed with the titrant solution to remove water (from washing) adhering to the inside of the buret. Leave the stopcock open while rinsing. Close the stopcock, and fill the buret with titrant, and adjust to zero. Examine the buret tip carefully for bubbles, or incomplete filling. To remove bubbles, drain rapidly and refill the buret. Be sure your eye is at the same level at the meniscus. After finishing titration, remove all the titrant from the buret; be especially sure not to let any alkali solution stay in it for a long time. Such solutions can slowly dissolve glass and can freeze the stopcock. After using the buret, rinse several times with distilled water and cap it to keep dust out, or store in the upside-down position. When using not aqueous solution in the buret, rinse with acetone and store dry, upside down. Automatic burets should be emptied after titration is completed.

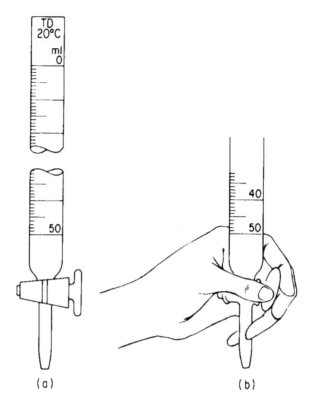

Figure 4.7
(a) Buret and (b) method of grasping stopcock.

4.2.4   Cleaning Volumetric Glassware

Laboratory glassware, especially volumetric glassware, must be kept spotlessly clean. Dirty glassware may destroy analytical work, and may cause false results.

Glassware must be soaked first in a 2% hot detergent solution for about 10 to 15 minutes. A longer soaking period is not recommended, because the detergent may cause a rough surface to develop on the glassware that may prevent uniform wetting. Do not use household detergent! Commercially available laboratory detergent, such as Alconox, Liquinox, etc. are available for laboratory glassware washing. For stronger cleaning, when needed, a commercial substitute of dichromate–sulfuric acid cleaning solutions such as Nichromix may be used. These have strong oxidizing agents that are added to concentrated sulfuric acid. Such substitutes do not have the toxicity of chromium and hence are safer to use. After 1 to 2 hours of soaking, the glassware should be rinsed thoroughly to remove all traces of concentrated sulfuric acid. Other treatment of glassware for special tests is discussed in the appropriate methodologies. Proper storage of glassware is also important to protect it from contamination, dust, breaking, and chipping.

4.3 THE ANALYTICAL BALANCE

Balances are mechanical devices used to determine the mass of objects. Analytical balances are delicate instruments; therefore comprehensive rules should be observed in caring for and using them.

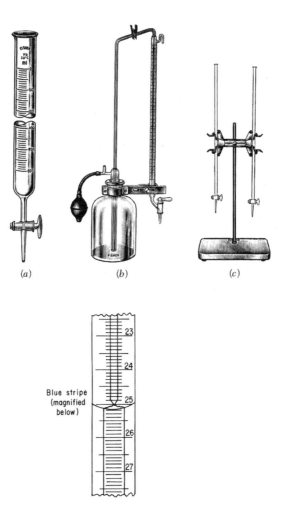

Figure 4.8
Different types of burets. (a) Single-dispensing buret with graduated etched scale, standard taper, and stopcock. (b) Automatic filling buret. Pumping the rubber bulb fills it to a precise 0.01 mL graduation, and overfill automatically returns to storage. Such burets are used when many titrations with the same solution are to be made. (c) Titration assembly and stand. (d) Schellbach burets are employed for more accurate work. They have a white background with a blue stripe and can be read at the point of least magnification.

1. Level the balance. With most balances the leveling method is built in, complete with a leveling bubble and vertically positionable legs.

2. Check the balance zero. The zero point changes if anything is spilled on the pan or if the doors are open. It drifts as a result of temperature changes as well.

3. Place the object to be weighed on the center of the pan.

4. Be certain the beam is locked before removing or changing weights or objects to be weighed.

5. Keep the balance scrupulously clean. Weigh chemicals in a weighing bottle or scoop or glazed weighing paper. Never place any moist objects or any chemicals directly on the pan. Remove any spilled chemicals immediately with a soft brush.

6. Work in front of the balance to avoid parallel errors.

7. Handle all weights and objects with forceps, never with fingers. The object placed on the pan must be kept free of fingerprints, or other interfering substances that could add weight and give an erroneous result.

Figure 4.9
Desiccator. The desiccator cover should slide smoothly on its ground-glass surface. This surface should be lightly greased with a light lubricant, such as petroleum jelly or silicon grease (never stopcock grease!).

8. The sample to be weighed should be brought to room temperature!
9. Release the locking mechanism slowly, avoiding jars.
10. Do not overload the balance.
11. After weighing, return all weights to zero, and make sure that the pan and the floor of the chamber are free from spilled chemicals, dust, or other materials. Be sure that both the right and the left sliding doors are closed.

4.4 DESICCATORS

A desiccator is a container, usually of heavy glass, plastic or, occasionally, metal used to keep samples and crucibles protected from the laboratory environment, and especially from moisture. The desiccator cover should slide smoothly on its ground-glass surface. This surface should be lightly greased with a light lubricant such as Vaseline (never stopcock grease!). A typical desiccator is shown in Figure 4.9. *Vacuum desiccators* are equipped with side arms, so that they may be connected to a vacuum and the contents will be subject to a vacuum rather than the dried air. A vacuum-type desiccator should be used to dry crystals that are wet with organic solvent and should not be used for substances that sublime readily.

A quantity of water-absorbing chemical, called *desiccant*, is placed in the bottom of the container and will absorb all the moisture inside the sealed container, thus providing a dry environment. A good commercially available desiccant is DRIERIT (anhydrous $CaSO_4$). It can be purchased in "indicating" form: when the original blue color of the material has changed to pink, it is saturated with moisture. It easily can be recharged by heating in an oven. Needless to say, the desiccator should be scrupulously clean and should never contain exhausted desiccant.

Chapter 5

CHEMICALS, REAGENTS, AND STANDARDS

5.1 CHEMICALS

Chemicals are manufactured in varying degrees of purity. Carefully select the degree of purity that meets the need of the work to be done. Always recheck the label of the chemical in use. Using the wrong chemical can ruin the analytical work and can cause explosion, fire, or an accident in the laboratory. Check all of the information on the chemical's label carefully! It contains the name, formula, formula weight, percentage of impurities, analytical grade, health hazards, and safety codes. (See Section 2.3.2).

5.1.1 Grades and Purity of Chemicals

Chemicals are graded according to purity as follows.

- *AR Primary Standard:* A specially manufactured analytical reagent of exceptional purity for standardizing volumetric solutions and preparing reference standards.
- *AR:* Analytical reagent grade for all general laboratory work.
- *ACS:* The chemical meets the requirements of the American Chemical Society Committee on Analytical Reagents.
- *USP:* A grade meeting the requirements of the U.S. Pharmacopeia.
- *NF:* The grade meeting the requirements of the National Formulary.
- *TAC/FCC:* Tested additive chemical/food chemical codex. Meets the requirement for food chemical codex and satisfactory for approved food uses.
- *Technical:* A grade suitable for general industrial uses.
- *AR Select:* High-purity acids for trace element analysis.
- *OR:* Organic laboratory chemicals of suitable purity for most research work and for most general laboratory purposes. It is the highest grade of the particular chemical generally available.
- *Certified:* Applies to stains certified by the Biological Stain Commission and bears their label of certification.
- *ChromAR:* Solvents specially purified for use of chromatography.
- *GenAR:* Used for biotechnology or genetic laboratories.

- *Nanograde*: Specially controlled for electron captured gas chromatographic (GC) techniques, such as pesticide residue analysis, etc.
- *ScintillAR*: Used in liquid scintillometry.
- *SilicAR*: Used for column and thin layer chromatography.
- *SpectrAR*: Used for Spectrophotometry.
- *StandAR*: A line of prepared solutions including various titrants and atomic absorption (AA) standards.
- *Mercury-free chemicals*: Used for mercury determination.
- *Nitrogen-free chemicals*: Used for nitrogen compounds analysis.

5.1.2 Proper Storage of Chemicals and Standards

TABLE 5.1

Storage of Chemicals, Reagents, and Standards

Chemical	Method of Storage
Acids (HCl, H_2SO_4, HNO_3, acetic acid)	Stored in original containers in specially designed and "Acids"-labeled cabinet, grouping by safety color code bottles
Flammable solvents	Stored in original containers, in specially designed and "Flammable"-labeled cabinet. Large quantities should be stored in metal safety cans, and outside the laboratory in marked "Flammable" storage area.
Solvents	Stored in separate solvent cabinet, in original containers, and in a well-ventilated area.
Chemicals used for analysis of VOCs	Stored in separate solvent cabinet, in original containers. No other chemicals stored in this area.
Chemicals	Stored in alphabetical order in chemical storage room, with records of "date of arrival" and "date opened" on each container.
Phenol Hydrogen peroxide (H_2O_2)	Stored in "Chemical Storage"-marked refrigerator in closely capped and sealed containers.
Turbidity standard, ammonia, nitrate, nitrite, phosphorus stock, and standard solutions	Stored in refrigerator, marked as "Refrigerator for Inorganics."
Silica stock solution	Stored in plastic bottle.
Oil and grease standard	Refrigerate in sealed containers, refrigerator marked as "Refrigerator for Organics."
Stocks and standards for trace organics	Stored in vials in specially marked freezer.
Metals, stock solutions	Stored at room temperature.
Metals, standards (100 to 10 ppm)	Stored at room temperature (each contains 0.5% acid as described in the method) in designated storage area in the laboratory.
pH, conductivity standards	Stored at room temperature in properly marked cabinet.
Microbiology (sample, media, reagents)	Stored in separate, marked refrigerator in the microbiology laboratory.

Note: All stocks and standards have to be marked by the date received or prepared, expiration date, and signature.

5.2 LABORATORY PURE WATER

One of the most important aspects of analysis is the quality of the laboratory pure water to be used for preparation of standard solutions, reagents, dilutions, and blanks. The American Society for Testing and Materials (ASTM) specifies different grades of laboratory pure water.

Type I water — It has no detectable concentration of the compound or element to be analyzed at the detection limit of the analytical method. Use Type I water in test methods requiring minimum interference and bias. It is prepared by distillation, deionization, or reverse osmosis treatment of feedwater followed by polishing with a mixed bed deionizer and passing through a 0.2 μm poresize membrane filter. A mixed bed deionizer typically adds small amounts of organic matter to the water. The regular quality check of the water by measurement of conductivity does not show organic, nonionized contaminants; therefore, the quality check of the Type I water always includes total organic carbon (TOC) analysis, silica (SiO_2) measurement, and bacterial content.

Type I water cannot be stored without significant degradation; therefore, it is best produced continuously and used immediately after processing.

Type II water — It can be used for laboratory procedures in which the presence of bacteria can be tolerated. It is produced by distillation or deionization. It may be stored, but keep storage to a minimum and provide quality consistent with the intended use. Store only in materials that protect the water from contamination, such as tetrafluoroethylene (Teflon) or glass for organic analysis, and plastic for analysis of metals or other inorganic parameters.

Type III water — Type III water may be used for glassware washing, preliminary rinsing of glassware, and as a feedwater for production of higher-grade waters. Select a storage container material to protect from contamination.

5.2.1 Methods for Preparation of Laboratory Pure Water

Distilled water — Distillation is the procedure in which the liquid is vaporized, recondensed, and collected. Distilled water quality depends on the type of still and the quality of the feedwater. Deionized feedwater is preferred. High-quality distilled water has a specific conductance less than 1 μmho/cm (which can also be expressed as microsiemens per centimeter (μS/cm) when using the SI unit for conductance), which corresponds to about 0.5 ppm of dissolved solids. Absolutely pure water has a conductivity of 0.55 μmho/cm at 25°C.

Demineralized or deionized (DI) water — This is water purified by a mixed-bed ion exchanger. Commercial resin purification trains can produce superior water quality. *Demineralizers* are used to remove mineral ions from water. Demineralizers contain several beds of resins and are packaged in cartridges; the water that is obtained this way is free from both mineral anions and cations. Demineralizers will not remove organic matter or nonelectrolyte from water. These must be removed first by distillation. Demineralized water cannot be substituted for distilled water in every case. Check all requirements!

Redistilled water — This is prepared by redistilling single distilled water from an all borosilicate-glass apparatus.

Reagent-grade water — Reagent-grade water is the highest purity that is available; it is even purer as triple distilled water. Reagent grade water is obtained from pretreated water, that is, water which has been distilled, deionized, or subjected to reverse osmosis; it is then passed through an activated carbon cartridge to remove the dissolved organic materials, through two deionizing cartridges to remove any dissolved inorganic substances,

TABLE 5.2

Reagent-Grade Water Specification

Quality Parameter	Type I	Type II	Type III
Bacteria (HPC), CFU/mL	<10	<1000	NS
pH	NS	NS	5–8
Conductivity, μmhos/cm at 25°C	<0.1	1	10
Silica (SiO_2), mg/L	<0.05	<0.1	<1
Particulate matter[a]	0.22 μm filter	NS	NS
Organic contaminants[b]	Activated carbon	NS	NS

Note: HPC = heterotrophic plate count; CFU = colony forming units; NS = not specified.

[a] Process specification; not measured by end user.
[b] Pretreatment, and possibly posttreatment for certain uses.

and through membrane filters to remove microorganisms and any particulate matter with a diameter of 0.22 μm. Specifications for the different types of laboratory pure water according to ASTI are shown in Table 5.2. The quality check of the laboratory pure water with monitoring frequency and limits is shown in Table 5.3.

5.3 REAGENTS AND SOLUTIONS

Solution is a homogeneous mixture of two or more substances, consisting of ions or molecules. When a solution is prepared, one substance is mixed with another substance in such a manner that, after mixing, only one physical state is observed. Solutions may exist in one of the three states of matter, that is, they may be gases, liquids, or solids. For example, air is a homogenous mixture of gases, soda water is a liquid solution where gas (CO_2) is dissolved in water. Ethanol in water is an example of a liquid solution of two completely miscible liquids, brine is a liquid solution of a solid (NaCl) dissolved in liquid (water), and alloy is a solution of two or more solid metals. Beer is a solution of a liquid (alcohol), a gas (carbon dioxide), and a solid (malt) that are dissolved in water.

In a solution, the substance in larger amount and whose physical state is observed is called the *solvent*. The substance in smaller amount that becomes incorporated into the solvent is called the *solute*. Solutes dissolve in solvents to yield solutions. The solubilities of substances vary widely. One substance may dissolve in one solvent but be nearly

TABLE 5.3

Quality Check of Laboratory Pure Water

Parameter	Monitoring Frequency	Limit
Conductivity	D	1–2 μmhos/cm
pH	D	5.5–7.5 unit
Total organic carbon	A	<1.0 mg/L
Trace metal single Cd, Cr, Cu, Ni, Pb, Zn	A	<0.05 mg/L
Trace metal total Cd, Cr, Cu, Ni, Pb, Zn	A	<1.0 mg/L
Ammonia, as NH_3-N	M	<0.1 mg/L
Free chlorine, Cl_2	M	<0.1 mg/L
Heterotrophic count		
Fresh water	M	<1000 cnt/mL
Stored water	M	<10,000 cnt/mL
Water suitability test	A	Ratio: 0.8–3.0

Note: A = annually; M = monthly; D = daily; cnt = count (bacterial).

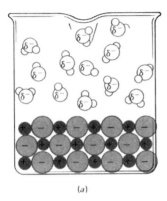 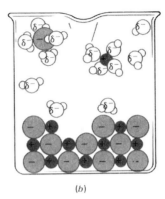

(a)   (b)

Figure 5.1

Ionic substances are soluble in water. (a) When ionic substances are dissolved in water, the very polar water molecules attract the anions and cations, weakening and then breaking the ionic bonds that hold the ions to the crystal lattice. (b) After breaking free, the ions are surrounded and bonded to water molecules, and then they diffuse away from the undissolved crystals.

insoluble in another. The general rule is "like dissolves like." The more similar two compounds are, the more likely it is that one is soluble in the other. When we say "like," we mostly mean alike in *polarity*. That is, polar compounds dissolve in polar compounds and nonpolar compounds dissolve in nonpolar compounds. Polar molecules dissolve in polar molecules because the positive end of one attracts the negative end of the other. The most important polar solvent is water, and most ionic compounds are soluble in water. Organic compounds are usually not polar molecules, so they dissolve in organic solvent and not in water. Some polar molecules such as sugars and alcohols are exceptions; they dissolve in water because they are alike in having OH^- groups through which strong hydrogen bonding attraction arises. However, as the hydrocarbon end of the molecule becomes the more prominent portion of the molecule, it becomes less like water. Now the forces of attraction between these molecules and water become weaker; therefore, the solubility of sugar and alcohol molecules in water decreases with increasing length of the hydrocarbon end. Attraction of water molecules to ions because of the ion–dipole force is shown in Figure 5.1. Ionic substances (for example, NaCl) dissolved in water yield an electric-conducting solution, called an *electrolyte*. A substance such as sucrose or table sugar that dissolves in water to yield a nonconducting or very poorly conducting solution is called a *nonelectrolyte*.

Solubility — Solubility is the measure of the amount of solute that dissolves in a given amount of solvent at a specific temperature. Generally, in analytical work, solubility is measured by finding the amount of solute that dissolves in 100 g of solvent. Three factors are most significant in predicting solubility: the nature of the solvent and solute, temperature, and pressure. For example, the solubility of NaCl is 36 g/100 mL at 20°C.

Solubility rules for common inorganic compounds

- NO_3^- All nitrates are soluble.
- $C_2H_3O_2^-$ All acetates are soluble. Silver acetate, $AgC_2H_3O_2$, is moderately soluble.
- Cl^- All chlorides are soluble, except AgCl, $PbCl_2$, and HgCl. $PbCl_2$ is soluble in hot water, but slightly soluble in cold water

- SO_4^{2-} All sulfates are soluble, except $BaSO_4$ and $PbSO_4$. Ag_2SO_4, Hg_2SO_4, and $CaSO_4$ are slightly soluble.

- CO_3^{2-}, PO_4^{3-}, CrO_4^{2-}, SiO_4^{2-} All carbonates, phosphates, chromates, and silicates are insoluble, except those of sodium, potassium and ammonium, and $MgCrO_4$.

- OH^- All hydroxides are insoluble, except sodium, potassium, and ammonium hydroxides.

- S^{2-} All sulfides are insoluble, except sodium, potassium, ammonium, magnesium, calcium, and barium sulfides. Aluminum and chromium sulfides are hydrolyzed and precipitate as hydroxides.

- Na^+, K^+, NH^+ All sodium, potassium, and ammonium salts are soluble.

- Ag^+ All silver salts are insoluble, except $AgNO_3$.

Saturated solution — Saturated solution is in equilibrium with respect to a given dissolved substance; for example, 36 g NaCl in 100 mL H_2O.

Unsaturated solution — Unsaturated solution is not in equilibrium with respect to a given dissolved substance, and more of the substance can dissolve; for example, 30 g NaCl in 100 mL H_2O.

Supersaturated solution — Supersaturated solution contains more dissolved substance than a saturated solution; for example, 40 g NaCl in 100 mL H_2O. The additional 4.0 g NaCl remains undissolved.

The effect of temperature on solubility — Generally, an increase in temperature is accompanied by an increase in the solubility. At higher temperatures, larger masses of solutes dissolve in a fixed mass of water than at lower temperature (Figure 5.2). However, this is not always the case. For a small number of ionic compounds, a decrease in solubility is observed with increasing temperature. Calcium sulfate, $CaSO_4$, and calcium hydroxide, $Ca(OH)_2$, are common examples. Cerium selenate, $Ce_2(SeO_4)_3$, and cerium sulfate non-ahydrate, $Ce(_2SO_4)_3 \cdot 9H_2O$, are very soluble at 0°C but much less soluble at higher temperature. Most gases become less soluble in water at higher temperature. Figure 5.3 shows a graph of temperature versus the solubility of selected solutes in water. Other substances, including sodium chloride (NaCl), have a relatively constant solubility with increasing temperature.

Heat can be released or absorbed when ionic substances are dissolved in water. In some cases the released heat is very noticeable, and should be considered when addressing safety in laboratory work. For example, when sodium hydroxide, NaOH, is dissolved in water, the solution becomes very hot (the solution process is called *exothermic*). The safety rule for diluting sulfuric acid, H_2SO_4, is to add the acid slowly to the water, and never add the water to the acid, because the acid may splatter from the exothermic reaction. On the other hand, when ammonium nitrate, NH_4NO_3, is dissolved in water, the solution becomes very cold (the solution process is *endothermic*).

5.3.1 Expressing Concentration of a Solution

The concentration of a solution is the amount of solute dissolved in a given quantity of solvent. Several different ways can be used to express the concentration of a solution.

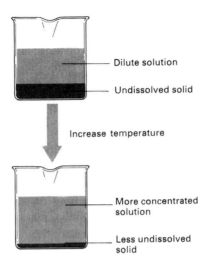

Figure 5.2
Temperature influences solubility. For many solutes, as the temperature of the solvent increases, the solubility of the solute increases. For solutes that give color to a solution, the increased solubility is recognized by a more intense color of the solute in the solution.

Percent (%) concentrations

Weight/volume (w/v) % solution — This is the mass of solute contained by 100 mL solvent. For example; 3% w/v NaCl solution contains 3 g of NaCl in 100 mL of solution.

Weight/weight (w/w) % solution — This is the mass of the solute contained in 100 g of solvent. For example 3% w/w NaCl solution contains 3 g NaCl in 100 g of solution (3 g NaCl + 97 g H_2O).

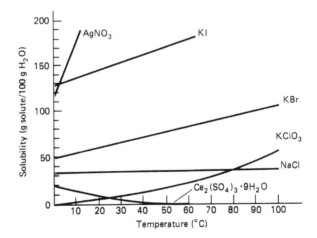

Figure 5.3
Temperature versus solubility of selected solutes in water. The solubility of a substance in water is usually measured in terms of the number of grams of solute that dissolve per 100 g of water. A solubility curve normally shows the solubility of a substance from 0 to 100°C. The solubilities of most ionic solids rise with increasing temperature. However, the solubility of sodium chloride (NaCl) increases only slightly and the solubility of cerium selenate ($Ce_2(Se)_4)_3$) falls with increasing temperature.

Volume/volume (v/v) % solution — This is the volume of the solute in 100 mL solution. For example, 5% v/v acetic acid (CH_3COOH) solution (vinegar) contains 5 mL of concentrated (conc.) acetic acid in 100 mL solution (5 mL acetic acid + 95 mL H_2O).

Molar concentration, molarity (M) — The molarity of the solution is the moles of solute that are in 1 liter of solution. For example, 1 *M* NaOH solution contains 1 mol or 40 g NaOH in 1 L of solution. Accordingly, 0.1 *M* NaOH contains 4.0 g NaOH in 1 L solution, and 0.01 *M* solution, 0.40 g NaOH in 1 L solution.

A mole (abbreviated mol) is a counting unit that allows us to keep track of the number and the mass of atoms, molecules, and ions. A mol is the formula weight of a molecule, element, or ion expressed in grams. For example, 1 mol of H_2O is 18.01 g, 1 mol of Na is 22.99 g, 1 mol of Cl is 35.45 g, 1 mol of Cl_2 is 70.90 g, 1 mol of Na_2SO_4 is 142.04 g, etc.

Molal concentration, molality (m) — The molality of a solution is the number of moles per 1 kg of solvent. For example: 2 *m* glucose ($C_6H_{12}O_6$) solution contains $2 \times 180 = 360$ g glucose and 640 g (1000 g – 360 g) H_2O, and 0.2 *m* glucose solution is $0.2 \times 180 = 36$ g glucose and 964 g H_2O.

Normal concentration, normality (N) — The normality of a solution is the number of equivalents of a substance dissolved in 1 L of solution. One *equivalent weight* depends on the reaction a substance undergoes.

In acid–base reactions — One equivalent is the number of grams of a substance that supplies or combines with 1 mol of H^+. Equivalent weight of any substance is the molecular weight divided by the number of H^+ (acids) or by the number of OH^- (bases).

In oxidation–reduction reaction — One equivalent is the number of grams of a substance that supplies or combines with 1 mol of electrons. Equivalent weight of any substance is the formula weight of substance divided by the electrons gained or lost in the reaction.

In precipitation and in complex formation reactions — One equivalent is the number of grams of a substance that supplies 1 mol of 1^+ ions, or the amount that reacts with 1 mol of 1^+ ions. Equivalent weight is the molecular weight divided by the charge on the ion.

CALCULATION OF EQUIVALENT WEIGHTS

$$\text{Equivalent weight of } \textit{elements} = \text{atomic weight/valency} \qquad (5.1)$$

$$\text{Equivalent weight of } \textit{acids} = \text{molecular weight/number of } H^+ \qquad (5.2)$$

$$\text{Equivalent weight of } \textit{bases} = \text{molecular weight/number of } OH^- \qquad (5.3)$$

$$\text{Equivalent weight of } \textit{salts} = \text{molecular weight/valency of the metal ion} \qquad (5.4)$$

$$\text{Equivalent weight of } \textit{any substance in oxidation–reduction} = \\ \text{formula weight of the substance/electrons gained or lost} \qquad (5.5)$$

EXAMPLES for calculating equivalent weight —

$$\text{Eqw HCl} = 36/1 = 36 \text{ (acid)}$$

$$\text{Eqw } H_2SO_4 = 98/2 = 49 \text{ (acid)}$$

$$\text{Eqw } H_3PO_4 = 98/3 = 32.6 \text{ (acid)}$$

$$\text{Eqw NaOH} = 40/1 = 40 \text{ (base)}$$

$$\text{Eqw } Ca(OH)_2 = 74/2 = 37 \text{ (base)}$$

$$\text{Eqw } Hg(NO_3)_2 = 324/2 = 162 \text{ (salt)}$$

$$\text{Eqw } KMnO_4 \ (+ 5 \ e^-) = 158/5 = 31.6 \text{ (oxidation–reduction)}$$

$$\text{Eqw } Fe^{2+} \rightarrow Fe^{3+} \ (1 \ e^-) = 56/1 = 56 \text{ (oxidation–reduction)}$$

$$\text{Eqw } Sn^{2+} \rightarrow Sn^{4+} \ (2 \ e^-) = 118/2 = 59 \text{ (oxidation–reduction)}$$

$$\text{Eqw } Cr_2O_7^{2-} \rightarrow 2 \ Cr^{3+} = 224/6 = 37.3 \text{ (oxidation–reduction)}$$

Milliequivalent per liter (meq/L) — The unit milliequivalent (Eq/1000) is commonly used in place of equivalent. Milligram equivalent per liter or milliequivalent per liter (meq/L) can be valuable for making water treatment calculations and checking analysis by anion–cation balance.

Parts per thousands (ppt) — This is also expressed as grams per kilogram (g/kg) or ‰. A 1 ppt solution contains one part of solute in 1000 g solvent. Commonly used to express salinity of certain industrial wastes, natural waters, and seawater.

Terms associated with salinity are *chlorinity* (g/L chloride at 20°C), and *chlorosity*, which is the chlorinity multiplied by the water density measured at 20°C.

$$\text{Salinity \textperthousand} = 0.03 + (1.805 \times Cl \text{ g/L at } 20°C \times \text{density})$$

In recent years, the conductivity and density (hydrometric measurement) methods have been used, because of their high sensitivity and precision. Although conductivity has the greatest precision, it responds only to ionic solutes. Density is less precise, but responds to all dissolved solutes. The densities and the corresponding salinities will be given in table form in Appendix B.

Parts per million (ppm) — This is also expressed as milligrams per liter (mg/L) or milligrams per kilogram (mg/kg). This unit is most commonly used in chemical analysis to express concentration. This unit was devised by Lord Kelvin during the British beer poisoning epidemic in 1900. Kelvin headed a Royal Commission that attempted to set the world's first measurement of tolerance to a poison (arsenic) in the unit of ppm. When dealing with solids, the ppm must be used in terms of milligrams constituent per kilogram of solid. If the result is higher than 10,000 ppm, the concentration is expressed as a percentage. If the concentration is lower than 0.1 ppm, it must be expressed in parts per billion, or ppb.

Parts per billion (ppb) — This is also expressed as micrograms per liter (μg/L) or micrograms per kilograms (μg/kg). It is the same as ppm, but instead of milligrams of solute, we use microgram (μg) of solute (1 mg = 1000 μg).

Concentrated (conc) solutions — In commercially available solutions, the term concentrated refers to the maximum concentration available. For example, concentrated aqueous ammonia solution contains about 28% NH_3 by mass, or concentrated hydrochloric acid is 37%. Table 5.4 summarizes the concentration units and the preparation of solutions. Preparation of standard acids is shown in Table 5.5. Preparation of the alkaline solutions sodium hydroxide (NaOH) and ammonium hydroxide (NH_4OH) is shown in Tables 5.6 and 5.7, respectively.

5.3.2   Diluting Solutions

The preparation of a lower concentration of solution from a higher concentration of solution is frequently accomplished in laboratory work by using the following formula:

$$C_1V_1 = C_2V_2 \qquad\qquad (5.6)$$

where C_1 = higher concentration, C_2 = desired lower concentration, V_1 = volume needed from the higher-concentration solution, and V_2 = desired volume of the lower-concentration solution. For example, there is a need to prepare 500 mL of 0.012 N NaCl solution by dilution from 0.1 N NaCl solution. Using formula (5.6), with C_1 = 0.10 N, V_1 = ?, C_2 = 0.012 N, and V_2 = 500 mL,

$$0.10V_1 = 0.012 \times 500$$

$$V_1 = (0.012 \times 500)/0.10$$

$$V_1 = 60$$

So, 60 mL of 0.1 N NaCl solution must dilute to 500 mL to prepare 0.012 N NaCl solution.

Exercises (Answers begin on Page 65)

1. How many grams of sodium chloride (NaCl) are needed to prepare 500 mL of 0.2% NaCl solution?
2. How many grams of copper sulfate ($CuSO_4$) are needed to prepare 1 L of 2% $CuSO_4$ solution?
3. How many grams of copper sulfate pentahydrate ($CuSO_4 \cdot 5\ H_2O$) are needed to prepare 1 L of 2% $CuSO_4$ solution?
4. How many grams of sodium hydroxide (NaOH) are needed to prepare 250 mL of 0.1 M NaOH solution?
5. How many grams of $Ca(OH)_2$ are needed to prepare 2 L of 0.25 N solution?
6. How many milliliters of concentrated sulfuric acid (H_2SO_4) are needed to prepare 1 L of 0.1 M H_2SO_4 solution? The density of the concentrated H_2SO_4 is 1.834 g/mL.
7. How many milliliters of concentrated H_2SO_4 is needed for preparation of 1 L 0.1 N solution? Density of the acid is 1.934 g/mL.
8. How many milliliters of 0.1 N H_2SO_4 should be diluted to 1 L to prepare 0.02 N H_2SO_4 solution?

5.3.3   Handling Reagents and Solutions

General rules — For successful analytical work, the availability of reagents and solutions of established purity is of prime importance. Following the rules below will successfully prevent contamination of the reagents and help to protect analytical results.

TABLE 5.4

Concentration Units of Solutions

Unit	Symbol	Definition
Percentage by mass	% w/w	(g of solute/g of solution) $\times$ 100
Percentage by mass volume	% w/v	(g of solute/mL of solution) $\times$ 100
Percentage by volume	% v/v	(mL of solute/mL of solution) $\times$ 100
Molarity	M	mol of solute/1 L
Molality	m	mol of solute/1 kg
Normality	N	equivalent of solute/1 L
Parts per thousand	ppt	g/L or g/kg
Parts per million	ppm	mg/L or mg/kg
Parts per billion	ppb	μg/L or μg/kg

TABLE 5.5

Preparation of Common Acid Solutions

Desired Concentration	HCl	H_2SO_4	HNO_3
Specific gravity at 20°C	1.174–1.189	1.834–1.836	1.409–1.1418
Percentage of active ingredient in conc reagent	36–37	96–98	69–70
Normality of conc reagent	11–12	36	15–16
Volume (mL) of conc reagent to prepare 1 L of			
18 N	—	500	—
6 N	500	167	380
1 N	83	28	64
0.1 N	8.3	2.8	6.4
Volume (mL) of 6 N reagent to prepare 1 L of 0.1 N	17	17	17
Volume (mL) of 1 N reagent to prepare 1 L of 0.02 N	20	20	20

TABLE 5.6

Preparation of Standard Alkaline Solution: Sodium Hydroxide (NaOH)

Normality of NaOH Solution	Weight of NaOH to Prepare 1 L Solution (g)	Volume of 15 N NaOH to Prepare 1 L Solution (mL)
15	625	—
6	240	400
1	40	67
0.1	4	6.4

Note: For preparation of solutions, use CO_2-free distilled water, by boiling distilled water for 15 minutes and cooling rapidly to room temperature. Cap flask (in which water boiled) with a slightly oversized inverted beaker, to minimize entry of atmospheric CO_2 during this cooling process. Store NaOH solutions in polyethylene (rigid, heavy-type) bottles with polyethylene screw caps or paraffin-coated bottles with rubber or neoprene stoppers. Check the normality of the solutions periodically.

TABLE 5.7

Preparation of Standard Alkaline Solution: Ammonium Hydroxide (NH_4OH)

Desired Concentration	NH_4OH, conc
Specific gravity at 20°C	0.90
Percentage of active ingredient	29
Normality of conc reagent	15
Volume (mL) of conc reagent to prepare 1 L of	
5 N	333
3 N	200
0.2 N	13

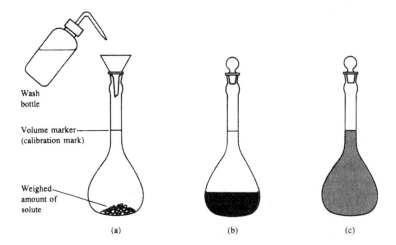

Figure 5.4
Steps involved in the preparation of standard solutions. (a) Insert a powder funnel into the flask and slowly pour the weighed amount of solute down the neck of the funnel. Rinse the funnel four times with DI water before removing it. Then rinse the neck of the flask with DI water to wash any solid into the body of the flask. Shake well. (b) Add a small quantity of DI water and dissolve the solid by gently swirling the flask (with the stopper in place). (c) Without contacting any surface above the mark, add DI water near to the mark, then add dropwise to fill exactly to the mark. With flasks of 100 mL or less, it may be desirable to use a Pasteur pipet or a long-barreled medicine dropper to match the bottom of the meniscus to the mark. Stopper the flask well, invert, and shake to mix thoroughly.

- Select the best available grade of chemicals suitable for the analytical work.
- Replace the top of the container immediately after removal of the chemical or reagent.
- Stoppers should never be set on the desk top. Hold between the fingers, or place in a clean paper towel or laboratory tissue paper laying the stopper upside down.
- Never return any excess reagent or solution to the original bottle to prevent the risk of contaminating the entire bottle.
- Do not insert pipets into a reagent bottle. Instead, shake the bottle with the cap in place, then pour out the desired volume.
- Keep the reagent shelf and laboratory balances clean! Immediately clean up any chemical spill!
- Always wear protective safety glasses, aprons, shoes, etc. when handling chemicals.
- Always clean the outside of acid bottles with water and dry them well before using. Wet bottles are slippery!

Pouring liquids from bottles and other containers — The correct procedures for pouring liquids from bottles, beakers, and other containers are illustrated in Figure 5.5.

Transferring solutions by pipet — The accepted method for transferring solution by pipet is shown in Figure 5.6.

Proper labeling of solutions and reagents — Prepared solutions and reagents should be properly labeled. The label should contain the name and concentration of the solution, date of preparation or date received, date container was opened, and the initials of the scientist who prepared the solution or opened the container. Special storage recommendations (for example, store in refrigerator, store in dark, etc.) and preparation frequency

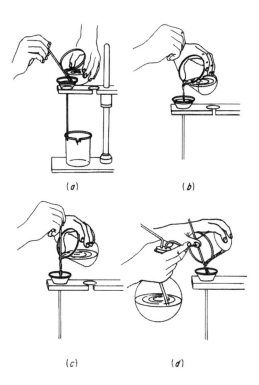

(a)   (b)

(c)   (d)

Figure 5.5
Pouring liquids from beakers and other containers. Use a stirring rod to guide liquid. Hold the stirring rod against the lip of the beaker. Tilt the container, allowing liquid to flow around the stirring rod which guides the liquid to the receiver. When the desired amount of liquid has been poured, position the pouring beaker vertically, allowing the last liquid to drain off the lip and down the rod.

are also important parts of a reagent bottle label. Label formats for reagents and solutions are shown in Figure 5.7. *CAUTIONS:* Mix thoroughly and completely when making solutions or dilutions; the most common errors are caused by failure to mix completely. Always read the label twice before using any solution!

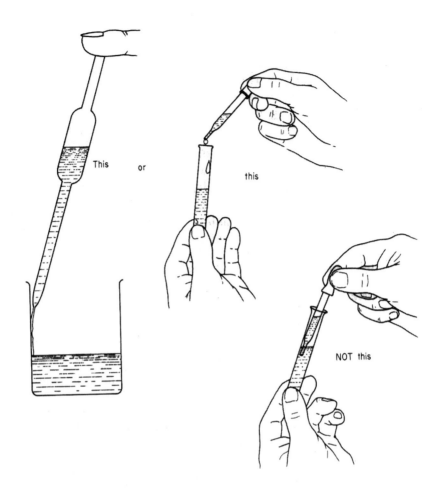

Figure 5.6
Transferring solutions into containers from pipets. Fill the pipet with the solution. Hold it above the solution to which it is to be added. Do not immerse it in the solution, because it will then become contaminated.

REAGENTS:

```
IDENTITY: _____
CONCENTRATION: _____
DATE OPENED: _____ INITIALS: ____
EXPIRATION DATE: _____
STORAGE CONDITIONS: RT/R/F/Other: _____
```

SOLUTIONS/MIXTURES:

```
IDENTITY: _____
CONCENTRATION: _____
PREP. DATE: _____ INITIALS: ____
EXPIRATION DATE: _____
STORAGE CONDITIONS: RT/R/F/Other: _____
```

Figure 5.7
Label formats for reagents and solutions (RT = room temperature, R = refrigerator, F = freezer).

Answers to exercises —

1. 0.2% NaCl solution = 0.2 g NaCl in 100 mL solution. *For 500 mL 0.2% NaCl, solution needs 0.2 × 5 = 1.0 g NaCl.*

2. 2% $CuSO_4$ solution = 2 g $CuSO_4$ in 100 mL solution. For 1000 mL 2% $CuSO_4$ solution needs 2 × 10 = 20 g $CuSO_4$. *20 g $CuSO_4$ is needed to prepare 1 L of 2% solution.*

3. As we calculated in question 2 above, we used 20 g of anhydrous $CuSO_4$ for the preparation of 1 L 2% solution. If we want to use $CuSO_4$ in the hydrated form, as $CuSO_4 \cdot 5$ H_2O we have to consider the water content of the hydrated form. First calculate the molecular weight (Mw) of the anhydrous and the hydrated forms: Mw of $CuSO_4$ anhydrous is 159.5 g, and 249.5 g for $CuSO_4 \cdot 5$ H_2O. For the anhydrous $CuSO_4$ we need 20 g (according to problem 2 above), so to determine how many grams are needed of the hydrated $CuSO_4 \cdot 5$ H_2O we use the following:

$$\frac{159.5}{20} = \frac{249.5}{x}$$

$$159.5\,x = (20)(249.5)$$

$$x = (20)(249.5)/159.5$$

$$x = 4990/159.5$$

$$x = 31.285 = 31.3$$

31.3 g of $CuSO_4 \cdot 5$ H_2O are needed to prepare 1 L of 2% $CuSO_4$ solution.

4. One mol NaOH is 40 g. 1 M NaOH contains 1 mol NaOH per liter, so 40 g per liter. 0.1 M contains 0.1 mol NaOH per liter, so 4.0 g per liter, and 4.0/4 = 1.0 g per 250 mL. *1.0 g NaOH are needed to prepare 250 mL of 0.1 M NaOH solution.*

5. $Ca(OH)_2$ is a base, so the equivalent weight is the molecular weight divided by the number of the OH^- ions. Equivalent weight of $Ca(OH)_2$ is 74/2 = 37 g. 1 N solution contains 37 g per liter. 0.25 N solution contains 37 × 0.25 = 9.25 g per liter, and 9.25 × 2 = 18.5 g per 2 liters of solution. *18.5 g $Ca(OH)_2$ are needed to prepare 2 L of 0.25 N solution.*

6. 1 L of 1 M H_2SO_4 contains 1 mol or 98 g of H_2SO_4. 1 L of 0.1 M H_2SO_4 contains 0.1 mol or 9.8 g of H_2SO_4. The calculated grams are converted to milliliters by using the density of the acid (grams/density = milliliters). 9.8/1.834 = *5.4 mL conc H_2SO_4 diluted to 1 L gives 0.1 M H_2SO_4 solution.*

7. Equivalent weight of H_2SO_4 is Mw/2, so 98/2 = 49 g. 1 L of 1 N H_2SO_4 solution contains 49 g H_2SO_4. 1 L of 0.1 N H_2SO_4 solution contains 4.9 g H_2SO_4. The grams converted to milliliters because of the density of the acid. 4.9/1.834 = *2.70 mL conc H_2SO_4 diluted to 1 L.*

8. Use the formula $c_1v_1 = c_2v_2$, where $c_1 = 0.1$, $v_1 = ?$, $c_2 = 0.02$, and $v_2 = 1000$ mL (1 L).

$$0.1\,v_1 = 0.02 \times 1000$$

$$v_1 = 20/0.1$$

$$v_1 = 200 \text{ mL}$$

200 mL of 0.1 N H_2SO_4 solution diluted to 1 L produces 0.02 N H_2SO_4 solution.

Chapter **6**

STANDARD SOLUTIONS

A *standard solution* is a solution whose concentration has been accurately determined. Standard solutions are prepared from highly pure chemicals, and the exact concentration is determined by a process called *standardization*. The substance available for standardization of a solution is highly pure and is called a *primary standard*.

Standard solutions are used in titrimetric analytical procedures, where an exact volume of a standard solution is needed to react with an unknown quantity of a second substance. The standard solution is called the *titrant*. If the volume and the concentration of the titrant are known, the unknown quantity of the substance titrated can be calculated.

Detailed theoretical treatments and actual methods of titration are given in a later chapter.

6.1 STOCK STANDARD SOLUTIONS

Stock standard solutions are high concentrations of standard solutions, used to prepare standard solutions by dilution. The analyst withdraws a portion of the stock solution using a volumetric pipet and dilutes it to the desired volume. The portion withdrawn by the pipet is called an *aliquot*. The quantity of the aliquot depends on the concentration of the stock solution and is determined as described earlier in Section 5.3.2. Tables 5.5 and 5.6 list the most commonly used concentrations and preparations of standard acids and bases.

6.2 STANDARDIZATION OF SOLUTIONS

It has been previously mentioned that the process by which the concentration of a solution is accurately determined is called standardization. Relatively few chemical reagents can be obtained in sufficiently pure form to meet the analyst's demand for accuracy. The few substances which are adequate in this regard are called primary standards. More commonly, a solution is standardized by a titration in which it reacts with a weighed portion of a primary standard. The primary standard should have the following characteristics:

- The material should be of known composition and highly pure; preferably 100% pure. The total amount of impurities should not exceed 0.01% to 0.02%.

- The substance should be stable and easy to dry, and it should not absorb water or carbon dioxide from the atmosphere.

- It should undergo a rapid and stoichiometric chemical reaction with the solution being standardized.

- It is desirable that the primary standard have a reasonably high equivalent weight in order to minimize the consequences of errors in weighing.

For *acid–base titration,* a widely used primary standard for base solution is the compound potassium hydrogen phthalate ($KHC_8H_4O_4$), abbreviated KHP. Sulfamic acid (HSO_3NH_2) and potassium hydrogen iodate ($KH(IO_3)_2$) are both strong acids, and are excellent primary standards. A common primary standard for standardizing strong acids is sodium carbonate (Na_2CO_3). For acid–base titration, it is customary to prepare solutions of an acid and base of the desired concentration and then to standardize one of the solutions against a primary standard. The solution thus standardized can be used as a *secondary standard* to obtain the concentration of the other solution. For highly accurate work it is preferable to standardize both the acid and base independently against a primary standard.

For *precipitation and complex formation titration,* pure salts are usually employed as primary standards. Sodium or potassium chloride can be used for standardization of silver nitrate ($AgNO_3$) or mercuric nitrate ($Hg(NO_3)_2$) solutions. Calcium carbonate ($CaCO_3$) is used as a primary standard for solutions of the complexing agent ethylenediaminotetraacetic acid (EDTA).

Many primary standards are available *for redox reagents.*

For standardization of potassium permanganate ($KMnO_4$) use sodium oxalate ($Na_2C_2O_4$), for sodium thiosulfate ($Na_2S_2O_3$) use potassium dichromate ($K_2Cr_2O_7$), and for iodine solution use arsenic oxide (As_2O_3).

6.3 PREPARATION AND STANDARDIZATION OF COMMON STANDARD SOLUTIONS

6.3.1 Sulfuric Acid (H_2SO_4) Solution 0.1 N

Caution: When diluting acid solutions, keep in mind the safety rule: never pour water into acid, always add acid to water!

Normal (N) solutions contain equivalent weight substances per liter. For equivalent weight of acids, the molecular weight is divided by the number of the hydrogen ions (see Section 5.3.1); therefore, the equivalent weight of sulfuric acid is the molecular weight divided by 2. 1 N H_2SO_4 solution has $98/2 = 49$ g acid in 1 L of solution. 0.1 N H_2SO_4 solution has $49/10 = 4.9$ g acid in 1 L of solution. Convert grams of acid to milliliters by using the specific gravity (grams divided by the specific gravity = milliliters). The specific gravity of the concentrated acid is 1.836; therefore, $4.9/1.836 = 2.4$ mL concentrated acid per 1 L of solution gives the 0.1 N concentration.

Standardization — Sulfuric acid is standardized with the sodium carbonate (Na_2CO_3) primary standard as follows:

- Dry Na_2CO_3 at 250°C for 4 hours in oven. Cool in desiccator.
- Weigh three clean and dry Erlenmeyer flasks (W_1).
- Transfer into each flask about 0.1 g of dried Na_2CO_3 and reweigh the flasks (W_2). The differences in weights ($W_2 - W_1$) are the weights of the Na_2CO_3 portions.
- Dissolve each sample in about 80 mL of DI water.

- Add methylorange or mixed bromcresol green–methyl red indicator (see determination of alkalinity) and titrate with 0.1 N H_2SO_4 until the end point (yellow to orange in the case of using methylorange and blue to pink by bromcresol green–methyl red indicator). Record the milliliters of acid used for each of the three Na_2CO_3 standards titration. Always titrate to the same color!

- Calculation:

$$N = [\text{weight of } Na_2CO_3 \text{ (g)} \times 1000]/[\text{mL of acid used} \times 53] \qquad \textbf{(6.1)}$$

Note that 53 = equivalent weight of Na_2CO_3.

The exact normality of the acid solution is given by the average of the three calculated normalities. For example:

Weight of the Na_2CO_3 (g)	mL Acid used	Normality
0.1238	23.2	0.1007
0.1096	20.5	0.1009
0.1646	31.1	0.0998

Normality is calculated by using formula (6.1):

$$(0.1238)(1000)/(23.2)(53) = 0.1007$$
$$(0.1096)(1000)/(20.5)(53) = 0.1009$$
$$(0.1646)(1000)/(31.1)(53) = 0.0998$$

The average of the three normalities is 0.1005; therefore, the actual normality of the H_2SO_4 solution is 0.1005.

6.3.2   Hydrochloric Acid (HCl) Solution 0.1 N

Equivalent weight of the HCl is 36 g (36/1); therefore, a 0.1 N solution contains 3.6 g HCl per 1 L of solution. The specific gravity of the concentrated HCl is 1.180, so measure 3.6/1.180 = 4.25 mL of concentrated HCl in a liter volumetric flask and dilute to the mark with DI water and mix thoroughly.

Standardization — Standardize the solution against sodium carbonate (Na_2CO_3) in the manner described in Section 6.3.1 for the standardization of sulfuric acid. Alternatively, both sulfuric and hydrochloric acid may be standardized with a known concentration of Na_2CO_3 solution, as follows:

- Prepare 0.05 N Na_2CO_3 solution by drying the primary standard Na_2CO_3 at 250°C for 4 hours and cooling in a desiccator. Weigh 2.5 ± 0.2 g (to the nearest mg) of chemical, dissolve in a 1-L volumetric flask, fill up to the mark, and mix. Do not keep this solution longer than 1 week.

- Measure 40 mL of 0.05 N Na_2CO_3 solution into a beaker and titrate potentiometrically with the acid solution to pH of about 5.00.

- Lift out electrode and rinse with DI water into the same beaker. Boil gently about 3 to 5 minutes under a glass watch cover.

- Cool to room temperature.

- Rinse cover glass with DI water into beaker, and finish titrating to the pH inflection point.
- Calculate the normality of the acid:

$$N = [A \times B]/[53 \times C] \tag{6.2}$$

where A = grams of Na_2CO_3 weighed into the 1-L volumetric flask; B = milliliters of Na_2CO_3 solution taken for titration; C = milliliters of acid used; and 53 = equivalent weight of Na_2CO_3.

6.3.3   Sodium Hydroxide (NaOH) Solution 0.1 N

Caution: Do not forget that the reaction between sodium hydroxide and water liberates heat!

Sodium hydroxide is a base, so calculation of the equivalent weight is the molecular weight divided by the number of hydroxide, (OH^-) ions; therefore, 40/1, or 40 g. So, the 0.1 N solution is prepared by dissolving 4.0 g of NaOH in DI water and diluting to 1 L.

Standardization — The solution is standardized against primary standard potassium hydrogen phthalate, or potassium biphthalate ($KHC_8H_4O_4$) as follows:

- Dry potassium hydrogen phthalate (KHP) at 120°C for 2 hours.
- Cool in desiccator.
- Weigh three Erlenmeyer flasks and record the weights (W_1).
- Place about 0.50 g of dried KHP into each of the preweighed Erlenmeyer flasks.
- Weigh each flask with the KHP in it (W_2).
- Calculate the mass of the KHP by subtracting W_1 from W_2, and record the exact weight of KHP in each flask.
- Add about 100 mL of DI water and stir gently for complete dissolution.
- Add a few drops of phenolphthalein indicator and titrate with the 0.1 N NaOH until a very faint pink color appears and persists.
- Calculation of the normality of the NaOH solution:

$$N = [\text{weight of KHP} \times 1000]/[\text{mL of NaOH} \times 204.3] \tag{6.3}$$

Note that 204.23 is the equivalent weight of KHP.

The exact normality is given by averaging the three normalities derived from the three parallel determinations, as shown previously in the sulfuric acid standardization.

One other choice is to use a potassium hydrogen phthalate (KHP) solution with known normality. Standardize 0.1 N NaOH solution with 0.05 N KHP solution as follows:

- Dry primary standard KHP at 120°C for 2 hours.
- Cool in desiccator.
- Weigh 10 ± 0.5 g, transfer into 1 L volumetric flask, and dilute to 1,000 mL.
- Measure 40 mL of 0.05 N KHP solution into a beaker and titrate with 0.1 N NaOH to the inflection point, which should be close to pH 8.7.
- Calculate the normality of the NaOH:

$$N = [A \times B] / [204.3 \times C] \tag{6.4}$$

where A = grams of KHP, B = milliliters of KHP solution taken for titration, C = milliliters of NaOH solution used, and 204.2 = equivalent weight of KHP.

6.3.4 Iodine (I_2) Solution 0.1 N

The equivalent weight of the iodine is atomic weight, 127, divided by the valency of iodine, which is 1. Therefore, to prepare 0.1 N iodine solution, dissolve 40 g potassium iodide (KI) in 25 mL DI water, add 12.7 g primary standard resublimed iodine, I_2, and stir until complete dissolution, then dilute to 1 L. Mix thoroughly.

Caution: Because I_2 may volatize and lose volume from both the solid and solution, transfer the solid immediately to KI, as specified above. Never let the solution stand in open containers for extended periods!

Standardization

- Dry arsenic oxide (As_2O_3) at 105°C for 1 hour.
- Cool in desiccator.
- Weigh three clean and dry Erlenmeyer flasks, W_1.
- Add about 0.20 g of dried As_2O_3 into each of these Erlenmeyer flasks and weigh. Record the weights as W_2. Calculate the exact weight of the As_2O_3 by subtracting W_1 (weight of the flask) from W_2 (flask + As_2O_3).
- Add 10 mL of 1 N NaOH solution and swirl to dissolve. When dissolution is complete, add 100 mL DI water and 10 mL of 1:35 diluted sulfuric acid (H_2SO_4).
- Slowly add sodium bicarbonate ($NaHCO_3$) crystals until effervescence ceases, then add 2 g in excess and stir until dissolved.
- Add 2 mL of starch indicator.
- Titrate with iodine solution to the first permanent blue color.
- Calculation:

$$N = [\text{weight of } As_2O_3 \text{ in g} \times 1000]/[\text{mL iodine} \times 49.45] \qquad (6.5)$$

Note that 49.45 is the equivalent weight of As_2O_3, 198/4.

Also, a standard 0.1 N arsenite solution can be used for determining the exact normality of the iodine solution, as follows:

1. Preparation and standardization of arsenite solution:
   - Accurately weigh a stoppered weighing bottle containing approximately 4.95 g arsenic trioxide (As_2O_3).
   - Transfer without loss to a 1-L volumetric flask and again weigh bottle. Do not attempt to brush out adhering arsenic oxide.
   - Moisten As_2O_3 with water and add 15 g NaOH and 100 mL DI water. Swirl flask contents gently to dissolve, dilute to mark, stopper, and mix thoroughly.
   - Calculate the normality of the As_2O_3 solution:

$$N = [\text{grams } As_2O_3]/[49.455] \qquad (6.6)$$

This solution will preserve its normality almost indefinitely!

2. Standardization of iodine solution:

- Accurately measure 40–50 mL of arsenic solution into a flask.
- Add starch indicator solution.
- Titrate with 0.1 N iodine solution until the first permanent blue color.
- Record the volume of the used iodine solution.
- Calculate the normality of the iodine solution:

$$N = [A \times B]/[49.455 \times C] \tag{6.7}$$

where A = grams of As_2O_3, B = milliliters of As_2O_3 solution taken for titration, C = milliliters of iodine solution used, and 49.455 = equivalent weight of As_2O_3.

CAUTION: As_2O_3 is toxic: avoid ingestion! Cancer suspect agent!

6.3.5 Sodium Thiosulfate ($Na_2S_2O_3$) Solution 0.1 N

The equivalent weight of crystallized sodium thiosulfate ($Na_2S_2O_3 \cdot 5H_2O$) is 248.12. For 0.1 N solution, dissolve 24.812 g sodium thiosulfate pentahydrate and make up to 1 L. Store in a tightly closed glass bottle and standardize against potassium dichromate or potassium biiodate after at least 2 weeks storage. This initial storage is necessary to allow oxidation of any bisulfite ion present. Add a few milliliters chloroform, $CHCl_3$, to minimize bacterial decomposition.

Standardization

- Dry potassium dichromate ($K_2Cr_2O_7$) at 120°C for 4 hours in desiccator.
- Dissolve 4.904 g anhydrous $K_2Cr_2O_7$ of primary standard quality in DI water and dilute to 1000 mL to yield a 0.1 N solution. Store in a glass-stoppered bottle.
- To 80 mL DI water, add with constant stirring, 1 mL concentrated H_2SO_4, 10 mL 0.1 N $K_2Cr_2O_7$ solution, and 1 g KI.
- Let the reaction mixture stand 6 minutes in the dark. It will have a yellow color from the liberated iodine.
- Titrate with 0.1 N $Na_2S_2O_3$ solution until the yellow color is almost discharged.
- Add 1 mL starch indicator. The color of the solution turns blue.
- Continue titration until the blue color disappears.
- Calculate the normality of the sodium thiosulfate solution:

$$N = [\text{milliliters } K_2Cr_2O_7 \times 0.1]/[\text{milliliters } Na_2S_2O_3 \text{ consumed}] \tag{6.8}$$

Alternatively, one can use the iodate method for standardization of sodium thiosulfate, as follows.

- Dissolve 3.249 g anhydrous potassium biiodate ($KH(IO_3)_2$) primary standard quality, or 3.567 g potassium iodate, (KIO_3) dried at 103 to 105°C for 1 hour and dilute to 1000 mL to yield a 0.1 N solution. Store in a glass-stoppered bottle.
- To 80 mL DI water, add, with constant stirring, 1 mL concentrated H_2SO_4, 10 mL 0.1 N $KH(IO_3)_2$, and 1 g KI.
- Titrate immediately with 0.1 N $Na_2S_2O_3$ titrant until the yellow color of the liberated iodine almost is discharged.
- Add 1 mL starch indicator solution, the color becomes blue.
- Continue titration until the blue color disappears.

- Calculate the normality of sodium thiosulfate:

$$N = [\text{milliliters KH(IO}_3)_2 \times 0.1]/\text{milliliters Na}_2\text{S}_2\text{O}_3 \text{ consumed} \qquad \textbf{(6.9)}$$

Sodium thiosulfate has to be restandardized weekly.

6.3.6 Potassium Permanganate ($KMnO_4$) Solution 0.1 *N*

Dissolve 3.2 g $KMnO_4$ in about 100 mL DI water and dilute to 1000 mL. Allow the solution to stand in the dark a few days and then filter through a fine-porosity sintered glass crucible. Do not use filter paper! Store the solution in a glass-stoppered, amber-colored glass bottle.

Standardization of KMnO₄ solution

- Dry sodium oxalate ($Na_2C_2O_4$) at 105°C for 1 hour.
- Cool in desiccator.
- Weigh three clean and perfectly dry Erlenmeyer flasks (W_1).
- Add about 0.15 to 0.17 g $Na_2C_2O_4$ into each flask and weigh (W_2).
- Calculate the exact weight of the $Na_2C_2O_4$ by subtracting W_1 from W_2.
- Add 20–25 mL DI water to each flask.
- Add 10 mL 20% H_2SO_4 and mix well.
- Heat to 70–80°C.
- Titrate with $KMnO_4$ solution (still hot) until a faint pink color persists for about 30 seconds.
- Always titrate a blank and subtract from the result.
- Calculate the normality of the $KMnO_4$ solution:

$$N = [\text{weight of Na}_2(\text{COO})_2 \text{ in grams}]/[\text{milliliters KMnO}_4 \times 0.06701] \qquad \textbf{(6.10)}$$

where 0.06701 = milliequivalent weight of $Na_2(COO)_2$. Equivalent weight = MW/2 = 134/2 = 67.01.

$KMnO_4$ solution must be standardized weekly!

Chapter **7**

COMMON LABORATORY TECHNIQUES

7.1 FILTRATION

Filtration is the process of removing material from a substrate in which it is suspended. Filtration is accomplished by passing the mixture to be processed through one of the many available sieves, called filter media. In the laboratory, filtration is generally used to separate solid impurities from a liquid or a solution or to collect a solid substance from the liquid or solution from which it was precipitated or recrystallized.

7.1.1 Filtration Methods

There are two general methods of filtration: gravity and vacuum (or suction). During *gravity filtration* the filtrate passes through the filter medium under the combined forces of gravity and capillary attraction between the liquid and the funnel stem. In *vacuum filtration*, a pressure differential is maintained across the filter medium by evacuating the space below the filter medium. Vacuum filtration adds the force of atmospheric pressure on the solution to that of gravity, with a resultant increase in the rate of filtration. A typical gravimetric filtration setup is shown in Figure 7.1 and one complete vacuum filtration assembly is shown in Figure 7.2.

7.1.2 Filter Media

Filter papers — There are several varieties or grades of filter paper for special purposes.

Qualitative-grade filter papers — These filters leave an appreciable amount of ash upon ignition and are therefore unsuitable for applications in quantitative analysis where precipitates are to be ignited on the paper and weighed.

Ashless quantitative filter papers — These filters can be ignited without leaving an ash.

Hardened-grade papers — These filters are designed to be used in vacuum filtration. They are processed to have great wet strength and a hard, lintless surface. They are available in low-ash, ashless, and regular grades.

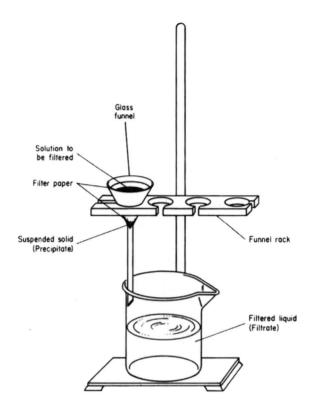

Figure 7.1
Gravity filtration setup.

Glassfiber filter papers — These filters are manufactured from very fine borosilicate glass and are used in Gooch, Buchner, or similar filtering apparatus. Glassfiber filter papers have very fine retention, very rapid filtration, and inertness to the action of most reagents. All grades of filter paper are available in a variety of sizes and in several degrees of porosity. The commonly used filter papers, with porosity and filtration speed, are listed in Table 7.1, and the properties of Whatman filter papers are summarized in Table 7.2.

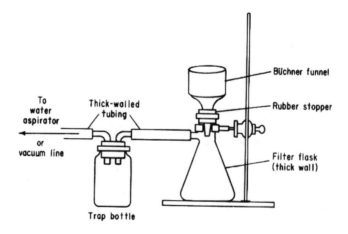

Figure 7.2
Vacuum filtration.

TABLE 7.1

Commonly Used Filter Papers

W	S&S	RA	Porosity	Speed	Use
Qualitative or Regular Grade Papers					
4	604	202	Coarse	Very rapid	Gelatinous
1	595	271	Medium	Medium	Crystalline
3	602	201	Medium	Slow	Fine (used with Buchner funnel)
Quantitative or Ashless Filter Papers (Less Than 0.1 mg Ash)					
41 Blue ribbon	589	—	Coarse	Very rapid	Gelatinous
40 White ribbon	589	—	Medium	Rapid	Crystalline
42 Black ribbon	589	—	Fine	Slow	Fine crystal

Note: W = Whatman filter papers; S&S = Schleicher & Schuell filter papers; RA = Rieve Angel filter papers.

Membrane filters — Membrane filters are thin polymeric structures with extraordinarily fine pores. They are composed of pure, biologically inert cellulose esters or other polymeric materials. With proper filter selection, they yield a filtrate that is ultra clean and/or sterile. Membrane filters are available in a wide variety of pore sizes in a number of different materials. The quality and performance of membrane filters vary with the manufacturer, type, brand, and lot.

Well-known, high-quality membrane filters are "Millipore" and "Gelman," with pore sizes of 0.025, 0.05, 0.1, 0.22, 0.45, 1.2, 3.0, 5.0, and 14 μm.

Fritted glassware — Some of the problems encountered in using filter paper are minimized by using fritted-glass equipment. Fritted or sintered glass is available in different porosities

TABLE 7.2

Properties of Whatman Filter Papers

	No. 4	No. 1	No. 2	No. 3	No. 5
Qualitative					
Porosity	Coarse	Medium	Medium-fine	Medium-fine	Fine
Flow rate	Fast	Medium fast	Medium slow	Medium slow	Slow
Particles retention	20–25 μm	11 μm	8 μm	6 μm	2.4 μm
Application	Air pollution	General	General	Good for suction	Useful for clearing cloudy suspension

	No. 41	No. 40	No. 44	No. 42	
Quantitative (Ashless)					
Porosity	Coarse	Medium	Fine	Fine	
Flow rate	Rapid	Medium	Slow	Slow	
Particles retention	20–25 μm	8 μm	3 μm	2.5 μm	
Application	Gelatinous precipitate; iron and aluminum hydroxides; air pollution monitoring	Widely used gravimetric crystalline precipitate	Retains fine particles faster than No. 42	For very fine precipitate	

Designation	Maximum Pore Size (μm)
Extra coarse	170–220
Coarse	40–60
Medium	10–15
Fine	4–5.5
Very fine	2–2.5
Ultra fine	0.9–1.4

Figure 7.3
Grades of fritted glassware.

and grades, as shown in Figure 7.3. To maintain optimum performance of fritted ware, it is best to follow the instruction of the manufacturer. Clean all fritted filters immediately after use. Many precipitates can be removed from the filter surface simply by rinsing from the reverse side with water under pressure not exceeding 15 lb/in. Some precipitates tend to clog the fritted filter pores, and special cleaning solutions are required, such as hot acid, hot detergent, organic solvent, etc. For example, oil and grease residues clean with organic solvent, organic matter with hot cleaning solution (see Section 4.2.4), mercury (Hg) residues with hot nitric acid (HNO_3), and silver chloride (AgCl) residue with ammonia solution.

Sintered-glass crucibles — These are glass crucibles with fritted-glass disks sealed permanently into the bottom end. They are available in fine (F), medium (M), and coarse (C) porosities and may be heated to a temperature as high as 500°C. The fine porosity must be used for fine crystalline precipitates only, the medium porosity is suitable for easily reduced silver halides and other such precipitates, and the coarse porosity is suitable for coarse crystals. Sintered glass crucibles are always dried in an oven, never over a flame.

Porcelain crucibles with porous ceramic disks — The ceramic disks are permanently sealed in the bottom, and the crucibles are used in the same way as sintered-glass crucibles, except they may be ignited at extremely high temperatures. They are excellent for filtering fine precipitates.

Gooch crucible — This is a porcelain thimble with a perforated base. A Gooch porcelain crucible is shown in Figure 7.4. Filtering crucibles must be used with a crucible holder and suction system, as shown in Figure 7.5.

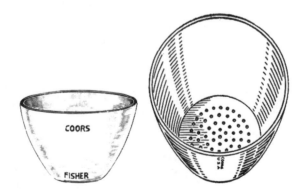

Figure 7.4
Gooch crucible with perforated bottom.

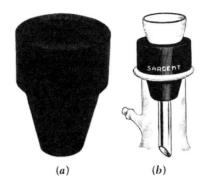

(a)   *(b)*

Figure 7.5
(a) Rubber adapter for suction filtration. (b) Adapter in use with a crucible in the suction flask.

Buchner funnels — Buchner funnels are used in vacuum filtration. They are not conical in shape, but have a flat, perforated bottom. A filter paper in a size sufficient to cover the perforated bottom area is placed on the flat bottom, moistened, and tightly sealed against the bottom by applying a slight vacuum. Plain Buchner funnels with removable plates are also available, as shown in Figure 7.6. A funnel with a fritted glass disk is used in vacuum filtration where the paper filter in a Buchner funnel would be attached, as shown in Figure 7.7.

Hirsch funnels — Hirsch funnels are used for collecting small amounts of solids. They are usually made of porcelain. The inside bottom of the funnel is a flat plate with holes in it which supports the filter paper. Buchner and Hirsch funnels can also be obtained in glass with sintered-glass disks.

Other funnels used like the Hirsch are the *Witt funnel* and the *filter nail* (see Figure 7.8). When using these funnels, a rubber ring forms the seal between the funnel and the filter flask, which is connected to the vacuum line or to an aspirator. A rubber stopper or cork can also be used instead of the rubber ring to fit the funnel to the filter flask.

7.1.3 Gravity Filtration

During gravity filtration, the filtrate passes through the filter medium under the forces of gravity and capillary attraction between the liquid and the funnel stem. The procedure

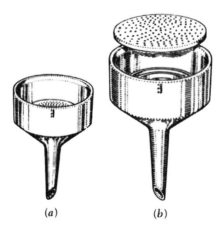

(a)   *(b)*

Figure 7.6
Buchner funnels. (a) Buchner suction funnel. (b) Buchner funnel with a removable plate, available in various sizes (14.5 to 308 mm).

Figure 7.7
A funnel with a fritted-glass disk is used for vacuum filtration where the paper filter in the Buchner funnel would be attached.

is slow, but it is highly favored for gravimetric analysis over the more rapid vacuum filtration because there is better retention of fine particles and less rupturing or tearing of the filter paper.

Optimum filtering speed is achieved in gravity filtration by proper folding and positioning of the paper in the funnel (Figure 7.9). Filter paper is folded so as to provide a space between the paper and the funnel, except at the top of the paper, which should fit snugly to the glass. The second fold is made so that the ends fail to match by about 1/8 in. Then the paper is opened into a cone. Place the cone in the funnel and add DI water to wet the paper. A clean finger is applied cautiously to smooth the paper and obtain a tight seal of paper to glass at the top. Air does not enter the liquid channel with a properly fitted paper, and thus the drainage from the stem of the funnel establishes a gentle suction that facilitates the filtration. A malfunctioning filter can seriously delay the analysis; it is preferred by far to reject such a filter and prepare a new one.

Filter paper circles are available in various diameters. The paper and the funnel should match in size. It is important that the paper does not extend above the edge of the glass funnel, but comes within 1 or 2 cm of the edge. The nonfilterable residue should occupy about one third of the filter paper cone, and never more than one half. The stem of the funnel should extend well into the vessel receiving the filtrate, and the tip of the stem should touch the inner surface of the vessel to prevent spattering of the filtrate. The varied shapes and sizes of glass funnels are shown in Figure 7.10, and the best way to pour the supernatant liquid into the filter is shown in Figure 7.11.

7.1.4 Vacuum Filtration

Vacuum filtration is a very convenient way to speed up the filtration process, but the filter medium must retain the very fine particles without clogging up. The vacuum is

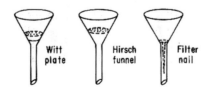

Figure 7.8
The Hirsch and other funnels for collecting small quantities of solids.

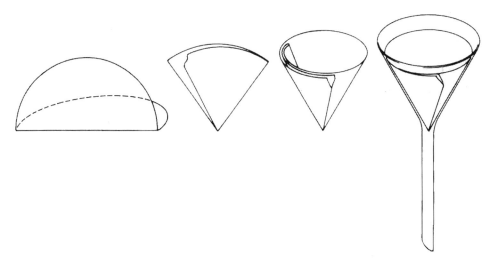

Figure 7.9
Folding filter paper. The second fold is not exactly at the right angle. Note the tear, which makes the filter paper stick to the funnel better. The paper and the funnel should match in size, and the paper should not extend above the edge of the glass funnel, but should come within 1 or 2 cm below the edge.

normally provided by a water aspirator, although a vacuum pump, protected by suitable traps, can be used. Because of the inherent dangers of flask collapse from the reduced pressure, thick-walled filter flasks should be used, and the technician should be always on the alert for the possibility of an implosion. Wear protective glasses when the assembly is under reduced pressure. Be careful that the liquid level in the "safety trap bottle" is never as high as the inlet tube. A typical setup for vacuum filtration is shown in Figure 7.2. Vacuum filtration can be performed with filter paper as well as with various crucibles. The common conical funnel paper is easily ruptured at the apex of the cone when under a vacuum. To straighten the cone of the apex, a small metal liner is often inserted, as shown in Figure 7.12.

7.2 CENTRIFUGATION

The principle of centrifugation is that the rate of settling of a precipitate or the rate of separation of two immiscible liquids is increased manifold by the application of a

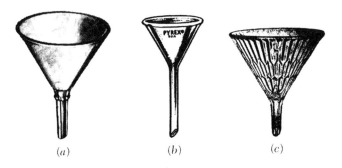

(a)   (b)   (c)

Figure 7.10
Varied shapes of glass funnels. (a, b) Varied size and stem length. (c) Funnel with heavy ribbed construction; raised ribs on inner surface facilitate rapid filtration.

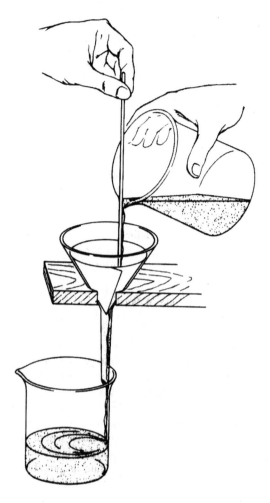

Figure 7.11
Technique of filtration with filter paper. The stem of the funnel should extend well into the vessel receiving the filtrate, and the tip of the stem should touch the inner surface of the vessel to prevent spattering of the filtrate. All transfers into the funnel should be made with the aid of a stirring rod. Pour the liquid onto the paper, not onto the glass.

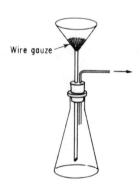

Figure 7.12
Funnel with a wire gauze cone to be used as a support for the filter paper.

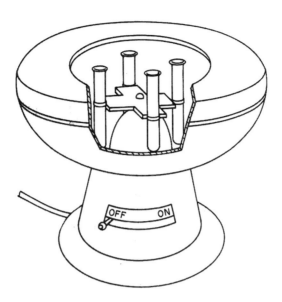

Figure 7.13
Common laboratory centrifuge.

centrifugal force thousands of times that of gravity. A *centrifuge* is a device in which solid or liquid particles of different densities are separated by rotating them in a tube in a horizontal circle. The denser particles tend to move along the length of the tube to a greater radius of rotation, displacing the lighter particles to the other end. The force exerted on rotating particles in a liquid is described in terms of *relative centrifugal force* (RCF). Centrifuges are rated by their RCFs, which range from less than 10 to almost 750,000. The speed of the centrifuge is expressed as *rpm* (revolutions per minute). Special centrifuges rotate at high speeds up to 70,000 rpm, and may have an RCF of 750,000. They may be refrigerated because the rotating head generates heat as a result of friction with air. An ultracentrifuge is a high-speed centrifuge used to measure the rate of sedimentation of colloidal particles or to separate macromolecules, such as proteins or nucleic acids, from solutions. Figure 7.13 shows a common laboratory centrifuge.

7.2.1 Using A Common Laboratory Centrifuge

Centrifuges rotate at high speeds, and therefore the head and tubes containing the liquid must be balanced to avoid excessive vibration, which can damage the equipment. Special centrifuge tubes must be used, because the tubes must be able to withstand the very high centrifugal forces. Never use ordinary test tubes in a centrifuge; they will break.

- Divide the liquid to be centrifuged among the minimum number of centrifuge tubes. Use an equal number of tubes, or fill one with a counterbalancing solution. All tubes and contained liquid should be massed and adjusted to the same mass.
- Insert the tubes equally spaced from each other.
- Start the motor, and run the centrifuge until the objective has been achieved.
- Turn off the switch, and allow the rotating centrifuge assembly to come to rest. Do not attempt to stop the rotation manually when the centrifuge is rotating at high speed; use the brake.

Figure 7.14
Decanting off supernatant liquid.

- If the solid is firmly packed against the bottom of the tube, the liquid can be decanted from the solid by pouring off the liquid or withdrawing it with a pipet fitted with a rubber bulb (for example, Pasteur pipet), as shown in Figures 7.14 and 7.15.
- To wash centrifuged precipitates, add sufficient water, mix well, centrifuge again, and discard the washing. Several small-volume washes are more effective than one large wash.

7.2.2 Centrifuge Safety Rules

- Always use balanced pairs of centrifuge tubes which are placed opposite each other on the rotating head. If necessary, use a counterbalance tube containing water for balancing.
- Always use rubber cushions to prevent tube breakage.
- Always use centrifuge tubes (Figure 7.16) — others may break.
- Always close the centrifuge cover before starting the motor, and open it only after the centrifuge has stopped.
- Always turn off the centrifuge if it starts to vibrate.
- Always use the break, never your hands to stop the centrifuge.

7.3 DISTILLATION

Distillation is a process that depends on differences in the ease of vaporization (how readily liquids become gases) of a component.

7.3.1 Simple Distillation

In simple distillation, the liquid mixture is heated. The most volatile component vaporizes at the lower temperature, and the vapor passes through a cooled tube (condenser), where it condenses back into its liquid state. The simple, one-stage distillation apparatus shown in Figure 7.17 works very well when only one component of the mixture is volatile. For example, from a solution of sodium chloride (NaCl) and water, sodium chloride is easily separated from the water and at the same time can collect the purified water. The sodium chloride–water mixture is placed in the distillation flask and heated

Figure 7.15
Withdrawing liquid after centrifugation.

Figure 7.16
Centrifuge tubes of general and special designs.

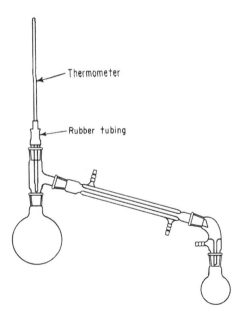

Figure 7.17
Apparatus for simple distillation.

to boiling. As the solution boils, water in the form of vapor fills the flask and travels toward the distilling head. From there, the hot water vapor travels down the double-jacketed, water-cooled condenser. As the vapor travels down within the inner walls of the condenser, it is cooled by tap water running between the outer walls of the condenser. The water vapor *condenses*, or changes back to its liquid form. The condensed liquid, known as the *distillate*, is collected in the receiving flask. Because only the water, and not the sodium chloride, vaporizes and passes through the condenser, the distillate contains only pure water. The sodium chloride remains behind in the distillation flask.

Setup for simple distillation

- The distilling flask should accommodate twice the volume of the liquid to be distilled.
- The thermometer bulb should be slightly below the sidearm opening of the flask. The boiling point of the corresponding distillate is normally accepted as the temperature of the vapor. If the thermometer is not positioned correctly, the temperature reading will not be accurate. If the entire bulb of the thermometer is placed too high, above the side arm leading to the condenser, the entire bulb will not be heated by the vapor of the distillate, and the temperature reading will be too low. If the bulb placed too low, too near the surface of the boiling liquid, there may be a condition of superheating, and the thermometer will show a too-high temperature.
- All glass-to-glass or glass-to-cork connections should be firm and tight.
- The flask, condenser, and receiver should be clamped independently in their proper relative positions on a steady base.
- The upper outlet for the cooling water exiting from the condenser should point upward to keep the condenser full of water.

Simple distillation procedure

- Pour the liquid into the distilling flask with a funnel which extends below the side arm.
- Add a few boiling stones to prevent bumping.
- Insert the thermometer.
- Open the water valve for condenser cooling.
- Heat the distilling flask until boiling begins; adjust the heat input so that the rate of the distillate is a steady 2 to 3 drops per second.
- Collect the distillate into the receiver.
- Continue distillation until only a small residue remains. Do not distill to dryness.

7.3.2   Fractional Distillation

Two or more miscible liquids that have different boiling points may also be separated by the process of distillation. For example, when a mixture of ethyl alcohol (boiling point 78.5°C) and water (boiling point 100°C) is heated, the component with the lower boiling point (ethyl alcohol) vaporizes more readily than the component with the higher boiling point (water). Therefore, the distillate is richer in ethyl alcohol than in water. And it can be distilled again to yield even purer ethyl alcohol. If this process is repeated enough times, a distillate of essentially pure ethyl alcohol is obtained. This technique is called *fractional distillation*. It consists essentially in the systematic redistillation of distillates (fractions of increasing purity).

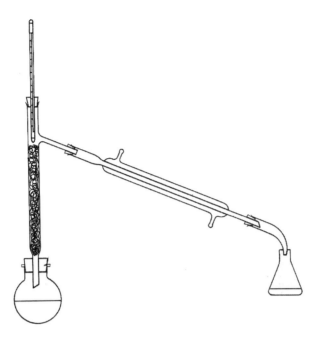

Figure 7.18
Fractional distillation apparatus. Column should be packed.

This technique is extremely useful in petroleum refining. The different components of crude petroleum all have different boiling points, so they can be separated by distillation. Petroleum companies use huge distilling columns, known as *fractionating columns*, to separate the various petroleum products. When the crude oil is heated, the components (or fractions) with lower boiling points rise to the top of the column, those with higher boiling points rise to the middle of the column, and solids remain at the bottom of the column. As the components rise up the column and condense, they are drawn off, to be purified further if necessary. A fractional distillation apparatus is shown in Figure 7.18.

7.3.3   Vacuum Distillation

Many substances cannot be distilled satisfactorily at atmospheric pressure because they are sensitive to heat and decompose before the boiling point is reached. Vacuum distillation is a distillation under reduced pressure and it makes distillation possible at lower temperatures.

Setup for vacuum distillation

- Source of vacuum: water pumps, aspirators. Use rubber pressure tubing.
- Lubricate all joints and connections.
- Apparatus should be airtight and free from leaks.
- Allow air into the system very carefully by using the pressure gauge (manometer).
- Safety trap to protect manometer and vacuum source from overflow liquid contamination.
- Maintain constant pressure by using the pressure regulator (manostat).
- Use capillary air inlet (Figure 7.19).

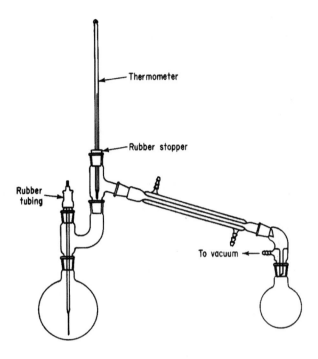

Figure 7.19
Vacuum distillation with gas-capillary bubbler.

- Special vacuum distillation flasks are used to minimize contamination of the distillate caused by frothing of the boiling solution.
- Use heating baths and electric mantles.

Vacuum distillation procedure

- Fill the distillation flask about one third full.
- Apply vacuum; adjust the capillary air inlet with the pinch clamp.
- Heat bath to about 20°C higher than the temperature at which the material will distill.
- Open cooling water.
- Check safety trap connections.
- When distillation is completed, remove the heating bath and allow the flask to cool. Remove the capillary clinch clamp. Cut off the cooling water. Turn off the vacuum pump.

7.4 REFLUXING

Reflux procedure allows a reaction mixture to be heated for an extended period of time without a loss of solvent. The condenser, which is fixed in a vertical position directly above the heated flask, condenses all vapors to liquid. Because none of the vapors escape, the volume of the liquid remains constant. It is used for carrying out reactions over long periods in organic synthesis.

Reflux setup

- The water inlet to the condenser is the lower one. The water outlet to the condenser is the upper one.

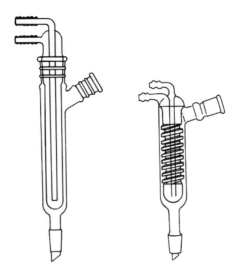

Figure 7.20
The two most common reflux condensers.

- Fill the heating flask no more than half full and add a few boiling stones.
- Turn on the cooling water.
- Heat the reflux for the desired period of time.

Commonly used reflux condensers are shown in Figure 7.20.

7.5 ION EXCHANGE

The exchange of ions of the same charge between a solution (usually aqueous) and a solid contact with it is called *ion exchange*. The process occurs widely in nature, especially in absorption and retention of water-soluble fertilizers by soils. For example, if potassium is dissolved in water and applied to soil, potassium ions are absorbed by the soil and sodium and calcium ions are released from it. A wide variety of materials, organic and inorganic, exhibit ion exchange behavior, but in laboratories the preferred ion exchangers are synthetic materials known as ion exchange resins. Ion exchange resins are useful and flexible tools. They commonly are in the form of sodium or hydrogen counter-ions (attached to the matrix) for cation exchangers and of chloride, formate, acetate, and hydroxide counter-ions for anion exchangers. In water analysis, ion exchangers can be applied to remove interfering ions, determine total ion content, concentrate trace quantities of cations, and separate anions from cations. The resin is agitated with the sample for a given time, after which the resin is removed by filtration. The column method is more efficient in that it provides continuous contact between the sample and the resin, thereby enabling the exchange reaction to go to completion. In this modification, the solution passes slowly through the resin bed, and ions are removed quantitatively from the sample. Elution of the resin permits recovery of the exchanged substances. Use resins that are specifically manufactured for a given analytical application. Prepare the ion exchanger by rinsing the resin several times with good-quality distilled water to remove any fine and colored matter and other leachable material that might interfere with the analytical method.

7.6 DRYING AND ASHING SAMPLES

7.6.1   Drying Samples

Samples often contain water, either as chemically combined hydrates or as occluded or surface absorbed water. The water content of different materials varies within wide limits. The moisture content of most samples is not constant, but varies with changes in the atmospheric humidity. The amount of moisture absorbed also varies with particle size, since samples of smaller particle size offer more surface for absorption. To afford reproducible analytical results, samples are usually dried before analysis, and the percentage composition of the sample is then calculated on a dry basis.

The goal of drying is either to make the sample anhydrous or to remove absorbed moisture but retain chemically combined water. Ordinarily, samples are dried in an oven at 100–110°C for 1 or 2 hours. However, some samples are partially decomposed under these conditions and should be dried at a lower temperature or not at all. Drying of heat-sensitive samples may be done in a desiccator containing a drying agent or in a vacuum desiccator.

Sometimes it is advantageous to analyze samples on an "as-received" or "wet basis". While the sample is being analyzed, the moisture content of another portion is determined. The analysis of the sample on a dry basis can be calculated from the data obtained.

Moisture — Moisture of any solid is determined by drying a known quantity aliquot of a well mixed sample at 100–105°C in a laboratory oven. Cool the dried sample in a desiccator and measure its weight, then calculate percentage of moisture as follows (values are in grams):

$$[\text{(original sample} - \text{dried sample)}/\text{original sample}] \times 100 \qquad (7.1)$$

Converting "on wet basis" result to "on dry basis" reporting form — Solid samples are analyzed on their wet bases and reported on their dry bases. To convert the wet basis result to this reportable form, it should be divided with the decimal fraction of the dry weight of the sample. For example, a sample is analyzed "as received" on wet basis, and found to contain 25 mg/kg barium and 12% moisture. The dry weight is 100% – 12% = 88%. The barium content of the sample on dry basis is 25/0.88 = 28.4 mg/kg.

Karl Fisher method to determine water content — Another way to determine water content is the Karl Fisher method. It is used primarily to determine the water content in organic compounds. The Karl Fisher method is a titration of water with an anhydrous methyl alcohol solution containing iodine, sulfur dioxide, and excess pyridine. The titration is based on the reaction between iodine and sulfur dioxide that will only occur if water is present. The sample is titrated with the Karl Fisher reagent until a permanent iodine color is observed. Because of other reaction products, the color change is usually from yellow to a brownish color that may be difficult to detect visually. A much sharper end point can be obtained using the electrometric "dead stop" end point.

7.6.2   Ashing

The ash content of any solid is determined by igniting a known quantity aliquot of the well-mixed sample at 1000°C in a muffle furnace. By knowing the weight of the original sample and the weight of the remaining ash, and using the formula below, the percentage of ash content of the sample is determined and calculated.

$$\% \text{ ash} = (\text{ash grams}/\text{original sample grams}) \times 100 \qquad \textbf{(7.2)}$$

Ash is considered as inorganic, and the difference between the moisture and the ash gives the percentage of the volatile and organic compounds. The sum of the percentages of moisture, organics, and ash should be a total of 100%.

Dry ashing — This is the technique used for liquid samples. A suitable volume of the well mixed sample is transferred into a platinum or high-silica glass (Cycor, a product of Corning Glass Work, Corning, New York, or equivalent). Evaporate to dryness on a steam bath. Transfer dish to a muffle furnace and heat sample to a white ash. If volatile elements are to be determined, keep temperature at 400 to 450°C. If sodium (Na) only is to be determined, keep temperature up to 600°C. Dissolve the ash in a minimum quantity of concentrated HNO_3 and warm water. Filter the diluted sample if necessary and dilute to a known volume, preferably so that the final HNO_3 concentration is about 1%. Take a portion of this solution to metal determination.

7.7 DISSOLVING THE SAMPLE

A solid sample usually must be put into solution before an analysis can be completed. Often dissolution is one of the most time-consuming steps in a quantitative analysis.

7.7.1   Decomposition Under Pressure in a Closed System

Often the increased pressure and higher temperature that can be realized in a closed system will greatly speed the dissolution process. For example, by placing the sample and solvent in a hard glass tube, which is then put inside a closed steel cylinder, temperatures up to 300°C and pressures above 100 atm are possible. Another device is a small steel autoclave that has a Teflon lining to prevent attack of the container itself.

7.7.2   Decomposition in an Open System

This is the more commonly used dissolution method. The sample is covered with a suitable solvent and usually is gently heated to make it dissolve faster. The solvent used should dissolve the sample completely in as short a time as possible. A solvent should be chosen that will not interfere in the subsequent steps of the analysis.

Many inorganic salts and some organic compounds dissolve readily in DI water. Occasionally, a small amount of acid is added to prevent hydrolysis and partial precipitation of certain metal ions. Dissolve organic compounds prior to analysis using common organic solvents. Concentrated or slightly diluted acids will dissolve most metals and metal alloys and many oxides, carbonates, and sulfides. However, some of them require fusion. Also, some phosphates are soluble in acids, but many require fusion with an alkali flux. Most silicate samples dissolve in hydrofluoric acid, but fusion is used extensively.

7.7.3   Digestion

Samples containing particulates or organic materials generally require pretreatment prior to the actual analysis. The digestion procedure breaks down an organically bound substance and converts the substance to the analyzed form by using liquid oxidizing agents such as sulfuric acid, nitric acid, perchloric acid, or hydrochloric acid, or using oxidizing mixtures such as "Aqua Regia" (a 3:1 mixture of concentrated HCl and concentrated HNO_3). Addition of bromine (Br_2) and hydrogen peroxide (H_2O_2) to mineral acids often increases their solvent action and hastens the oxidation of organic materials in the sample. More details about digestion procedures are specified in individual methods, such as those for metals, total phosphorus, total Kjeldahl nitrogen (TKN), etc. Hydrofluoric acid (HF) is primarily used for the decomposition of silicates (never use glass laboratory wear with HF because of its dissolving effect on glasses). Microwave-assisted digestion for metal analysis is also available. Only laboratory-grade microwave equipment and closed digestion containers with pressure relief that are specifically designed for hot acids may be used.

7.7.4   Fusion

Samples that fail to dissolve in any other way can usually be brought into solution by fusing them with a high-temperature acid such as potassium pyrosulphate ($K_2S_2O_7$), a base such as sodium carbonate (Na_2CO_3), or an oxidant such as sodium peroxide (Na_2O_2). A finely ground sample is mixed intimately with the granular solid flux, and the mixture is melted in a crucible. The melted flux attacks and dissolves the sample. Then the crucible is cooled, and the solidified melt is dissolved in dilute, aqueous acid, or water alone.

7.8   EXTRACTION

Substances have different solubilities in different solvents, and the process of selectively removing a solute from a mixture with a solvent is called *extraction*. The solvent used for the extraction process may be water, a water-miscible solvent, or a water-immiscible solvent. The selection of the solvent used depends upon the solute and upon the requirements of the procedure. Extraction procedures are used to separate, purify, and analyze substances.

Water is a polar solvent, and polar substances are soluble in it. For example, inorganic salts, salts of organic acids, strong acids and bases, carboxylic acids, alcohols, amines, and carbohydrates are water soluble.

A *dilute aqueous acid solution* (5 to 10% HCl) will extract basic substances. A *dilute aqueous basic solution* (5 to 10% NaOH, 5% $NaHCO_3$) will extract acidic substances from immiscible organic solvents. The acidic substances are converted to the sodium salt form which is soluble in water.

Organic solvents tend to dissolve organic substances. Common organic solvents lighter than water are diethylether, benzene, petroleum ether, and hexane. Organic solvents heavier than water are chloroform, ethylene dichloride, methylene chloride, and tetrachloromethane.

When a solute dissolved in a solvent such as water is shaken with a second immiscible solvent, there is a competition between the two solvents for the solute. What determines then whether the solutes end up primarily in one solvent phase or the other? Why do some solutes prefer an organic phase while others stay in aqueous phase? The general answer is that a solute will go into a phase in which it can form the more stable chemical species. A rule that often works is that "like dissolves like" (see Section 5.3).

7.8.1   Liquid–Liquid Extraction by Separatory Funnel

Many organic liquids are not miscible with water. When such a liquid is added to water, two layers are formed. Whether the organic liquid is the upper or lower layer depends on whether the density of the liquid is less or greater than that of water. Suppose there is an aqueous solution containing two dissolved solutes, A and B. An immiscible organic solvent is added to the solution and shaken vigorously and the mixture allowed to stand until the two solvent layers settle out. If the organic liquid has a much greater affinity for one of the dissolved solutes than does water, the solute will pass from the aqueous phase to the organic liquid phase. This solute is extracted. If the other solute prefers the aqueous phase to the organic, it will not be extracted. If extraction is carried out in a separatory funnel, the lower liquid can be carefully drained off and the two solutes will be physically separated.

The vapors of organic solvents are toxic to varying degrees. For this reason, solvent extraction should be performed in a hood and the use of more toxic solvents should be avoided.

Techniques by using separatory funnel

1. Pour the solution to be extracted into the separatory funnel. The funnel should be large enough to hold at least twice the total volume of the solution and the extraction solvent.
2. Pour in the extraction solvent.
3. Close the funnel with the stopper.
4. Shake the funnel gently.
5. When the funnel is shaken, pressure inside the funnel increases as a result of the added partial vapor pressures of the two solvents. The high pressure is reduced by holding the inverted funnel in both hands and opening the stopcock slowly to relieve the pressure. This is called the *venting procedure*. The venting procedure is repeated as many times as is necessary until no further pressure can be detected (i.e., if there is no more sound of escaping vapors).
6. Close the stopcock while the funnel is inverted and shake.
7. Repeat steps 5 and 6.
8. Place the funnel in a ring stand support and allow the two layers to separate.
9. Remove the stopper. Open the stopcock slowly and drain off the bottom layer.
10. Repeat the extraction with fresh extraction solvent as many times as desired (starting at step 2).
11. Combine the lower layers which have been drained off.

A separatory funnel with a ring stand is shown in Figure 7.21, and how to hold, shake, and vent a separatory funnel is shown in Figure 7.22.

Frequently, when aqueous solutions are extracted with organic solvents, *emulsions* form instead of two separate phases. An emulsion is a colloid in which small particles of one liquid are dispersed in another liquid. It may happen when the separatory funnel is shaken especially vigorously. Once emulsions form, it may take a very long time to separate. Emulsion formation can be minimized by gently swirling of the separatory funnel instead of shaking it vigorously, or the funnel may be gently inverted many times to achieve extraction. Emulsions caused by too small a difference in the densities of the water and organic layer can be broken by the addition of a high-density organic solvent tetrachloromethane (carbon tetrachloride). One or more of the following techniques may also be of value in breaking emulsions: add a few drops of silicone defoamer; add a few drops of dilute acid (if permissible); add a few drops of detergent solution; allow the emulsion to stand for a time; place the emulsion in a freezer; place the emulsion in a suitable centrifuge tube and centrifuge until the emulsion is broken; use a glass stirring rod to break the emulsion.

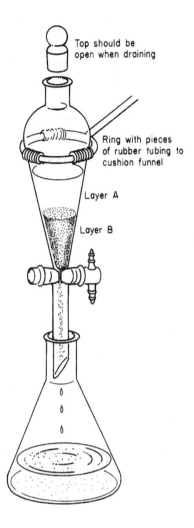

Figure 7.21
Separating immiscible liquids by means of a separatory funnel. After completion of the extraction, allow the two layers to separate. Remove stopper closure. Open the stopcock slowly and drain off the bottom layer.

7.8.2 Continuous Liquid–Liquid Extraction

When a solute is to be transferred from one solvent into another, requiring many extractions with a large volume of solvent due to its relative insolubility, the procedure is called *continuous extraction*. Specially designed laboratory glassware is required, because the extracting solvent is continually reused as the condensate from the total reflux. The solvents must be immiscible.

Higher density solvent extraction — In this extraction method the solvent has a higher density than the solution being extracted. The condensate from the higher density solvent passes through the extractable solution and returns to the boiling flask. Continuous heating vaporizes the solvent, and the process is continued as long as necessary.

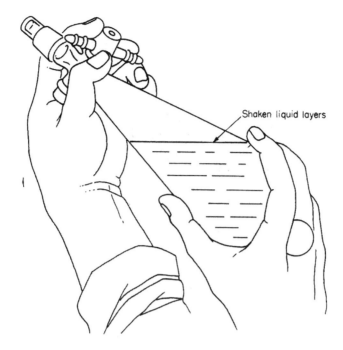

Shaken liquid layers

Figure 7.22
How to use a separatory funnel. Pour the extraction solvent into the solution to be extracted. After gentle shaking, invert the funnel and open the stopcock slowly to relieve the pressure built up (vent the separatory funnel). Close the stopcock and shake again several times.

Lower density solvent extraction — The extracting solvent has a lower density than the solution being extracted. The condensate of the extracting solvent is caught in a tube from the increased pressure, forcing it through the higher density solution, extracting the solute, and flowing back to the boiling flask. Continuous heating vaporizes the solvent, and the process is continued as long as necessary.

7.8.3 Sohxlet Extraction

A Sohxlet extractor (Figure 7.23) can be used to extract solutes from solids. Any volatile solvent can be used. The solvent is vaporized, and in the condensed phase dropped onto the solid substance placed in a thimble. When the liquid level fills the body of the extractor, it automatically siphons back into the flask. This process continues repeatedly.

Procedure setup

- Put the known-weight solid substance in the porous thimble and place it in the Sohxlet inner tube.
- Fill the flask one half full of the extracting solvent.
- Assemble the unit.
- Turn on the cooling water. Heat.
- When the extraction is complete, turn off the heat and cooling water.

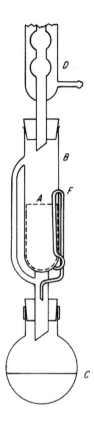

Figure 7.23
Sohxlet extraction. D = reflux condenser, B = body of extractor, A = extraction thimble, F = liquid return siphon, and C = extracting solvent.

- Dismantle the apparatus, and pour the extraction solvent containing the solute into a beaker. Isolate the extracted component by evaporating the solvent.
- Determine the component and calculate its concentration according to the method used.

Chapter **8**

GRAVIMETRIC METHODS OF ANALYSIS

8.1 GENERAL PRINCIPLES

Quantitative analytical chemistry is divided into two groups, *wet chemistry* and *instrumental methods*. Detailed discussion of these methods can be found in Chapter 1. The first of the two wet chemical methods we will consider is the *gravimetric method of analysis* (the second, titrimetric or volumetric analysis will be discussed in Chapter 9). The measurement step in a gravimetric method is weighing. A weighed sample is dissolved, after which an excess of precipitating reagent is added. The precipitate is filtered, washed, dried or ignited, and weighed. From the weight and known composition of the precipitate, the amount of the analyte can be calculated. From this and the weight of sample taken, the percentage of the desired substance in the original sample can be calculated.

A precipitate suitable for analytical procedures should be relatively insoluble, readily filterable, reasonably pure, and stable. It should be nonhygroscopic and have a known chemical composition.

8.1.1 Types of Precipitates

The three types of analytical precipitates are the curdy, gelatinous, and crystalline precipitates.

Curdy and gelatinous precipitates — Both form in the same way: the cation and anion react to form a soluble colloid, and then coagulate to filterable-sized particles. Heating is necessary to produce a precipitate. In contrast to the curdy precipitate, gelatinous precipitates trap a great deal of water and many more impurities.

The most useful curdy precipitates are the silver halides compounds. For example, when an aqueous solution of sodium chloride reacts with aqueous silver nitrate solution, the product will be aqueous solution of sodium nitrate and solid silver chloride precipitation, according to the following equation:

$$NaCl(aq) + AgNO_3(aq) \rightarrow NaNO_3(aq) + AgCl(s)$$

The best examples of gelatinous precipitates are metal hydroxides, such as $Fe(OH)_3$.

Crystalline precipitates — The precipitates of alkaline earth sulfates, such as barium sulfate, are good examples of crystalline precipitates. They precipitate as regularly shaped, discrete particles. Since these crystals are not pure and are often too small to filter effectively, they are heated to achieve larger, purer crystals. For example, sulfate ion is precipitated by adding an excess barium chloride ($BaCl_2$):

$$Na_2SO_4(aq) + BaCl_2(aq) \rightarrow 2\ NaCl(aq) + BaSO_4(s)$$

8.1.2 Aging and Digestion of Precipitates

Aging and digestion of precipitates frequently help to make them suitable for analytical procedures. Freshly formed precipitates are aged by leaving them in contact with the supernatant liquid at room temperature for a period of time. Heating during the aging process is called digestion. The precipitate is kept in contact with the supernatant at a temperature near boiling for a period of time.

8.1.3 Manipulations Associated With the Filtration Process

Whether one uses gravity or vacuum filtration (Sections 7.1.3 and 7.1.4), three operations must be performed: decantation, washing, and transfer.

Decantation — Carefully decant the clear liquid above the precipitate through the filter paper or filtering crucible (see Chapter 7), keeping as much as possible of the precipitate in the beaker. Use a stirring rod to direct the liquid into the filter or filtering crucible instead of down the outside of the beaker. This process is called decantation. The decantation process and the transfer of the precipitate to the filter are shown in Figure 8.1.

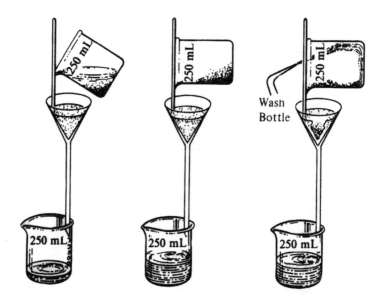

Figure 8.1
Decanting and transferring precipitate to filter paper (left). Decanting, keeping most of the precipitate in the beaker (center). Transferring the bulk of the precipitate down a stirring rod (right). Washing the remainder of the precipitate into the funnel with the wash bottle (a rubber policeman may also be needed in this step).

Washing — Washing the precipitate is necessary to remove interfering substances. Wash the precipitate down from the beaker's wall with a stream of washing solution. Allow the precipitate to settle and pour the supernatant into the filter. Repeat two or three times. Make sure that the stem of the funnel is well inside the beaker receiving the filtrate and that its tip touches the wall of the beaker.

Transferring the precipitate to the filter — Remove the bulk of the precipitate from the beaker with the help of a stream of the wash liquid. Use a stirring rod to direct the flow of liquid into the filtering medium. The last traces of precipitate are removed from the walls of the beaker by scrubbing the surfaces with a rubber policeman attached to the stirring rod.

Drying the precipitate — Some precipitates can be dried sufficiently for analytical determination without resorting to high temperatures. Some precipitates are collected in sintered glass crucibles and oven dried at 120°C to constant mass. Others are ignited and ashed before being brought to constant mass.

Silver chloride does not adsorb water strongly and is normally dried in this manner for ordinary analytical work.

Transfer of paper and precipitate to a crucible — When filtration and washing of the precipitate is completed, transfer the filter paper and the precipitate to a crucible.

If any type of filtering crucible must be ignited over a flame, it should be an ordinary porcelain crucible. Platinum crucibles also are ideal for this job, but are expensive.

Ignition of filter paper in a crucible — Ignition at high temperature is required for complete removal of water that is occluded or very strongly adsorbed, and for complete conversion of some precipitates to the desired compound. Gelatinous precipitates such as the hydrous oxides adsorb water quite strongly and must be heated to very high temperature to remove water completely.

If time allows, the best result is obtained when the filter paper in the crucible is left to dry overnight prior to ignition.

The steps of the ashing process are:

Charring the paper — Cover the crucible containing the filter paper and char the paper with gentle heat until the organic material gasifies and either escapes or condenses as a tar on the inner surface of the cover.

Burning off carbon residue — Remove the lid and increase the temperature slowly until all of the black carbon residue is burned away. When the carbon is gone, ignite for the recommended time at the highest temperature of the burner with the bottom of the crucible just touching the top of the blue cone of the flame. Ignition of a precipitate can be seen in Figure 8.2.

Cooling — Allow the crucible to cool until the red glow disappears, and then place the crucible in a desiccator. The lid should not be put on the crucible at that point, but doing so reduces the possibility of mechanical loss of the precipitate. Cool to constant weight.

Errors during ignition — Errors other than incomplete removal of water or volatile electrolytes can occur during ignition. One of the most serious is reduction of the precipitate

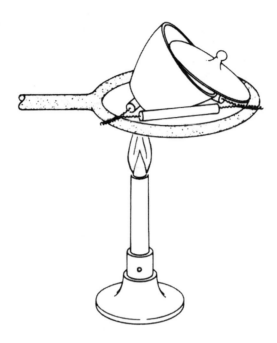

Figure 8.2
Ignition of a precipitate.

by carbon when filter paper is employed. Precipitates also can be overignited, leading to decomposition to substances of indefinite composition. Errors can result from reabsorption of water or carbon dioxide by an ignited precipitate upon cooling. Crucibles should be properly covered and kept in a desiccator as they cool.

8.2 GRAVIMETRIC CALCULATION

In the usual gravimetric procedures, from the measurement of the original sample (solid samples by weight, liquid samples by volume) and the weight of the precipitate, the percentage content of the analyte in the sample is calculated.

A common example of a gravimetric analysis is the determination of sulfate (SO_4) content of a sample by precipitation of the sulfate in the form of barium sulfate ($BaSO_4$). From the weight of the barium sulfate and the weight of the original sample, the actual percentage of the sulfates in the sample is calculated. For example,

- The original weight of the sample is 2.4386 g.
- It is dissolved in 100 mL DI water.
- To the acidified sample solution, barium chloride ($BaCl_2$) solution is added in excess with mixing.
- The solution is then heated and left to cool overnight (aging the precipitate).
- The filtered, ignited and cooled pure barium sulfate, $BaSO_4$, is found to weigh 0.9825 g.
- Calculate the sulfate ion (SO_4^{-2}) content of the precipitated $BaSO_4$ (0.9825 g) as follows:
  1. The weight of 1 mol of $BaSO_4$ is 233.389 g.
  2. The weight of SO_4^{-2} ion in 1 mol of $BaSO_4$ is 96.062 g.

3. If 233.389 g of $BaSO_4$ contains 96.062 g of SO_4^{-2}, how many grams of SO_4^{-2} will be in the precipitated 0.9825 g of $BaSO_4$? The problem can be written as a ratio:

$$\% \text{ substance} = \frac{\text{we}}{}$$

After cross multiplication and solving for x,

$$233.389 \, x = (96.062)(0.9825)$$

$$x = (96.062)(0.9825) / 233.389$$

$$x = 0.4044$$

Therefore, 0.4044 g of SO_4^{-2} is in the precipitated 0.9825 g $BaSO_4$.

- The next step is to calculate the percent sulfate content of the original sample:
  1. 0.9825 g $BaSO_4$ contains 0.4044 g SO_4^{-2}.
  2. 100 g $BaSO_4$ contains $\times$ g SO_4^{-2}.

$$\frac{0.9825}{0.4044} = \frac{100}{x}$$

After cross multiplication and solving for x,

$$x = (0.4044)(100) / 0.9825$$

$$x = 16.5$$

So, the sulfate (SO_4^{2-}) content of the sample is 16.5%.

8.2.1   Utilization of the Gravimetric Factor

By using the gravimetric factor, the calculation is much simpler and faster.

Gravimetric factor

The gravimetric factor is calculated by dividing the weight of 1 mol of the substance of interest with the weight of 1 mol of the precipitated compound.

$$\text{factor} = \frac{\text{weight of 1 mol of substance}}{\text{weight of 1 mol of the precipitated compound}} \tag{8.1}$$

Calculate the percentage of analyte with the gravimetric factor

For solid samples —

$$\% \text{ substance} = \frac{\text{weight of the precipitate } x \text{ factor} \times 100}{\text{weight of original sample}} \tag{8.2}$$

For liquid samples —

$$\% \text{ substance} = \frac{\text{weight of the precipitate } x \text{ factor} \times 100}{\text{volume of original sample}} \qquad (8.3)$$

Applying the gravimetric factor calculation for the above sulfate determination

$$\text{factor} = \frac{\text{weight of 1 mol SO}_4}{\text{weight of 1 mol BaSO}_4} = \frac{96.062}{233.389} = 0.4116$$

Substituting the values into formula (8.2),

$$\% \text{ SO}_4^{2-} = \frac{0.9825 \times 0.4116 \times 100}{2.4386} = 16.5$$

The sulfate (SO_4) content of the sample is 16.5%.
Selected gravimetric factors are shown in Appendix C, Table C–4.

Exercises

1. A sample that weighs 0.9516 g is analyzed for phosphorus (P) content by precipitating the phosphorus in the form of magnesium pyrophosphate ($Mg_2P_2O_7$). If the precipitate weighs 0.6294 g, what is the %P in the sample?
2. What will be the gravimetric factor if the substance sought is chloride (Cl) and the precipitated compound is lead chloride ($PbCl_2$)?
3. Calculate the gravimetric factor if the substance sought is chlorine (Cl_2) and the precipitated compound is aluminum chloride ($AlCl_3$).

8.3 GRAVIMETRIC ANALYSIS AND MODERN ANALYTICAL CHEMISTRY

Although it is true that the gravimetric technique has been displaced in its routine aspects by instrumental analysis, it is still of great importance in the field of analytical chemistry. In general, where only a few determinations are required, a gravimetric procedure may actually be faster and more accurate than an instrumental method which requires extensive calibration or standardization.

Gravimetric methods compare favorably with other analytical techniques in terms of accuracy attained. If the analyte is a major constituent of the sample, the gravimetric method offers great accuracy. If the analyte is present in minor or trace amounts, a gravimetric method is generally not employed.

Answers to exercises

1.

$$\text{factor} = \frac{\text{weight of 2 mol P}}{\text{weight of 1 mol Mg}_2\text{P}_2\text{O}_7} = \frac{61.9476}{222.5674} = 0.2783$$

$$\%\text{P} = \frac{0.6294 \times 0.2783 \times 100}{0.9516} = 18.4\%$$

The sample has 18.4% of phosphorus.

2. $$\text{factor} = \frac{2\,\text{Cl}}{\text{PbCl}_2} = \frac{2(35.453)}{278.1} = 0.2550$$

3. $$\text{factor} = \frac{3\,\text{Cl}_2}{2\,\text{AlCl}_3} = \frac{3(70.906)}{2(133.34)} = 0.7977$$

Chapter 9

TITRIMETRIC ANALYSIS

9.1 GENERAL PRINCIPLES

The second wet chemistry method of analysis is the method known as *titrimetric* or *volumetric analysis*. In gravimetric analysis (discussed in Chapter 8), the sample is weighed and dissolved, then the constituent is converted to a form that is isolated and weighed. In the titrimetric procedure, a substance that reacts with the constituent is added, and the exact amount of the added reagent needed to consume the constituent is measured.

Titrimetric analysis is one of the major divisions of analytical chemistry, and the calculations involved are based on the simple stoichiometric relationships of chemical reactions.

Titration is a quick, accurate, and widely used way of measuring the amount of a substance in solution. A titration is performed by adding exactly the volume of a *standard solution* (a solution of exactly known concentration) needed to react with an unknown quantity of a second substance. The concentration of the standard solution is determined by a process called *standardization* (see Section 6.2). The standard solution is called the *titrant*; the volume of titrant needed for the titration is carefully measured by means of a *buret*. In order to know when to stop the addition of the titrant, the chemist may use a chemical substance, called the *indicator*, which responds to the appearance of excess titrant by changing color. If the volume and concentration of the titrant are known, the unknown quantity of the substance titrated can be calculated.

9.1.1 Chief Requirements for a Titration

- The reaction should be stoichiometric; that is, there should be a definite whole-number ratio between the number of the moles in the reaction.
- The rate of the reaction should be rapid so that the titration can be carried out quickly.
- The reaction should be quantitative. For the usual analytical accuracy, it must be at least 99.9% complete when a stoichiometric amount of titrant has been added.
- Some method should be available for determining the point in the titration at which a stoichiometric amount of titrant has been added and the reaction is completed. When this point is determined experimentally by an indicator color change or by some change in the electrochemical or physical properties of the solution, it is called the *end point* of the titration. The point at which the theoretical amount of titrant has been added is called the *equivalence point* of the titration. Ideally, the end point and the equivalence point coincide, but for various reasons there is frequently some difference between the two.

Choosing indicators to make the two points coincide or correcting for the difference between the two is one of the important aspects of titrimetric analysis.

9.1.2 Indicators

Indicators are usually weak organic acids or bases that change color when the hydrogen ion concentration changes. Not all indicators change color at the same pH. The choice of indicator for a particular titration depends on the pH at which the equivalence point is expected to occur.

Acid–base indicators — These are usually weak acids in which the un-ionized form HA has a different color from the negative ion A⁻. In solution, the indicator dissociates slightly:

$$HA \Leftrightarrow H^+ + A^-$$

In acid solution the concentration of H^+ is high, and the indicator is largely undissociated HA. In alkaline solutions the equilibrium is displaced to the right and A^- is formed.

Acid–base indicators and selection of the proper indicator is discussed in Section 10.3.1.

Other types of indicators — Indicators can be used for other reactions. *Starch*, for example, is used in iodine titration, because of the deep blue complex it forms. *Oxidation–reduction indicators* are substances that show a reversible color change between oxidized and reduced forms. These titrations are detailed in Section 11.2.2.

9.1.3 Volumetric Glassware

Volumetric glassware used for titrimetric procedures is discussed in Section 4.2.

9.1.4 Reaction Used for Titration

Any of the following types of chemical reaction can serve as the basis of a titration.

Precipitation titration — The end point of this titration can be detected by any of several visual-indicator methods (See Section 11.3).

Acid–base titration — Hundreds of compounds, both organic and inorganic, can be determined by a titration based on their acidic or basic properties. Acid is titrated with a base, and base is titrated with an acid. The end point is usually detected by adding an indicator, or by potentiometric titration using a pH meter.

Complex formation used in titration — Most metal ions can be accurately determined by titration with a standard solution of an organic complexing agent such as ethylenediaminotetraacetic acid (EDTA). This reagent reacts with the most metal cations to form a very stable, water soluble complex. The added indicator solution forms a highly colored complex with the metal ion. The color remains as long as some metal ion remains untitrated. When a stoichiometric amount of EDTA titrant has been added, the complex of the metal ion and indicator dissociates, causing a color change and marking the end point as discussed in Section 11.1.

Oxidation–reduction titration — Several elements that have more than one oxidation state can be determined by titration with a standard oxidizing or reducing agent. The end point is determined by addition of an indicator or by a potentiometric titration as discussed in Section 11.2.2.

9.1.5 Performing a Titration

- Rinse the buret three times with the solution to be used, draining completely each time.
- Fill the buret above the zero graduation.
- Drain slowly until the tip is free of air bubbles and completely filled with liquid and the meniscus of the liquid is at the zero graduation.
- Add the titrant solution slowly to the known volume titrating solution, swirling the flask with the right hand until the end point is obtained. To avoid error, with the tip well within the titration vessel, introduce solution from the buret in increments of a milliliter or so. Swirl (or stir) the sample constantly to assure efficient mixing. Reduce the volume of the additions as the titration progresses; in the immediate vicinity of the end point, the titrant should be added a drop at a time. Near the end point, the trail of the color from each drop is quite long!
- When it is judged that only a few more drops are needed, rinse down the wall of the titration vessel. Allow a minute or so to elapse between the last addition of reagent and the reading of the buret. For precision work, volume of less than one drop can be rinsed off the tip of the buret with analyte-free water.
- The end point is reached when the color change does not disappear after 30 s.
- Allow the buret to stand for 30 s, then read the final position of the meniscus.

CAUTION: Never allow reagent to remain in a buret overnight. The stopcock may "freeze" because of prolonged contact, especially with bases, such as sodium and potassium hydroxide (NaOH, KOH).

To prevent a stopcock from freezing, discard the titrant (but do not return it to the original container), and rinse the buret several times with DI water. If possible, fill it with DI water and cap it to keep dust out. Otherwise, store the buret upside down.

9.2 CALCULATING THE RESULT OF A TITRATION

To calculate the result of a titration, it is necessary to know the volume and the concentration of the titrant. The concentration of the titrant commonly used is molarity (*M*) or normality (*N*). As previously discussed, molarity of a solution is equal to the moles in 1 L of solution, and normality can be expressed by the equivalents in 1 L of the solution (Section 5.3.1).

The result of a titration may be expressed in different concentrations, such as molarity (*M*), normality (*N*), percent (%), parts per million (ppm), and parts per thousand (ppt).

9.2.1 Results Given in the Same Concentration as the Titrant

$$c_1 v_1 = c_2 v_2 \tag{9.1}$$

where c_1 = concentration of the titrant, c_2 = concentration of the unknown, v_1 = volume of the titrant used (mL), and v_2 = volume of the unknown sample used for titration.

For example, 100 mL of unknown concentration of HCl solution are titrated with 16.8 mL of 0.1 N NaOH. What is the normality of the HCl solution?

$$(x)(100) = (0.1)(16.8)$$

$$x = (0.1)(16.8)/100$$

$$x = 0.0168$$

The concentration of the HCl solution is 0.0168 N.

9.2.2   Results Given in Different Concentrations, Using Normality

$$\% \text{ w/w} = (\text{mL titrant} \times N \times \text{Eqw of unknown} \times 100)/\text{mg sample} \qquad \textbf{(9.2)}$$

$$\% \text{ w/v} = (\text{mL titrant} \times N \times \text{mEqw of unknown} \times 100)/\text{mL sample} \qquad \textbf{(9.3)}$$

$$\text{g/L} = (\text{mL titrant} \times N \times \text{mEqw of unknown} \times 1000)/\text{mL sample} \qquad \textbf{(9.4)}$$

$$\text{ppm (mg/L)} = (\text{mL titrant} \times N \times \text{Eqw of unknown} \times 1000)/\text{mL sample} \qquad \textbf{(9.5)}$$

$$\text{ppm (mg/kg)} = (\text{mL titrant} \times N \times \text{Eqw of unknown} \times 1000)/\text{g sample} \qquad \textbf{(9.6)}$$

$$\text{Normality } (N) = \text{g/L/Eqw of substance} \qquad \textbf{(9.7)}$$

where Eqw = equivalent weight and mEqw = milliequivalent weight.

9.3 BACK-TITRATION

Sometimes the rate of a chemical reaction is too low for a titration to be carried out directly by adding the titrant to the unknown solution until the reaction is completed. Instead, a measured volume of titrant is added that is in excess of that theoretically needed for the reaction. Then the excess titrant is back-titrated with a reagent (back-titrant) that will react rapidly with it.

The calculation method is to obtain a number of milliliters of titrant used to combine with the substance titrated, and then to calculate the results as was done for a direct titration.

$$\text{titrant added} - \text{titrant excess} = \text{net titrant}$$

To calculate the result as ppm,

$$[(\text{mL titrant} \times N) - (\text{mL back-titrant} \times N) \times \text{Eqw unknown} \times 1000]/\text{mL sample} \quad \textbf{(9.8)}$$

To calculate the result as a percentage,

$$[(\text{mL titrant} \times N) - (\text{mL back-titrant} \times N) \times \text{Eqw unknown} \times 100]/\text{mg sample} \quad \textbf{(9.9)}$$

Exercises

1. 100 mL HCl was titrated with 16.8 mL 0.1 N NaOH solution. Calculate the normality and the percentage concentration of the HCl solution.

2. Determine the calcium hydroxide, $Ca(OH)_2$, content of an unknown concentration solution. Calcium hydroxide is a base, therefore it is titrated with a known concentration of acid. It is determined that 25 mL of the unknown solution consumes 24.6 mL of 0.02 N HCl solution. Calculate the $Ca(OH)_2$ concentration and express in percentage and in ppm.

3. 5.00 g of an unknown material is analyzed for its chloride content. The sample was dissolved in 50 mL of analyte-free water and titrated with 18.2 mL of 0.141 N mercuric nitrate, $Hg(NO_3)_2$, standard solution. What is the percentage of (% w/w) chloride content of the material?

4. A water sample is analyzed for sulfide (S^{2-}) content by using the iodometric back-titration method. To the sample, add an amount of iodine solution estimated to be an excess over the amount of sulfide present. If the iodine color disappears, add more iodine so that the color remains. Back-titrate with sodium thiosulfate, $Na_2S_2O_3$ solution, adding a few drops of starch indicator as the end point is approached, and continuing until blue color disappears.

   For a 200 mL water sample, add 20 mL of 0.0250 N iodine solution, and back-titrate with 18.2 mL of 0.025 N sodium thiosulfate solution. What is the sulfide content of the water sample in ppm?

Answers to exercises

1. Using formulas (9.3), (9.4), and (9.7):

$$\% \ (w/v) \ HCl = (16.8 \times 0.1 \times 0.03646 \times 100)/100 = 0.06125\% \ HCl$$

$$g/L \ HCl = (16.8 \times 0.1 \times 0.03646 \times 1000)/100 = 0.6125 \ g/L \ HCl$$

$$N \ HCl = 0.6125/36.461 = 0.0168 \ N \ HCl$$

The solution is 0.06125% HCl solution. It contains 0.6125 g HCl per 1 L solution and its normality is 0.0168 N.

To solve the same problem without using the formulas:

1 N NaOH contains 39.997 g/L or 39.997 mg/mL NaOH

equivalent to

1 N HCl contains 36.461 g/L or 36.461 mg/mL HCl

0.1 N NaOH contains 3.9997 g/L or 3.9997 mg/mL NaOH

equivalent to

0.1 N HCl contains 3.6461 g/L or 3.6461 mg/mL HCl.

In this exercise 100 mL of unknown concentration, HCl solution consumed 16.8 mL 0.1 N NaOH, or $16.8 \times 3.9997 = 67.1950$ mg NaOH. Therefore, the 100 mL unknown concentration solution contains 67.1950 mg NaOH equivalent quantity HCl. If 3.9997 mg NaOH is equivalent to 3.6461 mg HCl, the 67.1950 mg NaOH is equivalent to X mg HCl.

$$3.9997 \quad 67.1950$$
$$3.6461 \quad x$$

Solve the x by cross multiplication.

$$x = (36.460)(67.1950)/39.9997 = 61.2167 \text{ mg} = 0.06122 \text{ g HCl}$$

0.0612 g HCl is in 100 mL (*0.0612% HCl*), so 0.6120 g in 1000 mL. Therefore, the solution contains *0.6120 g/L HCl*.

To calculate the normal concentration of this particular HCl, 1 N HCl has 36.460 g HCl in 1 L; x N HCl has 0.6120 g HCl in 1 L.

$$\frac{1}{36.460} = \frac{x}{0.6120}$$

$$x = 0.6120/36.461 = \textit{0.0168 N HCl}$$

2. Using formulas (9.1) and (9.4),

$$\% \text{ Ca(OH)}_2 = (24.6 \times 0.02 \times 0.0370 \times 100)/25 = 0.0728\%$$

$$\text{ppm Ca(OH)}_2 = (24.6 \times 0.02 \times 37.0 \times 1000)/25 = 728 \text{ ppm (mg/L)}$$

The solution is 0.0728% and contains 728 ppm of $Ca(OH)_2$.

To solve the same problem without using the formulas.

The equivalent weight of $Ca(OH)_2$ = 74.092 g/2 = 37.046 g. The equivalent weight of HCl = 36.460 g/1 = 36.460. 1 N HCl has 36.460 g/L and it is equivalent to 37 g/L $Ca(OH)_2$. 0.02 N HCl has 0.729 g/L and it is equivalent to 0.74 g/L $Ca(OH)_2$.

Therefore, 1 mL of 0.02 N HCl contains 0.729 mg of HCl and it is equivalent to 0.740 mg of $Ca(OH)_2$. 24.6 mL of 0.02 N HCl contains 24.6×0.729 = 17.93 mg HCl. It is equivalent to 18.20 mg $Ca(OH)_2$, in 25 mL of the unknown solution. Therefore, 18.20×40 = 728 mg is in 1000 mL, so the solution contains *728 mg/L or ppm of Ca(OH)₂*. If 1000 mL contains 728 mg, 100 mL contains 72.8 mg or 0.0728 g. So, the solution is *0.0728% Ca(OH)₂*.

3. Using formula (9.2),

$$\% \text{ w/w Cl} = [18.2 \times 0.141 \times 0.0365 \times 100]/5 = 1.87$$

The material contains 1.87% chloride.

To solve the same problem without using the formula.

The equivalent weight of the $Hg(NO_3)_2$ is 324.6/2 = 162.3 g, 162.3×0.141 = 22.8 g/L or 22.8 mg/mL in the 0.141 N solution. If 162.3 g of $Hg(NO_3)_2$ is equivalent to 36.460 g of chloride, the 22.8 g is equivalent to 36.460×0.141 = 5.14 g Cl, so the 1 mL of the 0.141 N $Hg(NO_3)_2$ solution (22.8 mg) is equivalent to 5.14 mg Cl, and 18.2 mL is equivalent to 5.14×18.2 = 93.55 mg Cl. So, 93.55 mg is the chloride content of the 5 g sample material, and 93.55×20 = 1871 mg Cl = 1.871 g Cl is in 100 g of sample. In other words, *the chloride content of the material is 1.87% w/w.*

4. Substituting into the formula (9.8),

$$\text{ppm S}^{-2} = [(20)(0.025) - (18.2)(0.025) \times 16 \times 1000]/200 = 3.6$$

Note that 16 = equivalent weight of sulfide (32/2). *The water sample has 3.6 ppm of sulfide (S^{2-}) content.*

Chapter 10

Acid–Base Titration

Acid–base titration is a quick and accurate method of determining acidic or basic substances in analytical samples. Most commonly, the end point of an acid–base titration is detected by observing the color change of an indicator. However, a pH meter can also be used to follow the pH changes during an acid–base titration, called potentiometric titration.

10.1 ACIDS AND BASES

Long ago, scientists used simple tests to classify a substance as an acid or a base. Acids had a sour taste and turned blue litmus to red, and corroded metals. Aqueous solutions of bases generally had a bitter taste and turned red litmus to blue. They also had a soapy "feel." However, the taste test isn't very safe, and you are strongly urged not to apply it to anything!

Antoine Lavoisier was one of the first chemists to try to explain what makes a substance acidic. In 1777, he proposed that oxygen was an essential element in acids. (Oxygen, which he named, means "acid former" in Greek). But in 1808, Sir Humphry Davy showed that hydrogen chloride, which dissolves in water to give hydrochloric acid (HCl), contains only hydrogen and chlorine. Chemists then noted that hydrogen, not oxygen, must be an essential constituent of acids.

10.1.1 Arrhenius Theory of Acids and Bases

According to Svante Arrhenius, *an acid is a substance that, when dissolved in water, increases the concentration of hydrogen ion (H^+). A base is a substance that, when dissolved in water, increases the concentration of the hydroxide ions (OH^-)*. An acid which ionizes in dilute aqueous solution to produce many hydrogen ions (nearly complete ionization) is classified as a *strong acid*. An acid that ionizes slightly in water to produce few hydrogen ions is classified as a *weak acid*. A *strong base* ionizes in water nearly completely to produce many hydroxyl ions. A *weak base* ionizes slightly in water to produce a few hydroxide ions.

The hydrogen ion is actually hydrated and can be represented as $H^+ \cdot H_2O$ or H_3O^+, commonly known as hydronium ion. But for simplicity, instead we write H^+.

$$\text{Acid: } HX \Leftrightarrow H^+ + X^-$$
$$\text{Base: } BOH \Leftrightarrow OH^- + B^+$$

10.1.2 Bronsted–Lowry Concept of Acids and Bases

In 1923 the Danish chemist Johannes Bronsted and, independently, the British chemist Thomas Lowry pointed out that acid–base reactions can be seen as proton-transfer reactions and that acids and bases can be defined in terms of proton (H$^+$) transfer. According to the Bronsted–Lowry concept, *an acid is the species donating a proton in a proton-transfer reaction. A base is a species accepting the proton in a proton-transfer reaction.* Substances that can be either acids or bases depending on the other substance present are said to be *amphoteric.* Another term that is sometimes used is "amphiprotic," to stress that proton donating/accepting ability is of central concern (from the Greek amphoteros, "partly one and partly the other"). For example, the bicarbonate ion (HCO$_3^-$) found in baking soda can either donate a proton to a base or accept a proton from an acid. Toward OH$^-$, therefore, HCO$_3^-$ is an acid:

$$K = \frac{[C]^c[D]^d}{}$$

Toward H$^+$, however, HCO$_3^-$ is a base:

$$HCO_3^-(aq) + OH^+(aq) \rightarrow H_2CO_3(aq) + H_2O$$

10.2 BASIC CONCEPT OF CHEMICAL EQUILIBRIUM

Chemical reactions relating to chemical analysis are frequently reversible reactions. That is, a "reverse" reaction competes with the "forward" reaction. After a time, these two opposing reactions occur at the same rate, and the equilibrium is established. Such a reaction system is written with the double arrow ($\Leftrightarrow$) as in the following general reaction:

$$aA + bB \Leftrightarrow cC + dD \tag{10.1}$$

A and *B* represent reactant chemicals, and *C* and *D* represent the products, and *a, b, c, d* represent the coefficients required to balance the equation.

A mathematical expression, called the equilibrium constant expression, can be written as follows for this reaction:

$$K_{eq} = \frac{[C]^c[D]^d}{[A]^a[B]^b} \tag{10.2}$$

The brackets ([]) refer to the molar concentration of the chemicals. The value of K_{eq} is constant at a given temperature. The explanation of this constancy is based on Le Chatelier's Principle. It says that if a stress is placed on a chemical reaction system at equilibrium, the equilibrium shifts so as to relieve the stress. For example, in the case of addition of more reactant or product to a reaction system, the equilibrium will shift so as to maintain the constancy of K_{eq}.

10.2.1 Acid–Base Equilibrium

Acid–base equilibrium is an extremely important topic throughout chemistry and in other fields which utilize chemistry, such as biology, medicine, and agriculture. A strongly acidic species, H$^+$, and a strongly basic species, OH$^-$, exist in identical concentrations in pure water. Water is therefore neutral in an acid–base sense, meaning that it is neither acidic nor basic. When the molar concentration of H^+ exceeds that of OH^-, the solution is acidic; when it is less, the solution is basic. The relative concentrations of H^+ and OH^- thus enable us to define acidic and basic solutions.

Acidic solution $[H^+] > [OH]$

Basic solution $[H^+] < [OH^-]$

Neutral solution $[H^+] = [OH^-]$

10.2.2 Self-Ionization of Water

Because water, the most common solvent for acids and bases, is itself a (very) weak acid, we consider the self-ionization first. The self-ionization of water is written:

$$H_2O \Leftrightarrow H^+(aq) + OH^-(aq) \tag{10.3}$$

We can write the equilibrium constant:

$$K_{eq} = \frac{[H^+][OH^-]}{[H_2O]} \tag{10.4}$$

Water self-ionization constants do not change the value of $[H_2O]$. Initially, $[H_2O]$ equals 55.6 mol/L. If we take the density of water to be 1.00 g/mL (or 1000 g/L), the molar concentration of H_2O is $1000/18 = 55.6$ mol/L (18 g is the weight of 1 mol of water). Because the concentration of water (55.6 mol/L) is practically constant, it can be incorporated into the expression

$$K[H_2O] = [H^+][OH^-] \tag{10.5}$$

We can use a new constant, called the *ion-product constant of water*:

$$K_w = [H^+][OH^-] \tag{10.6}$$

In a neutral solution at 25°C, $[H^+] = [OH^-] = 1 \times 10^{-7}$; therefore,

$$K_w = (1 \times 10^{-7})(1 \times 10^{-7}) = 1 \times 10^{-14} \tag{10.7}$$

The hydrogen ion concentration expressed as a power of 10 is known as the pH. To simplify discussions of very small values of $[H^+]$, scientists use the negative of the logarithm of the numerical part of the molar concentration of H^+ and call this number the pH of the solution. By definition, the pH is equal to the negative logarithm of the hydrogen ion concentration and pOH is equal to the negative logarithm of the hydroxide ion concentration:

$$pH = -\log [H^+] \tag{10.8}$$

$$pOH = -\log [OH^-] \tag{10.9}$$

where p stands for the exponent or "power," and H and OH refer to H^+ and OH^-.

The sum of the pH and pOH is 14, because they originated from the hydrogen and hydroxide ion concentrations (expressed in moles per liter).

$$pH + pOH = 14.00 \text{ (at } 25°C)\qquad\qquad\textbf{(10.10)}$$

Because $[H^+] = 1 \times 10^{-7}$ mol/L in pure water at 25°C, the pH of pure water is 7.00.

$$pH = -\log(1 \times 10^{-7})$$
$$= -(-7)$$
$$= 7.00$$

An alternative expression for pH is

$$pH = \log\frac{1}{[H^+]}\qquad\qquad\textbf{(10.11)}$$

We have to be very careful to notice the reciprocal relationship between pH and $[H^+]$. Low values of pH mean relatively high values of $[H^+]$, and high values of pH mean relatively low values of $[H^+]$.

10.2.3 pH Calculations

We can calculate pH from either $[H^+]$ or $[OH^-]$, and we can find the value of $[H^+]$ or $[OH^-]$ from a given pH.

pH of strong acids and bases — Acids and bases are commonly classified as *strong* or *weak*, to indicate the approximate degree of ionization. The strong acids include hydrochloric (HCl), hydrobromic (HBr), hydroiodic (HI), nitric (HNO_3), perchloric ($HClO_4$), sulfuric (H_2SO_4), and organic sulfonic acids (RSO_3H). Acids in this group are assumed to be 100% ionized in dilute aqueous solution. If the concentration of a strong acid is known, the pH can be easily calculated. Strong bases are also considered to be 100% ionized in dilute aqueous solution, and they include alkali metal hydroxides, certain alkaline earth metal hydroxides, and quaternary ammonium hydroxides (R_4NOH). The pH of these solutions is found after first calculating pOH.

Examples

1. Because rain washes pollutants out of the air, the lakes in many parts of the world are undergoing changes in pH. The water in one lake was found to have $[H^+] = 3.2 \times 10^{-5}$ mol/L. Calculate the pH and pOH of the lake's water and decide whether it is acidic or basic.

   Given a value of $[H^+]$ we can calculate pH directly from equation (10.8):

   $$pH = -\log 3.2 \times 10^{-5}$$
   $$= -(-4.49485)$$
   $$= 4.49$$

Because the pH is less than 7.00, the water of the lake is acidic. Calculate the pOH from equation (10.10):

$$pOH = 14 - 4.49$$

$$= 9.51$$

The value of the [OH⁻] may be calculated by taking the antilog of the pOH:

$$\text{antilog } (-9.51) = 3.09 \times 10^{-10}$$

$$[OH^-] = 3.09 \times 10^{-10}$$

The lake water has a pH = 4.49, pOH = 9.51, [H⁺] = 3.2 × 10⁻⁵, and [OH⁻] = 3.09 × 10⁻¹⁰.

2. A water sample has [OH⁻] 1.47×10^{-9} mol/L. What is the pH of the water sample?

To calculate the pOH value, use equation (10.9).

$$pOH = -\log (1.47 \times 10^{-9})$$

$$= -(-8.8327)$$

$$= 8.83$$

Calculate pH using equation (10.10).

$$pH = 14 - pOH$$

$$= 14 - 8.83$$

$$= 5.17$$

Alternatively, one can calculate the [H⁺] by using equation (10.7).

$$1 \times 10^{-14} = [H^+][OH^-]$$

$$1 \times 10^{-14} = [H^+](1.47 \times 10^{-9})$$

$$[H^+] = (1 \times 10^{-14}) / (1.47 \times 10^{-9})$$

$$[H^+] = 6.80 \times 10^{-6}$$

Calculate pH according to equation (10.8).

$$pH = -\log 6.80 \times 10^{-6}$$

$$= 5.17$$

The pH of the sample is 5.17.

pH of weak acids and bases — Acids and bases are classified as weak acids or bases because they are slightly ionized in solution. The pH of a weak acid can be calculated from the ionization constant, K_a. Similarly, the pH of a weak base can be calculated from

TABLE 10.1

Acid Ionization Constants at 25°C

Substance	Formula	K_a
Acetic acid	$HC_2H_3O_2$	1.7×10^{-5}
Benzoic acid	$HC_7H_5O_2$	6.3×10^{-5}
Boric acid	H_3BO_3	5.9×10^{-10}
Carbonic acid	H_2CO_3	4.3×10^{-7}
Cyanic acid	$HCNO$	3.5×10^{-4}
Formic acid	$HCHO_2$	1.7×10^{-4}
Hydrocyanic acid	HCN	4.9×10^{-10}
Hydrofluoric acid	HF	6.8×10^{-4}
Hydrogen sulfide	H_2S	8.9×10^{-8}
Hypochlorous acid	$HClO$	3.5×10^{-8}
Nitrous acid	HNO_2	4.5×10^{-4}
Oxalic acid	$H_2C_2O_4$	5.6×10^{-2}
Phosphoric acid	H_3PO_4	6.9×10^{-3}
	H_2OP_{4-}	6.2×10^{-8}
	HPO_4^{2-}	4.8×10^{-13}
Propionic acid	$HC_3H_5O_2$	1.3×10^{-5}
Pyruvic acid	$HC_3H_3O_3$	1.4×10^{-4}
Sulfurous acid	H_2SO_3	1.3×10^{-2}
	HSO_3^-	6.3×10^{-8}

Note: The degree of ionization of a weak acid depends on both K_a and the concentration of the acid solution. For a given concentration, the larger the K_a, the greater the degree of ionization. For a given value of K_a, however, the more dilute the solution, the greater the ionization.

its K_b value ($K_a = [H^+][A^-]/[HA]$, $k_b = [HB_+][OH^-]/[B]$). Values of K_a and K_b for selected weak acids and bases are shown in Tables 10.1 and 10.2.

Exercises (Answers begin on Page 121)

1. What are the concentrations of hydrogen ion and acetate ion in a solution of 0.10M acetic acid, CH_3COOH? What is the pH of the solution?
2. What is the hydrogen ion concentration of an 0.20 M solution of ammonia solution?

TABLE 10.2

Basic Ionization Constants at 25°C

Substance	Formula	K_b
Ammonia	NH_3	1.8×10^{-5}
Aniline	$C_6H_5NH_2$	4.2×10^{-10}
Dimethylamine	$(CH_3)_2NH$	5.1×10^{-4}
Ethylamine	$C_2H_5NH_2$	4.7×10^{-4}
Hydrazine	N_2H_4	1.7×10^{-6}
Hydroxylamine	NH_2OH	1.1×10^{-8}
Methylamine	CH_3NH_2	4.4×10^{-4}
Pyridine	C_5H_5N	1.4×10^{-9}
Urea	NH_2CONH_2	1.5×10^{-14}

10.2.4 Measurement of pH

pH test papers — pH test papers are available for every value of pH; they cover both broad and narrow pH ranges. The paper is impregnated with one indicator. A strip of test paper is wet with liquid to be tested and then immediately compared with the standard color chart provided for each paper and range. The pH can be determined visually by comparison of the colors. This measurement is just an approximation of the pH value.

pH meter — A pH meter functions by measuring the electric potential between two electrodes that are immersed into the solution of interest. The basic principle is to determine the activity of the hydrogen ions by potentiometric measurement using a glass and a reference electrode. The most popular, called a *combination electrode* incorporates the glass and the reference electrode into a single probe. The *glass electrode* is sensitive to the hydrogen ions, and changes its electrical potential with the change of the hydrogen ion concentration. The *reference electrode* has a constant electric potential. The difference in potential of these electrodes, measured in millivolts (mV), is a linear function of the pH of the solution. The scale of the pH meters is designed so that the voltage can be read directly in terms of pH. The composition and the working theory of ion selective and reference electrodes as well as a discussion of the potentiometric methods of analysis can be found in Section 13.2.3.

10.2.5 Buffers

A buffer is a solution characterized by the ability to resist changes in pH when limited amounts of acid or base are added to it. It is composed of a weak acid and the salt of that acid or a weak base and the salt of that base. Therefore, buffers contain either a weak acid and its conjugate base or a weak base and its conjugate acid. Thus, a buffer solution contains both an acid species and a base species in equilibrium. Many chemical and biological processes are very sensitive to changes in the pH of a solution, and it is extremely important to maintain as constant a pH as possible. Buffer solutions are therefore of considerable interest in the chemical and biological sciences. The capacity of a buffer is a measure of its effectiveness in resisting changes in pH upon the addition of acid or base.

10.3 ACID–BASE TITRATION

An acid–base titration is a procedure for determining the amount of acid (or base) in a solution by determining the volume of base (or acid) of known concentration that will completely react with it. Several inorganic acids and bases and hundreds of organic compounds are sufficiently acidic or basic to be determined by acid–base titration.

Most commonly, the end point of an acid–base titration is detected by observing the color change of an indicator. However, a pH meter can be used to follow the pH changes that occur during an acid–base titration and locate the equivalence point. (Terms related to titrimetric analysis are discussed in Section 9.1).

10.3.1 Acid–Base Titration Curves

The acid–base titration curve is a plot of the pH of the solution of acid (or base) against the volume of added base (or acid). Such curves are used to gain insight into the titration

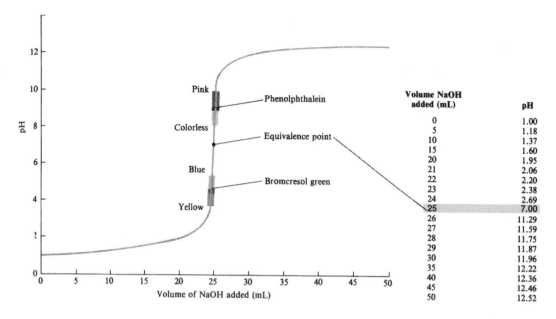

Volume NaOH added (mL)	pH
0	1.00
5	1.18
10	1.37
15	1.60
20	1.95
21	2.06
22	2.20
23	2.38
24	2.69
25	7.00
26	11.29
27	11.59
28	11.75
29	11.87
30	11.96
35	12.22
40	12.36
45	12.46
50	12.52

Figure 10.1

Curve for the titration of a strong acid by a strong base. 25 mL of 0.1 M HCl is titrated by 0.1 M NaOH. The portion of the curve where indicators bromcresol green and phenolphthalein change color are shown. Note that both indicators change color where the pH changes rapidly (the nearly vertical part of the curve). The equivalence point is at pH 7.00. To detect the equivalence point, the selected indicator color change should be within pH 3 to 11. Phenolphthalein can be used, because it changes from colorless to pink in the pH range of 8.2 to 10.0. The indicator bromcresol green, whose color changes in the pH range 3.8 to 5.4, would also work.

process. They can be used to choose an indicator that will show when the titration is complete. pH changes slowly at first until the titration is near to the equivalence point; however, it changes rapidly near the equivalence point. The rate of change (pH change per milliliter change of titrant) is greatest at the equivalence point.

Strong acid titrated with strong base; strong base titrated with strong acid — The titration curve for HCl titrated with NaOH is shown in Figure 10.1. At the equivalence point, a small increment of titrant causes a pH change of several units. The pH changes rapidly from pH 3.00 to about 11. Any indicator that has a color transition range within the nearly vertical portion of the titration curve is suitable for the titration. Methyl red, phenolphthalein, and several other indicators may be used for this type of titration.

Weak acid titrated with strong base — The titration of a weak acid by a strong base gives a somewhat different curve (Figure 10.2). As before, the pH changes slowly at first, then rapidly near the equivalence point. The pH range is shorter than that for the titration of a strong acid by a strong base. This means the choice of the indicator is more critical. The equivalence point of a weak acid-strong base titration occurs on the basic side; therefore, the optimum choice of indicator would be one that changes color over a range that includes the pH of the equivalence point. Phenolphthalein would work, because it changes color between pH 8.2 and 10.00.

Weak base titrated with a strong acid — The titration curve is similar to that obtained when a weak acid is titrated with a strong base. Figure 10.3 shows that the pH declines

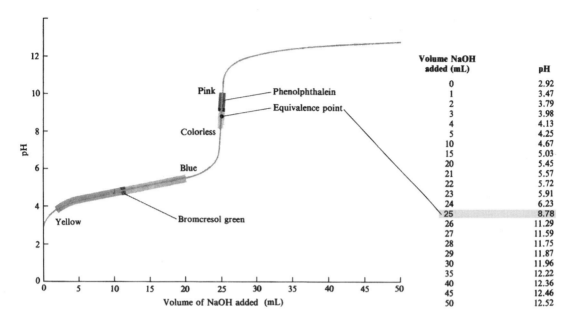

Volume NaOH added (mL)	pH
0	2.92
1	3.47
2	3.79
3	3.98
4	4.13
5	4.25
10	4.67
15	5.03
20	5.45
21	5.57
22	5.72
23	5.91
24	6.23
25	8.78
26	11.29
27	11.59
28	11.75
29	11.87
30	11.96
35	12.22
40	12.36
45	12.46
50	12.52

Figure 10.2
Curve for the titration of a weak acid by a strong base. Here 25 mL of 0.1 M nicotinic acid, a weak acid, is titrated by 0.1 M NaOH. Note, that bromcresol green changes color during the early part of the titration, well before the equivalence point. Phenolphthalein changes color where the pH changes rapidly (near to the equivalence point). Thus, phenolphthalein would work — it changes color in the range of 8.2 to 10.0. Bromcresol green changes color in the range 3.8 to 5.4; therefore, it would not work. The equivalence point is at the basic side, pH 8.87.

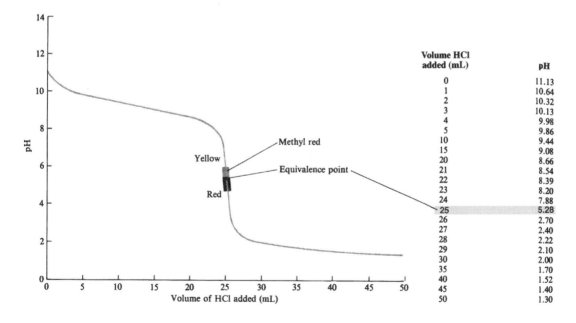

Volume HCl added (mL)	pH
0	11.13
1	10.64
2	10.32
3	10.13
4	9.98
5	9.86
10	9.44
15	9.08
20	8.66
21	8.54
22	8.39
23	8.20
24	7.88
25	5.28
26	2.70
27	2.40
28	2.22
29	2.10
30	2.00
35	1.70
40	1.52
45	1.40
50	1.30

Figure 10.3
Curve for the titration of a weak base by a strong acid. Here 25 mL of 0.1 M NH_3 (weak base) is titrated by 0.1 HCl. In this case the pH declines slowly at first, then falls abruptly from about pH 7 to pH 3. Methyl red, which changes color from yellow at pH 6.0 to red at pH 4.8 is a possible indicator for this titration.

slowly at first, than falls abruptly from about pH 7.00 to pH 3.00. So, the equivalence point of the weak base-strong acid titration occurs on the acidic side, and the selected indicator changes the color in this pH area. Note that phenolphthalein could not be used!

10.3.2 Acid–Base Indicators

Acid–base indicators are compounds that change color on going from acidic to basic solutions. They are usually weak acids in which the un-ionized form HA has a color different from the negative ion A⁻. In solution the indicator dissociates slightly:

$$HA \Leftrightarrow H^+ + A^-$$

In acid solution the concentration of H⁺ is high, and the indicator is largely undissociated HA. In alkaline solutions the equilibrium is displaced to the right and A⁻ is formed. Useful acid–base indicators show a sharp color change over a range of about 2 pH units. Most indicators are two-color indicators in which the acidic and basic forms have contrasting colors. There are also a few one-color indicators such as phenolphthalein, in which the acidic form is colorless and the basic form is magenta. An indicator changes color over a pH range and not at a single pH. Common acid–base indicators with their pH ranges and color changes are listed in Table 10.3.

Selection of proper indicator — When color change indicators are used to indicate the equivalence point of an acid–base titration, the selection of the correct indicator for that particular system is most important.

TABLE 10.3

Some Acid–Base Indicators

Indicator	Color Change With Increasing pH	pH Range
Picric acid	Colorless to yellow	0.1–0.8
Thymol blue	Red to yellow	1.2–2.8
2,6-Dinitriphenol	Colorless to yellow	2.0–4.0
Methyl yellow	Red to yellow	2.9–4.0
Bromphenol blue	Yellow to blue	3.0–4.6
Methyl orange	Red to yellow	3.1–4.4
Bromcresol green	Yellow to blue	3.8–5.4
Methyl red	Red to yellow	4.2–6.2
Litmus	Red to blue	5.0–8.0
Methyl purple	Purple to green	4.8–5.4
p-Nitrophenol	Colorless to yellow	5.6–7.6
Bromcresol purple	Yellow to purple	5.2–6.8
Bromthymol blue	Yellow to blue	6.0–7.6
Neutral red	Red to yellow	6.8–8.0
Phenol red	Yellow to blue	6.8–8.4
p-a-Naphtholphthalein	Yellow to blue	7.0–9.0
Phenolphthalein	Colorless to red	8.0–9.6
Thymolphthalein	Colorless to blue	9.3–10.6
Alizarin yellow R	Yellow to violet	10.1–12.0
1,3,5-Trinitrobenzene	Colorless to orange	12.0–14.0

Note: The ranges are roughly 1 to 2 pH units. It should also be noted that various indicators change color at widely different pH values. It is necessary for the analyst to select the proper indicator for the titration.

If we titrate *strong acid with a strong base*, for example hydrochloric acid (HCl) with sodium hydroxide (NaOH), the *equivalence point will be at pH 7.00*. In the titration of a *weak acid with a strong base*, for example acetic acid (CH_3COOH) with sodium hydroxide (NaOH), the *equivalence point will be at basic pH range*. For titration of a *weak base with a strong acid*, for example ammonia solution (NH_4OH) with hydrochloric acid (HCl), the *equivalence point is in acidic pH range*.

For example, when selecting a proper indicator for a weak acid-strong base titration, methyl red (pH interval 4.4 to 6.2) should not be used, because the equivalence point is between pH 7.00 and pH 10. Methyl orange (pH interval 3.1 to 4.4) should not be used. Bromthymol blue (pH interval 6.0 to 7.6) would also give a false indication before the equivalence point is reached. Phenolphthalein (pH interval 8.3 to 9.8) would be satisfactory. As a general rule, one should select an indicator that changes color at approximately the pH at the equivalence point of the titration. For weak acids, the pH at the equivalence point is above 7.00, and phenolphthalein is the usual choice. For weak bases, where the pH is below 7.00, methyl red (pH 4.2 to 6.2) or bromthymol blue are suitable.

In order to sharpen the color change of some indicators, mixtures of two indicators have been recommended. For example, the color change of mixed bromcresol green–methyl red (blue to pink) is more easily detected than the gradual change of methyl orange from yellow to red through a number of shades of orange.

Typical acid–base titrations with calculations are discussed in Section 9.4, Exercises 1 and 2.

Answers to exercises

1. K_a for acetic acid is 1.7×10^{-5}.

$$K_a = [H^+][CH_3COO^-]/[CH_3COOH]$$

Substitute the values:

$$1.7 \times 10^{-5} = [x][x]/0.1$$

$$x^2 = (1.7 \times 10^{-5})(0.1)$$

$$x^2 = 1.7 \times 10^{-6}$$

$$x = 1.3 \times 10^{-3}$$

$$pH = -\log 1.3 \times 10^{-3}$$

$$pH = 2.89$$

$[H^+] = [CH_3COO^-] = 1.3 \times 10^{-3}$; pH = 2.89.

2. K_b for ammonia is 1.8×10^{-5}.

$$K_b = [NH_4][OH^-]/[NH_3]$$

Substitute the known values:

$$1.8 \times 10^{-5} = x \times x / 0.2$$

$$x^2 = (1.8 \times 10^{-5})(0.2)$$

$$x = \sqrt{3.6 \times 10^{-6}}$$

$$x = [OH^-] = 1.9 \times 10^{-3}$$

$$pOH = -\log 1.9 \times 10^{-3} = 2.72$$

$$pH = 14 - 2.72 = 11.28$$

The pH of the ammonia solution is 11.28.

Chapter 11

COMPLEXOMETRIC, OXIDATION–REDUCTION, AND OTHER TITRATIONS

11.1 COMPLEXOMETRIC TITRATION

Acid–base (neutralization) reactions are only one type of many that are applicable to titrimetric analysis. There are reactions that involve the formation of a complex ion.

11.1.1 Complex Formation

Many metal ions, especially those of the transition elements, form coordinate covalent bonds with molecules or anions having lone pairs of electrons. Complexes are formed by the reactions of a metal ion, a cation, with an anion or neutral molecule. The metal ion in the complex is called a *central atom*, and the group attached to the central atom is called a *ligand*. Nearly all ligands have one thing is common: they possess a lone pair of electrons that may be shared with the metal cation in coordinate covalent bonding. The number of bonds formed by the central metal atom is called the *coordination number* of the metal. For example, in the reaction of silver ion with cyanide ion to form the very stable $Ag(CN)_2^-$ complex, the silver ion is the central atom with a coordination number of two, and cyanide is the ligand. Some typical complexes along with some of their properties are listed in Table 11.1.

When the formula of a metal complex is written, we usually indicate the species that are bonded to the metal by enclosing them and the metal ion within square brackets. For example, $[CoCl_6]^{3-}$. These brackets are not to be confused with those that we used earlier when we wished to denote molar concentration! Ligands such as Cl^-, or NH_3, which have one atom that can bond to a metal cation, are said to be *monodentate ligands*. Molecules or ions that have two atoms that may coordinate the metal ion, are said to be *bidentate ligands*. There are also more complex *polydentate ligands* containing three, four, or even more donor atoms.

Heterocyclic rings formed by the interaction of a metal ion with two or more functional groups in the same ligand are called *chelate rings*, the organic molecule is a *chelating agent*, and the complexes are named *chelates*, from the Greek "chela," meaning "claw". The ligand in this case bites the metal with two claws (donor atoms) much like a crab.

TABLE 11.1

Some Typical Complexes

Metal	Ligand	Complex	Coordination Number of Metal	Geometry
Ag^+	NH_3	$Ag(NH_3)_2^+$	2	Linear
Hg^+	Cl^-	$HgCl_2$	2	Linear
Cu^+	NH_3	$Cu(NH_3)_4^+$	4	Tetrahydral
Ni^{2+}	CN^-	$Ni(CN)_4^{2-}$	4	Square planar
Co^{2+}	H_2O	$Co(H_2O)_6^{2+}$	6	Octahedral
Co^{3+}	NH_3	$Co(NH_3)_6^{3+}$	6	Octahedral
Cr^{3+}	CN^-	$Cr(CN)_6^{3-}$	6	Octahedral
Fe^{3+}	CN^-	$Fe(CN)_6^{3-}$	6	Octahedral

A particularly important polydentate ligand is the anion of ethylenenediaminotetraacetic acid, abbreviated EDTA:

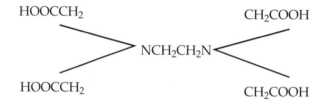

Its six coordinating atoms firmly attach themselves to free metal ions. EDTA has been used as an antidote in lead poisoning, because it binds to Pb^{2+}. The calcium disodium salt of EDTA is used in this treatment, because EDTA by itself would remove too much of the blood serum's calcium. In solution, EDTA has a greater tendency to complex with Pb^{2+} than with Ca^{2+}. As a result, the calcium is released and the lead tied up in the complex:

$$[CaEDTA]^{2-} + Pb^{2+} \rightarrow [PbEDTA]^- + Ca^{2+}$$

The lead chelate is then excreted in the urine. The structure of the chelate formed when the anion of EDTA envelopes a lead(II) ion is shown in Figure 11.1.

EDTA also binds to iron and calcium and is used as a water softener in products such as shampoos. EDTA is also added to foods to tie up metal ions that catalyze oxidation

Figure 11.1
The structure of the chelate formed when the anion of EDTA envelopes a lead(II) ion.

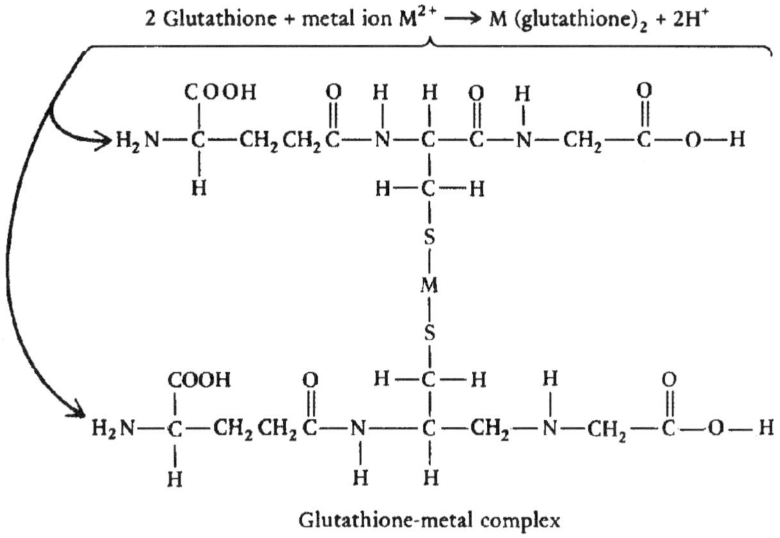

$$2 \text{ Glutathione} + \text{metal ion } M^{2+} \longrightarrow M \text{ (glutathione)}_2 + 2H^+$$

Figure 11.2
Glutathione reaction with a metal (M).

(and hence deterioration) of the food product. It has also been found that EDTA increases the storage life of whole blood by removing free Ca^{2+}, which promotes clotting.

Arsenic and heavy metals owe their toxicity primarily to their ability to react with and inhibit sulfhydryl (–SH) enzyme systems, such as those involved in the production of cellular energy. For example, glutathione (a tripeptide of glutamic acid, cystein, and glycine) occurs in most tissues; its behavior with metals illustrates the interaction of a metal with sulfhydryl groups. The metal replaces the hydrogen on two sulfhydryl groups on adjacent molecules (Figure 11.2), and the strong bond that results effectively eliminates the two glutathione molecules from further reaction. Glutathione is involved in maintaining red blood cells.

The problem of developing a compound to counteract lewisite, an arsenic-containing poison gas used in World War I, led to an understanding of how arsenic acts as a poison and subsequently to the development of an antidote. Once it was understood that lewisite poisoned people by the reaction of arsenic with protein sulfhydryl groups, British scientists set out to find a suitable compound that contained highly reactive sulfhydryl groups that could compete with sulfhydryl groups in the natural substrate for the arsenic, and thus render the poison ineffective. Out of this research came a compound now known as *British anti-lewisite* (BAL). BAL, which bonds to the metal at several sites, is a chelating agent.

With the arsenic or heavy metal ion tied up, the sulfhydryl groups in vital enzymes are freed and can resume their normal functions. BAL is a standard therapeutic item in a hospital's poison emergency center and is used routinely to treat heavy metal poisoning. BAL chelation of a heavy metal ion is shown in Figure 11.3.

Many interesting metal chelate compounds are found in biological systems. Oxygen is carried throughout our body tissues by hemoglobin, which is packaged in our red blood cells. Hemoglobin is a tetrameric protein of four polypeptide chains, each subunit of which binds a nonprotein entity called heme. Heme is the ferrous chelate of tetrapyrrole called protoporphyrin IX. The chelate is shown in Figure 11.4. Chlorophyll in green plants is the magnesium chelate of a tetrapyrrole, where a magnesium ion of coordination number 4 is chelated by the four electron donor nitrogen atoms of the four pyrrol rings, as shown

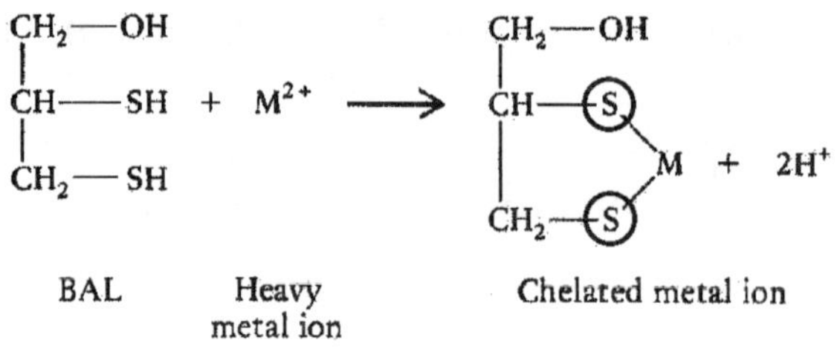

$$CH_2-OH \qquad\qquad CH_2-OH$$
$$CH-SH + M^{2+} \longrightarrow CH-\textcircled{S}$$
$$\qquad\qquad\qquad\qquad M + 2H^+$$
$$CH_2-SH \qquad\qquad CH_2-\textcircled{S}$$

BAL   Heavy   Chelated metal ion
metal ion

Figure 11.3
British anti-lewisite (BAL) chelation of a heavy metal ion. Lewisite is an arsenic-containing poison gas used in World War I. BAL is a chelating agent and makes the arsenic and other heavy metals ineffective. BAL is a standard therapeutic item in hospital's poison emergency centers and is used routinely to treat heavy metal poisoning.

in Figure 11.5. Another example is Vitamin B_{12}, which is a Co(III) chelate. It is similar to the heme and chlorophyll systems.

11.1.2 Metal Ion Determination

Many metal ions can be determined by titrating them with a reagent that complexes them in solution. The solution to be titrated is buffered at a suitable pH, an indicator is added, and the metal ion is titrated with a standard solution of the complexing agent. Usually, a sharp color change marks the end point of the titration. Titrations of this type are convenient and accurate; in many instances, they have replaced time-consuming gravimetric procedures. Except for the alkali metals, most metal cations can be determined by titration with a suitable complexing agent. For titration of a metal ion with a complexing ligand, the formation constant of the complex must be large so that the titration reaction

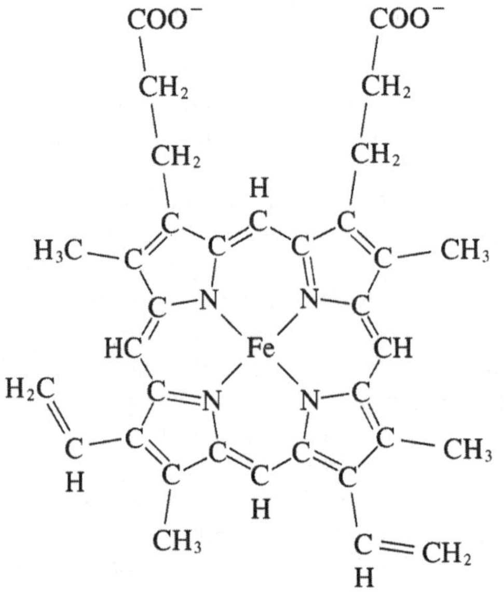

Figure 11.4
The structure of hemoglobin.

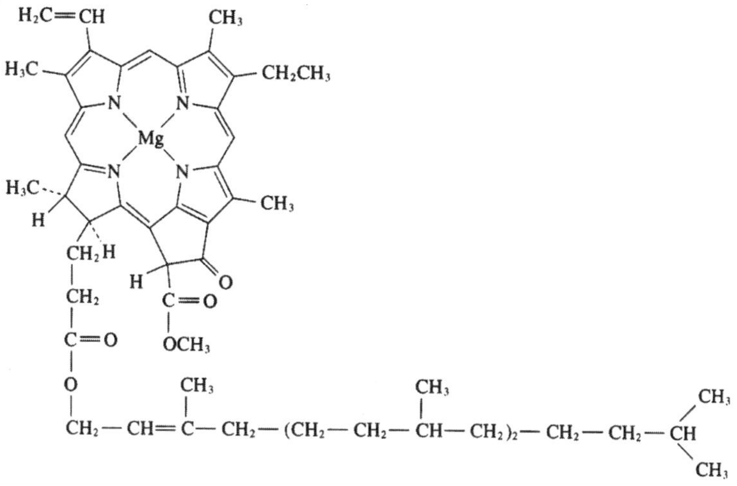

Figure 11.5
The structure of chlorophyll.

will be stoichiometric and quantitative. Complexing agents, such as EDTA, are suitable as titrants for metal ion determination.

A variety of substances, often called metallochromic indicators, are now available for this titration procedure. Basically, *organochromic indicators* are colored organic compounds which themselves form chelates with metal ions. The most common indicator used in complexometric titration is *Eriochrom Black T*. The molecule is represented as a tripotic acid, H_3In, which is dissociated in aqueous solution into H^+ and H_2In^-. This form of the indicator is red. The indicator forms stable complexes with a number of metal ions. Many EDTA titrations are performed in buffers of 8 to 10 pH range in which the predominant form of Eriochrom Black T is the blue HIn^{2-} form. Eriochrom Black T is, unfortunately, unstable in solution, and solutions must be freshly prepared in order to obtain the proper color change. Another indicator of similar structure is called *calmagite*. It is stable in aqueous solution and maybe substituted for Eriochrom Black T.

Applications of EDTA titrations — Hard water contains dissolved calcium and magnesium salts. Water heaters, faucets, and other objects that are in long contact with hard water often "lime up" with a deposit of precipitated metal salts. Calcium and magnesium salts of soaps are insoluble and form a scum called "bathtub ring." In a great many industries, the hardness of water used must be kept under strict control. Titration with standard EDTA is used throughout most of the world to determine the amount of hardness in water. Although both calcium and magnesium are titrated with the standard EDTA, total hardness is calculated as if it were all calcium carbonate ($CaCO_3$). Analysis of total hardness is outlined in Section 21.2.

11.2 OXIDATION–REDUCTION TITRATION

11.2.1 Theory of Oxidation–Reduction

Oxidation occurs when a substance loses electrons and reduction occurs when a substance gains electrons. Oxidation and reduction always take place at the same time. When a substance loses electrons (oxidation), another substance must gain these electrons (reduction). The reaction is called *redox reaction*. In other words, an oxidation–reduction reaction is one in which one or more of the reacting chemicals goes to a more positive oxidation state by losing electrons and one or more reactants is converted to a less positive oxidation state by gaining electrons. Therefore, *oxidation occurs when the oxidation number of a substance increases in a chemical reaction, and reduction occurs when the oxidation number of the substance decreases in a chemical reaction.*

Oxidation states (sometimes called oxidation numbers) — Oxidation numbers allows us to keep track of the electrons associated with each atom. To assign oxidation numbers, the following set of rules is used:

Rule 1. All pure elements are assigned the oxidation number of zero.

Rule 2. All monatomic ions are assigned oxidation numbers equal to their charges.

Rule 3. Certain elements usually possess a fixed oxidation number in compounds:

- The oxidation number of O in most compounds is –2.
- The oxidation number of H in most compounds is +1.
- The oxidation number of halogens in many, but not all, binary compounds is –1.
- The oxidation numbers of alkali metals and alkaline earth metals are +1 and +2, respectively.

Rule 4. The sum of all oxidation numbers in a compound equals zero, and the sum of oxidation numbers in a polyatomic ion equals the ion's charge.

For Na, Fe, H_2, and Cl_2, the oxidation number is 0. For Cl^-, Na^{1+}, Ca^{2+}, and O^{2-} the oxidation numbers are equal to the charge.

Exercises (Answers on Page 132)

1. What is the oxidation number for N in the compound NO?
2. What is the oxidation number of N in the compound N_2O?
3. What is the oxidation number of N in HNO_3?
4. What is the oxidation number of Cr in potassium dichromate, $K_2Cr_2O_7$?

Determining which substance goes through oxidation and which through reduction by using oxidation numbers — Let's consider the oxidation–reduction reaction in which metallic sodium combines with chlorine gas to produce sodium chloride:

$$2\ Na(s) + Cl_2(g) \rightarrow 2\ NaCl(s)$$

To determine which substance has undergone oxidation or reduction, it is necessary to assign oxidation numbers to each element in the equation and then look for those that have changed.

During the reaction, the oxidation number of sodium increases (0 to +1) and the oxidation number of chlorine decreases (0 to –1). If the oxidation number of a substance increases during the reaction, it indicates that it has lost electrons. Electrons are negative particles; consequently, when a negative particle is lost, the substance becomes more positive.

$$2\ Na \rightarrow 2\ Na^+ + 2\ e^- \qquad Oxidation$$

At the same time, the electrons lost by Na are picked up by Cl_2, resulting in the decrease of oxidation number of Cl_2.

$$2\ e^- + Cl_2 \rightarrow 2\ Cl^- \qquad Reduction$$

Therefore, oxidation and reduction can be redefined in terms of oxidation numbers:

1. Oxidation occurs when the oxidation number of a substance increases during a chemical reaction.
2. Reduction occurs when the oxidation number of a substance decreases during the reaction.
3. The reactant that gains electrons is the oxidizing agent and the reactant that releases electrons is the reducing agent.

An oxidizing agent takes electrons from another substance, resulting in the oxidation of that substance. A reducing agent gives up electrons to another substance, bringing about its reduction. In other words, *the substance undergoing reduction is the oxidizing agent, and the substance undergoing oxidation is the reducing agent.*

Leo the **germ** is a mnemonic that will help you to remember: *l*ose **e**lectrons **o**xidation, and **g**ain **e**lectrons **r**eduction. One "**m**" is added at the end to form an English word.

11.2.2 Oxidation–Reduction Titrations (Redox Titrations)

Chemical reactions involving oxidation–reduction are widely used in titrimetric analysis. The ions of many elements can exist in different oxidation states, resulting in the possibility of a very large number of redox reactions. For example, the determination of iron in solution by first reducing it to iron(II) and then titrating with standard potassium permanganate ($KMnO_4$) is one of the oldest of all titrating methods. The development of modern analytical methods such as spectroscopy has somewhat reduced our dependence on oxidation–reduction titrations for inorganic analysis. However, these methods continue to be useful, especially for determinations in which a high degree of accuracy is required.

Calculations with normality are based on the use of equivalent weight. As discussed previously in Section 5.3.1, the equivalent weight of a substance involved in redox reaction is the formula weight divided by the number of electrons transferred in the redox reaction.

There are several types of indicators that may be used in redox titrations. These indicators are called the *redox indicators*. The redox indicators are organic molecules that undergo structural changes upon being oxidized or reduced. *Diphenylamine* was one of the first redox indicators to be widely used in titrimetric analysis. The reduced form of this indicator is colorless, the oxidized form is deep violet. The indicator *ferroin* is the iron(II) complex of the organic compound 1,10-phenanthroline, with a blood-red color. It is prepared by mixing equivalent quantities of iron(II) sulfate and 1,10-phenanthroline. Its color change from red to light blue during oxidation is discussed below.

Titration with strong oxidizing agent — *Potassium permanganate*, $KMnO_4$, has been widely used as an oxidizing agent for over 100 years. It is a reagent that is readily available, inexpensive, and requires no indicator. Permanganate reacts rapidly with many reducing agents according to the reaction below:

$$MnO_4^- + 9\,H^+ + 5e^- \rightarrow Mn^{2+} + 4\,H_2O$$

The reaction takes place in a very acidic solution, and some substances require heating or the use of a catalyst to speed up the reaction. Preparation and standardization of the permanganate standard solution is discussed in Section 6.3.6.

Cerium(IV) titrant is a powerful oxidizing agent. It is prepared in a sulfuric acid solution. It is known that both Ce(IV) and Ce(III) ions form a stable complex with various anions. When Ce(IV) ion is used as a titrant, the compound ferroin is normally employed as an indicator. Ferroin is the iron(II) complex of the organic compound 1,10-phenantroline, with a blood-red color. The iron(II) ion can be oxidized to iron(III), and the latter ion also forms a complex with 1,10-phenantroline. The color of the iron(III) complex is light blue, and a sharp color change occurs when the iron(II) is oxidized to iron(III) in the presence of 1,10-phenantroline.

Potassium dichromate is a somewhat weaker oxidizing agent than permanganate and cerium(IV). It is used to titrate ferrous iron and certain other reducing agents.

Direct titrations with iodine — Elemental iodine, I_2, is a solid that is soluble in a number of organic solvents but insoluble in water. However, iodine is extremely soluble in an aqueous solution containing sodium or potassium iodide. The iodine reacts with the iodide ion and forms a soluble anionic complex, which is deep red-brown in color.

$$I_2 + I^- \rightarrow I_3^-$$

Iodine is a fairly weak oxidizing agent. Actually, the mild oxidizing agent action of iodine is often an advantage for titrations where a more powerful titrant might cause nonstoichiometric oxidations to occur. The indicator for such titrations is a *starch* suspension. The first trace of excess iodine marks the end point of the titration by forming an intense blue complex with the starch. Direct iodine titrations are usually carried out in neutral or acidic solution. At pH value higher than 11, the stoichiometry of iodine titrations is erratic.

Direct iodine titration is the *Karl Fisher method* for water. A quantitative determination of water is frequently needed, particularly in organic materials. The Karl Fisher method is a titration of water with an anhydrous methyl alcohol solution containing iodine, sulfur dioxide, and excess pyridine. Each mole of water requires 1 mol of I_2. The sample is titrated with the Karl Fisher reagent until a permanent iodine color is observed.

Indirect titrations involving iodine — An important general method for determining oxidizing chemicals involves iodine indirectly. An excess sodium or potassium iodide is added to the sample. The oxidizing chemical is reduced by the iodide, resulting in the formation of an equivalent amount of iodine, which is titrated with standard sodium thiosulfate ($Na_2S_2O_3$). The iodine formed in the first reaction is equivalent to the amount of oxidizing agent in the sample. Starch is the indicator in this titration. The disappearance of the blue starch–iodine complex marks the end point. It is important not to add the starch indicator until most of the iodine has been titrated. If the starch is added too soon, iodine is adsorbed onto the starch, making the end point very slow and hard to detect. The indirect iodine method works well for chlorine determination:

$$\overset{O}{Cl_2} + 2\,\overset{-}{KI} \to 2\,\overset{-}{KCl} + \overset{O}{I_2}$$

The formed iodine is then titrated with standard sodium thiosulfate, starch being the indicator:

$$I_2 + 2\,S_2O_3^{2-} \to 2\,I^- + S_4O_6^{2-}$$

This indirect iodine method works very well for determining either chlorine or bromine.

11.3 PRECIPITATION FORMATION TITRATIONS

The formation of a precipitate can be used as a basis of a titration, provided that there is a suitable way of determining when a stoichiometric amount of titrant has been added. The most important are the titration of halide ions with silver(I) and of sulfate with barium(II).

11.3.1 Mohr Method for Halides

The Mohr method for chloride was published more than a hundred years ago and is still used. Chloride is titrated with a standard silver nitrate ($AgNO_3$) solution; the indicator is chromate ion, (CrO_4^{2-}), the precipitate is brown silver chromate (Ag_2CrO_4). The first permanent appearance of the reddish-silver chromate precipitate is taken as the end point of the titration. The concentration of the chromate indicator is very important. If too much chromate is added, the end point occurs before the equivalence point; if not enough, the end point comes late.

The Mohr titration is limited to solutions with pH values from about 6 to 10. In more alkaline solutions, silver oxides precipitate.

11.3.2 Volhard Method

The Volhard method is a procedure for titrating silver(I) with standard potassium thiocyanate (KSCN). The titration is carried out in acidic solution. When the silver(I) has been precipitated as white silver thiocyanate, the first excess of titrant and the iron(III) indicator react and form a soluble red complex. The method is widely used for silver and chloride, because the titration can be done in acidic solution.

In the Volhard method for determining chloride and other anions, a measured volume of standard silver nitrate solution is added to the sample solution. The volume of silver nitrate is in excess of the amount needed to react with the halide. The excess silver(I) is then back-titrated with standard thiocyanate solution.

$$Ag^+ + SCN^- \to AgSCN\,(s)$$

$$Fe^{3+} + SCN^- \to FeSCN^{2+}\,(red)$$

11.3.3 The Use of Adsorption Indicators: The Fajans Method

Adsorption indicator method for halides — In the adsorption indicator method for halide ions, the end point reaction occurs on the surface of the silver halide precipitate. Adsorption of a colored organic compound on the surface of a precipitate may induce electronic shifts in the molecule that alter its color. The organic compounds thus employed are referred to as adsorption indicators. A commonly used indicator for chloride determination is *fluorescein*. When silver ions are in excess, the fluorescein ions (Fl^-) can be attracted to the surface of the positively charged particles:

$$(AgCl) \cdot Ag^+/Fl^-$$

The resulting aggregate is pink, and the color is sufficiently intense to serve as a visual indicator.

A list of some adsorption indicators is given in Table 11.1.

Adsorption indicator method for sulfate — This method for the determination of sulfate is very important. Where the gravimetric method is slow and time consuming, this method is quick and reasonably accurate. The precipitated barium sulfate ($BaSO_4$) is fluffy and highly adsorptive if the titration is carried out in a mixture of 50:50 water and methyl alcohol at a pH of 3.5. Alizarine red S is used as an indicator. The indicator is yellow in solution but forms a pink complex on the surface of the precipitate. The color change is sharp and distinct. Foreign cations as sulfates and foreign anions as barium salts may precipitate also with barium sulfate and cause errors. A few metal cations also interfere by forming a colored complex with the indicator. To avoid these errors, ion exchange is used to remove cations that would interfere with the end point or cause error by coprecipitation.

Answers to examples

1. According to rule 3 above, the oxidation number for oxygen is –2. Because –2 needs +2 to give 0, the oxidation number for N is +2.
2. Oxygen is –2, N should be +2, but we have two nitrogens, therefore N should have oxidation number of $2/2 = 1$. The oxidation number of N is +1.
3. Hydrogen +1, oxygen $-2 \times 3 = -6$. Make the sum of these numbers: $-6 + 1 = -5$, so the oxidation number for N should be +5.
4. $K = +1 \times 2 = +2$; $O = -2 \times 7 = -14$; $-14 + 2 = -12$. Because –12 needs +12 to give 0, Cr oxidation number is $+12/2 = +6$.

Chapter **12**

SPECTROPHOTOMETRY

12.1 FUNDAMENTALS OF SPECTROSCOPY

12.1.1 Early History of the Nature of Light

Since ancient times the nature of light, and particularly of color, had been a source of curiosity. The idea, from ancient Greece until the seventeenth century, was that colors consisted of a mixture of white light and darkness and could be changed by changing the mixture. This idea was radically changed by the work of Isaac Newton (1642–1727), who at age 24 started his research about light. A ray of sunlight passed through a hole into a darkened room and onto a screen. The prism, placed in the beam of the light, dispersed the light into a spectrum of colors ranging in the order of red, yellow, green, blue, and violet. Newton concluded that white light is a "confused aggregate of rays imbued with all sort of colors." The function of the prism is merely to analyze the light into its component colors. No more discoveries occurred until after Newton's death. In 1808 the British astronomer William Herschel discovered the infrared portion of the solar spectrum. Soon after, the ultraviolet part of the spectrum was found. It has been reported in 1802 that there are dark lines in the sun's spectrum, but without any satisfactory explanation. In 1817, Joseph von Fraunhofer, an optician and instrument maker, noted the same lines. With improved equipment, he proceeded to map the dark lines of the solar spectrum, and calculated the corresponding wavelengths. Until this day, these lines are called the Fraunhofer lines, which are "dark lines in the solar spectrum that result from the absorption by elements in the solar chromosphere of some of the wavelengths of the visible radiation emitted by the hot interior of the sun" (*A Concise Dictionary of Chemistry,* Oxford University Press, Second Edition, 1990, page 127).

During the first half of the nineteenth century, a good deal of experimentation took place with the colored flames produced by injecting various salts into a flame. When light passed through a slit and a prism onto a screen, bright discrete lines were seen against a dark background, the reverse of the solar spectrum. The connection between the two was not made for many years. Robert Bunsen, professor of chemistry at the University of Heidelberg (and designer of the laboratory "Bunsen gas burner") viewed the exhibited colored flames by different salts through spectroscope. He noted that the colors were linked to the element, not the compound in which it was bound. He realized that the bright lines in the visible region of the spectrum seen with a spectroscope were characteristic of specific elements, and that this technique could be used as an extremely sensitive and simple method of element identification. With this new method, Bunsen identified and isolated two new elements, cesium (Cs) and rubidium (Rb). Gustav Kirchhoff, professor of physics at Heidelberg University, became interested in Bunsen's work. Kirchhoff examined the dark lines of the spectrum and concluded that the appearance of these lines is due to a process of absorption as the emission rays passed through the cool outer layer of the sun's atmosphere, which causes them to show up dark against the bright background. This absorption spectrum is just as characteristic of a specific element as its emission spectrum.

The effectiveness of Bunsen and Kirchhoff's spectroscopy in chemical analysis was first used as a qualitative method. Modern quantitative methods did not begin until about the 1920s, when good commercial optical equipment began to appear. The method used was emission spectroscopy. In 1939, Woodson applied the absorption procedure first in quantitative elemental measurement by determining mercury. Atomic absorption spectroscopy had its birth in 1955, when two independently published papers both described the method.

12.1.2 Electromagnetic Radiation

One of the ways that energy travels through space is by electromagnetic radiation. The light from the sun, the energy used for cooking food in microwave ovens, the X rays used in the medical field, and the radiant heat from a fireplace are all examples of electromagnetic radiation. Although these forms of radiant energy seem quite different, they all exhibit the same type of wavelike behavior and travel at the speed of light in a vacuum. Waves have three primary characteristics: wavelength, frequency, and speed. *Wavelength* (symbolized by the Greek letter lambda, λ) is the distance between two consecutive peaks or through the wave, as shown in Figure 12.1. *Frequency* (symbolized by the Greek letter nu, ν) is defined as the number of waves (cycles) per second that pass a given point in space. Since all types of electromagnetic radiation travel at the speed of light, short wavelength radiation must have a high frequency. This implies an inverse relationship between wavelength and frequency, as

$$\nu \lambda = c \qquad\qquad (12.1)$$

where λ is the wavelength in meters, ν is the frequency in cycles per second (abbreviated s) in hertz (abbreviated Hz), and c is the speed of light (2.9979×10^8 m/s). The unit of Hz is $1/s$ or s^{-1}.

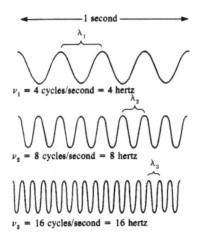

Figure 12.1
The nature of waves. Note that the radiation with the shortest wavelength has the highest frequency.

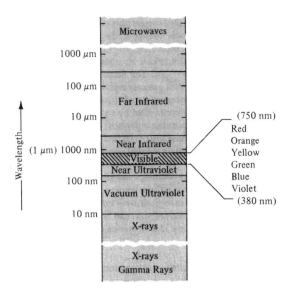

Figure 12.2
The electromagnetic spectrum.

Electromagnetic radiation is classified in Figure 12.2. Each portion of the spectrum has a popular name. For example, *radio waves* are electromagnetic radiation with low frequencies, and therefore very long wavelength. *Microwaves* also have low frequencies and are emitted by radar instruments. Microwaves are also absorbed by molecules in food, and the energy the molecules take on raises their temperature. This is why foods cook quickly in a microwave oven. *Infrared radiation* is emitted by hot objects and consists of the range of frequencies that can make molecules of most substances vibrate internally. We can't see infrared radiation, but we can feel how it is absorbed by our body by holding our hand near a hot radiator; the absorbed radiation makes our hand warm. Each substance absorbs a uniquely different set of infrared frequencies. A plot of frequencies absorbed versus the intensities of absorption is called an infrared absorption spectrum. It can be used to identify a compound, because each infrared spectrum is as unique as a set of fingerprints. *Gamma rays* are at the high frequency end of the electromagnetic spectrum. They are produced by some elements that are radioactive. *X rays* are much like gamma rays, but they are usually produced by special equipment. Both X rays and gamma rays penetrate living things easily. Our eyes are able to sense only a narrow band of wavelengths ranging from about 400 to 700 nm. This band is called the *visible spectrum* and consist of all the colors we can see, from red through orange, yellow, green, blue, and violet (see Figure 12.2). White light is composed of all these colors in roughly equal amounts, and they can be separated by focusing a beam of white light through a prism, which spreads the various wavelengths apart. Table 12.1 contains the wavelength region of each color.

The dual nature of the light — In 1901, the German physicist Max Planck proposed that electromagnetic radiation is emitted only in tiny *packets* or *quanta* of energy that were later called *photons*. The energy of one photon is called one quantum of energy.

$$E = h \nu \qquad\qquad (12.2)$$

where E is the energy of a photon, h is called the Planck's constant, and ν is the frequency of the electromagnetic radiation absorbed or emitted. The value of Planck's constant is

TABLE 12.1

Wavelength Region for Each Color

Wavelength Region	Color
380–450	Violet
450–495	Blue
495–550	Green
550–570	Yellow-green
570–590	Yellow
590–620	Orange
620–750	Red
—	Purple[a]

[a] Purple is seen by the eye when equal numbers of photons of blue and red lights strike the eye.

6.626×10^{-34} Js (units of energy, joules, multiplied by time, seconds). Each photon pulses with a frequency, ν, and travels with the speed of light. Planck proposed and Albert Einstein (1879–1955) confirmed that the energy of a photon of electromagnetic radiation is proportional to its frequency. According to Einstein's famous equation,

$$E = mc^2$$
$$m = E / c^2 = h / \lambda c$$

(12.3)

where m represents mass and c is the speed of light. The main significance of this equation is that energy has mass.

We can summarize the conclusion from Planck's and Einstein's work:

- Energy is quantized. It can occur in discrete units called quanta.
- Electromagnetic radiation, which was previously believed to exhibit only wave properties, was found to have certain characteristics of particulate matter. This is known as the dual nature of the light.

12.1.3 Continuous and Line Spectra

When the light from the sun or any other object is heated to a very high temperature (such as filament in a light bulb), it is split by a prism and displayed on a screen, a continuous spectrum form. It contains light of all colors. A rainbow after a summer shower is a *continuous spectrum*. In this case, the colors contained in the sunlight are spread out by tiny water droplets in the air. Water droplets act as a prism.

When a light given off by an electric discharge passes through a gas and is separated by a prism, a rather different spectrum is observed on the screen. The electric discharge in an electric current excites or energizes the atoms of the gas. The atoms absorb the energy, electrons will be promoted to higher energy levels, and when they return to the lower energy state, emit the absorbed energy in the form of light. When a narrow beam of this light is passed through a prism, only a few colors are observed as a series of discrete lines. This spectrum is called a *line spectrum*. Continuous and line spectrums can be seen in Figure 12.3.

According to the modern atomic theory, electrons exist at energy levels. By the addition of sufficient energy to the atoms, electrons can be promoted from a lower energy level to a higher, vacant energy level. When the light strikes an electron, causing it to be promoted

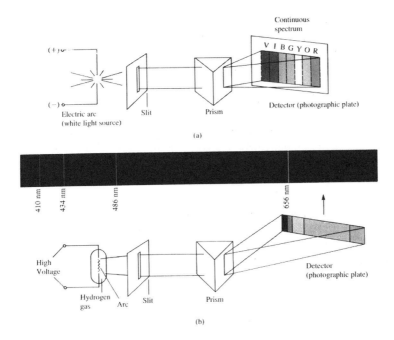

Figure 12.3
(a) Continuous spectrum obtaining all wavelengths of visible light (indicated by the initial letters of the colors of the rainbow) (b) the hydrogen line spectrum contains only a few discrete wavelengths.

to a higher energy level, the energy that once was light is now possessed by the electron. It is a less stable configuration, called the *excited state*. An important point concerning the process, however, is that the light coming in must be exactly the same energy as the energy difference between the two energy levels; otherwise it will not be absorbed at all. As this state is unstable, the electron drops back to the lower energy level and the atom will return spontaneously to its stable *ground state*. The radiant energy, equivalent to the energy that was absorbed, will be released and emitted in this state. Because only certain energy jumps can occur, only certain colors can appear in the spectrum. Figure 12.4 shows this electronic energy transfer. The wavelength of the radiant energy is directly related to the electronic transition that occurs. Either the energy absorbed in the excitation process or the energy emitted in the decay process can be measured and used in analytical purposes.

If light of the right wavelength reaches a ground state atom, the atom absorbing the light enters into the excited state. The quantity of the absorbance is measured. This is utilized in atomic absorption spectrophotometry.

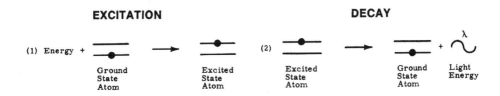

Figure 12.4
Electronic energy transition. Step (1) in the process, the excitation, is forced by supplying energy. The decay process in step (2), involving the emission of light, occurs spontaneously. Since every element has a unique electronic structure, the wavelength of light emitted is a unique property of each individual element.

In atomic emission, the sample is placed in one high thermal energy environment, and the atoms of the sample are excited and emit light. Measurement of the intensity of the emitted light is used in atomic emission spectrophotometry.

The other form of interaction between energy and electrons is vibrational energy. For vibration, less energy is needed. Atoms have no vibrational energy transition; all energy transfer is electronic, and only a limited number of wavelengths can be absorbed, so only those wavelengths show up in the spectrum. The result is line spectra. In molecular absorption, both electronic and vibrational transitions are possible, because all wavelengths have a chance of being absorbed to some degree. The result is a continuous spectra.

12.1.4 Beer's Law

The amount of light absorbed by a sample is proportional to the concentration of the absorbing species in the sample. So, there is a linear relationship between absorbance and concentration. This relationship is well defined in the Beer–Lambert Law, also called Beer's Law: the amount of light absorbed or transmitted by a solution is a function of concentration of the substance and the sample path length:

$$A = abc \tag{12.4}$$

where A = absorbance, a = absorptivity (sometimes called "extinction coefficient"): the ability of the absorbing species to absorb light. It depends on the electronic and vibrational transitions in a given species. The numerical value of a depends on the units used for expressing the concentration of the absorbing solution. Also in equation (12.4) above, b = diameter, or width of the cuvette, called "pathlength." A wider cuvette has more of the absorbing species, and therefore results in greater absorbance. Finally, c = concentration.

For example, a 2.00 ppm (mg/L) standard measured in a 1 cm cuvette shows absorbance of 0.246. What is the concentration of a sample, if the measured absorbance is 0.529, and measured also in a 1 cm cuvette?

$$c_1 = 2.00 \text{ ppm} \qquad c_2 = ?$$

$$b_1 = 1 \text{ cm} \qquad b_2 = 1 \text{ cm}$$

$$A_1 = 0.246 \qquad A_2 = 0.529$$

$$a = ?$$

To calculate the sample concentration from the known absorbance and pathlength, by using the Beer's Law, we need the value of the absorptivity of the species. From the knowledge of the values of the analyzed standard, we are able to calculate the numerical expression of the absorptivity.

$$A_1 = a \times b_1 \times c_1 \tag{12.5}$$

Substitute the values and solve for a:

$$0.246 = a \times 1 \times 2$$

$$a = 2 / 0.246$$

$$a = 0.123 \tag{12.6}$$

$$A_2 = a \times b_2 \times c_2$$

Substitute the values and solve for c_2:

$$0.529 = 0.123 \times 1 \times c_2$$

$$c_2 = 0.529 / 0.123$$

$$c_2 = 4.3 \text{ ppm}$$

The concentration of the sample is 4.3 ppm.

12.2 MOLECULAR SPECTROPHOTOMETRY

As discussed previously, there is a considerable difference in the absorption properties of atoms and molecules. The subject of the absorption of light by individual, nonbonded atoms must be considered separately from molecular absorption. Because of this, the techniques and instrument designs differ significantly and must be separately discussed. Absorption of light by molecules or ions causes two types of energy changes: electronic (change in the energy of the electrons of a molecule) and vibrational (change in the internuclear distance of two or more atoms in the molecule). Electronic transition needs more energy and thus occurs in the visible ultraviolet spectral region. Vibrational changes in a molecule result from the absorption of low-energy, infrared radiation.

12.2.1 Molecular Absorption and Color

The color of a molecule in solution depends on the wavelength of light it absorbs. Thus, when a sample solution of a molecule or ion is exposed to white light, certain wavelengths are absorbed, and the remaining wavelengths are transmitted to the eye. The color perceived by the eye is determined only by the wavelengths transmitted. The substance exhibits the color that is complementary to those absorbed. In simple terms, the color seen is the complementary color of the color absorbed. Table 12.2 shows the general relationship between wavelengths of visible light absorbed and the color observed. For example, if the color of the test solution is yellow, the selected wavelength should be 450 nm, or if the test solution is dark blue, the absorbance measured at 580 nm wavelength.

12.2.2 Molecular Absorption Spectrophotometry

The instrument used for measurement of absorbance is known generally as a spectrophotometer. A spectrophotometer is an instrument in which radiant energy of a very

TABLE 12.2

Visible Spectrum and Complementary Colors

Wavelength (nm)	Color	Complementary Color
400–435	Violet	Yellow-green
435–480	Blue	Yellow
480–490	Green-blue	Orange
490–500	Blue-green	Red
500–560	Green	Purple
560–580	Yellow-green	Violet
580–595	Yellow	Blue
595–610	Orange	Green-blue
610–675	Red	Blue-green

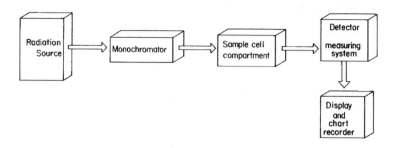

Figure 12.5
Basic construction of a simple single-beam spectrophotometer.

narrow wavelength range is selected from a source and passed through the sample solution. It is contained in a glass or quartz cell, called a cuvette. Some of the radiant energy is absorbed by the chemicals in the sample, and the rest passes on through. The quantitative basis of the spectrophotometry is that the amount of radiation absorbed (absorbance) at an appropriate wavelength is proportional to the concentration of the light-absorbing chemical in the sample. Spectrophotometry is a very fast and convenient method for quantitative analysis.

12.2.3 Basic Components of Spectrophotometers

The basic spectrophotometer consists of a light source, wavelength selector, sample holder or sample compartment, a detector, and a readout device, as shown in Figure 12.5.

Light source — The light source provides the light to be directed at the sample. The selection of the source depends on what region of the electromagnetic radiation is needed. There are considerable differences in technique and spectrum analysis between methods involving ultraviolet and visible (UV/Vis) light and infrared (IR) light. UV/Vis spectrophotometry is used mostly for quantitative analysis, IR spectrophotometry is mostly a qualitative technique, although quantitative applications can also be important. The light sources for visible and UV radiation are shown in Table 12.3. The tungsten filament lamp is the only common source for the visible region; it covers part of the UV region, but generally is not used below 320 to 330 nm. At wavelengths shorter than this, the deuterium lamp is used.

Wavelength selector or monochromator — The function of the monochromator is to select a beam of monochromatic (one-wavelength) radiation. The essential parts of the monochromator are (1) the entrance slit, which controls the intensity of the light; (2) a lens

TABLE 12.3

Ultraviolet and Visible Radiation Sources

Source	Wavelength Range	Intensity
Tungsten filament lamp	320–2500 nm	Weak below 400 nm Strong above 750 nm
Tungsten halogen lamp (quartz envelope)	250–2500 nm	—
Hydrogen discharge lamp	180–375 nm	Weak everywhere, but best in 200–325 nm region
Deuterium discharge lamp	180–400+ nm	Moderate

or mirror, which causes light to travel as parallel rays; (3) a dispersion device, which selects the light of different wavelengths, and can be a grating, prism, or various optical filters; (4) a focusing lens or mirror; and (5) an exit slit, which controls the color (or wavelength) of the light entering the sample compartment.

A diffraction grating is a surface with a large number of parallel grooves. Light striking the grating is diffracted so that different wavelengths come off at different angles. Rotating the grating allows radiation of the desired wavelength to be selected; this is done by turning the wavelength dial on the instrument. The *prism* disperses radiation by means of refraction. Radiation of different wavelengths is bent at different angles on entering and emerging from the prism. The simplest monochromators rely on *optical filters*. Absorption filters absorb certain parts of the spectrum and are made of colored glass. Absorption filters have a very large bandwidth. Interference filters reject unwanted wavelengths and transmit a narrow bandwidth. Instruments that use optical filters are inexpensive, portable, and can be designed for specific analysis in the field.

Sample holder, or sample compartment — The sample holder is a tight box where the sample is irradiated by the light emerging from the monochromator. The sample, in the form of solution, is contained in an optically transparent cell, called a *cuvette*, with a known width and optical length. These cells are made of optical glass. Some inexpensive spectrophotometers use a circular test tube cuvette. Cells used in the visible region are made of optical-quality borosilicate glass. These can be used down to about 320 nm, at which point the glass begins to absorb most of the radiant energy. For lower wavelengths, it is necessary to use more expensive cuvettes made of quartz or some other form of silica. Of course, these cells can also be used above 320 nm. *Matched cuvettes* are identical with respect to pathlength and reflective and refractive properties in the area where the light beam passes. If the pathlengths were different or if the wall of one cuvette reflected more or less light than another cuvette, then the absorbance measurement could be different for that reason, and not because of the concentration difference. Therefore, the cuvettes must be placed in the instrument exactly the same way each time, since pathlength and refractive properties can change by rotating the cuvette. A vertical line on the cuvette, lined up with similar line on the cuvette holder, helps to avoid the above-mentioned source of error.

Protect the cuvette from scratches. Avoid the use of abrasive cleaning agents; clean the cuvettes well with soft cloths, avoid finger marks and lint or dirt, and handle them by the top edge when inserting them into the instrument. Any liquid or fingerprints adhering to the outside wall of the cuvette must be removed with a soft cloth or soft tissue prior to measurement. Because of additional reflection from air-to-glass surfaces, empty cuvettes transmit less radiation than do cells filled with reference standards, blanks, or deionized water. *DO NOT USE AN EMPTY CUVETTE TO ZERO THE INSTRUMENT.* To avoid errors caused by removing a cuvette and then replacing it with another, a spectrophotometer is available with a fixed "flow-through" cell. The solution to be measured needs about 30 to 60 seconds to flow through the cell. The reading can be taken after the first few seconds, which are needed to wash out the cell and fill it with the new sample.

Detector — Transmitted radiation passing through the solution is picked up by a photosensitive detector. Detectors are phototubes, which convert light energy into electrical energy. A typical phototube consists of a half cylinder cathode and a wire anode in a sealed evacuated glass tube. Because the cathode emits electrons when photons strike it, the phototube is termed a "photoemissive tube." The response of the phototube to different wavelengths depends on the composition of the cathode coating. A schematic diagram of a phototube is shown in Figure 12.6.

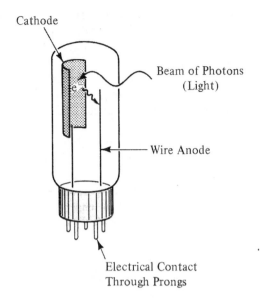

Cathode

Beam of Photons
(Light)

Wire Anode

Electrical Contact
Through Prongs

Figure 12.6
Schematic diagram of a phototube showing the emission of an electron from the cathode to the anode. A typical phototube consists of a half cylinder cathode and a wire anode in a sealed evacuated glass tube. A beam of photons passes through the sample and strikes the inner surface of the cathode and ejects electrons from the cathode. The electrons migrate through the vacuum to the positive wire anode and produce a current.

Readout device — When radiations strikes the detector, it generates electric current, which is amplified with an amplifier and is then transmitted to a recorder or displayed on the spectrophotometer via a digital or scale readout. Digital displays are now used except on lower-cost instruments. The readout can be either transmittance or absorbance. In using a conventional spectrophotometer, the measured absorbance is used to calculate the concentration of the measured sample component or to prepare a Beer's Law plot. Manipulation of data requires more time than the actual measurements. A modern spectrophotometer with a built-in microprocessor or microcomputer can do rapid computations, store information for later use, and control many operations of the meter. In such instruments, the operator inserts the cuvette into the instrument, but uses the keyboard for operation. Calibration plots are made by linear regression (which will be discussed later) or may be graphically displayed.

12.2.4 Single-Beam and Double-Beam Spectrophotometer

There are basically two general types of spectrophotometry instruments that are available. These are called "single-beam" and "double-beam" instruments.

Single-beam instrument — In the single-beam instruments, all measurements are based on the varying intensity of a single-beam of light. All of the energy from the light source can be directed through the sample cell. A schematic diagram of a single-beam optical system is shown in Figure 12.7. The disadvantage of the single-beam instrument is that the light intensity can change due to fluctuations occurring in the line voltage, the power source, or the light bulb itself. Thus, an error could result in the sample reading. Single-beam problems with lamp intensity drift have been controlled by designing more stable light sources, lamp power supplies, and the prewarming of light sources.

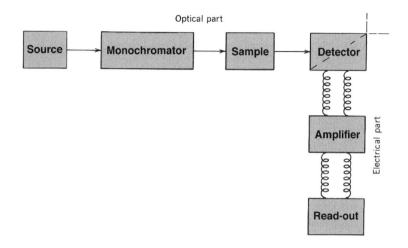

Figure 12.7
Block diagram showing components of a single-beam spectrophotometer. The optical part and the electrical part of the instrument meet at the detector, which converts radiant energy into electrical energy.

Double-beam instrument — The double-beam instrument uses additional optics to divide the light from the lamp into a *sample beam* (directed through the sample cell) and a *reference beam* (directed through the blank). A schematic diagram for a typical double-beam instrument is shown in Figure 12.8. The light coming from the monochromator is directed in either one of two paths with a rotating half-mirror, called a *chopper*. At one moment, the light passes through the sample, while at the next moment it passes through the blank. Both beams are joined again with a second rotating half-mirror prior to entering the detector. In any case, the detector sees alternating light intensities and automatically compensates for fluctuations, usually by automatically widening or narrowing the

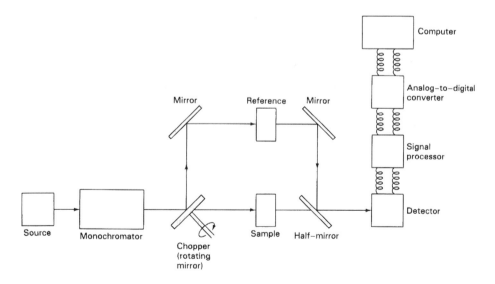

Figure 12.8
Schematic diagram of a double-beam spectrophotometer. The light coming through the monochromator is directed along either one of two paths with the use of a "chopper" or rotating half-mirror. At one moment the light passes through the sample, while at the next moment it passes through the blank. Both beams are joined again with a second rotating half-mirror prior to entering the detector.

entrance slit to the monochromator. If the beam becomes less intense, the slit is opened; if the beam becomes more intense, the slit is narrowed. Thus, the signal relayed to the readout device is free of the effects of intensity fluctuations from the source.

12.2.5 UV, Visible, and Infrared Spectrophotometers

Spectrophotometric instruments very greatly in price, performance, and sophistication.

Visible spectrophotometer — These instruments have inexpensive optical glass components, and operate from 325 nm to 900 to 1000 nm wavelength range. Older-type instruments select wavelengths mechanically through a wavelength knob, whereas the modern digital-readout instruments offer electronic wavelength selection via a keyboard. Older instruments rely on blue- and red-sensitive phototubes.

Ultraviolet–Visible (UV/Vis) spectrophotometer — UV/Vis spectrophotometer is designed for measurements in the ultraviolet as well as the visible region. Such an instrument measures absorption in the 200 to 1000 nm region. For measurements below 320 nm, the spectrophotometer must be equipped with an ultraviolet source of radiation. The most common source of radiation in the visible region is the tungsten filament lamp, and in the UV region, the deuterium discharge lamp. In some spectrophotometers, the tungsten halogen lamp can be used as low as 250 or 220 nm. Light sources for UV/Vis radiation are listed in Table 12.3.

Spectrophotometers with microprocessor or microcomputer — Modern spectrophotometers with a built-in microprocessor or microcomputer can perform rapid data processing and control many of the spectrophotometer operations. In such instruments the operator still inserts one or more cuvettes into the instrument, but uses the keyboard to program the necessary operating instructions. Calibration plots are made from standards by linear regression or may be graphically displayed.

Comparison of UV/VIS and IR spectrophotometers — There are considerable differences between methods involving UV/Vis light and IR light.

- UV/Vis spectrophotometry is mostly a technique for quantitative analysis, IR is mostly a qualitative technique, although both techniques may be utilized.
- For UV/Vis technique, the absorption spectra is recorded, while the transmittance spectra is used in IR technique.

TABLE 12.4

Ultraviolet and Visible Radiation Detectors

Detector	Useful Range
Human eye	380–750 nm
Blue-sensitive phototube (vacuum photodiode)	330–625 nm
Red-sensitive phototube (vacuum photodiode)	600–975 nm
Wide-range phototube (vacuum photodiode)	400–800 nm
Silicon photodiode (semiconductor photodiode)	350–1170 nm
Silicon photodiode, UV-enhanced	200–1170 nm
Photomultiplier tube	200 or 300–1100 nm

TABLE 12.5

Summary of Spectrophotometry Techniques

Region	Range	Source	Cuvette Material	Detector
UV	180–400 nm	Hydrogen discharge tube	Quartz	Phototube
Vis	400–750 nm	Tungsten filament light bulb	Glass or clear plastic	Phototube
IR	750–15,000 nm	Globar, Nernst glower, incandescent nichrome wire	Inorganic salt crystals	Thermocouple

- UV/Vis spectra are created from electronic transitions. IR spectra arise from molecular vibrational transitions, so they give more information about the molecular structure.

- UV/Vis and IR instruments are different in design, in cuvette material, and in sample preparation technique.

Infrared (IR) spectrophotometers — IR spectrophotometers have the same basic components as UV/Vis instruments, but the radiation source used in the optical systems, sample cells, and detectors are different. For Vis spectrophotometers, the light source is a tungsten-filament lamp; for UV light, a hydrogen discharge lamp is the most common. For IR, a source of heat is needed. The two most important infrared sources are the glowing silicon carbide rods (Globar) and rods made of rare earth oxides, such as zirconium and yttrium oxides (Nernst glower). Incandescent nichrome wires are also common. The monochromator system of an IR spectrophotometer is the same as that of a UV/Vis spectrophotometer; the only dispersing device used for wavelength selection in infrared is the diffraction grating.

When using visible light, any clear, colorless, transparent material can be used for a cuvette, including glass and plastic. A cuvette used for measuring in the UV region must be made of quartz glass. Both glass and quartz absorb infrared radiation; therefore, the monochromator optics and the cells must be made from ionic materials. Large, polished sodium chloride (NaCl) crystals are most often used. Cells of lithium fluoride (LiF) or calcium fluoride (CaF_2) provide better resolution at lower wavelength, and cells of potassium bromide (KBr) or cesium iodide (CsI) are more useful at higher wavelengths. The salt crystals are placed in a fixture of some type, such as a *demountable cell* (Figure 12.9). A small amount of liquid sample, introduced through a sample port with a syringe, is held within a gasketed space inside the cell and in the path of the light beam when placed in the instrument. Of course, since salts are highly water soluble, water cannot be a solvent for the sample. Usually, solvents such as carbon tetrachloride (CCl_4) or methylene chloride

Figure 12.9
A demountable cell.

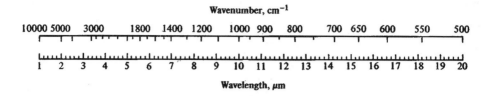

Figure 12.10
Interconversion of wavelength and wavenumber.

(CH_2Cl_2) are used, because their spectra show very few or no absorption lines in the IR region.

Infrared radiation may be measured by detecting the temperature change of a material in the infrared beam; this type of detector is a *thermal detector*. Because the radiant power of infrared radiation is so weak, the response of most thermal detectors is quite low. A preamplifier is usually necessary to obtain a good signal-to-noise ratio in the amplifier. Another problem is the heat radiated from objects in the room. To minimize this source of error, the detector must be housed in a vacuum or shielded from direct exposure to heat.

In ultraviolet and visible spectra, absorbance and transmittance is plotted against wavelength. In infrared spectra the preferred method uses wavenumbers instead of wavelengths. *Wavenumber* is the reciprocal of wavelength expressed in centimeters, and therefore has a unit of cm^{-1}. Conversion of wavelength and wavenumber is shown in Figure 12.10. The IR region of the spectrum is usually considered to start near the red end of the visible spectrum, at the point where the eye no longer responds to dispersed radiation (infrared = below the red). The fundamental IR region extends from 3600 cm^{-1} (wavenumber) or 2.8 μm (wavelength). The analytical useful IR region extends from 3600 cm^{-1} to somewhere around 300 cm^{-1} or 33 μm). A sample of an infrared spectrum is shown in Figure 12.11.

Infrared spectrophotometers are generally double-beam instruments. The sample cell and reference cell (solvent) are exposed to equivalent beams from the same infrared source. A rotating half-circle mirror is used to direct an equivalent beam alternately through the two cells many times a second. Thus, any condition that affects the sample beam will also affect the reference beam equally, so the condition will be cancelled out in the readout. Double-beam instruments are discussed above, and the schematic diagram of the operation is demonstrated in Figure 12.8.

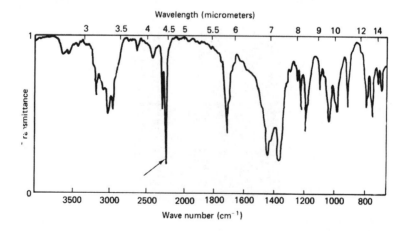

Figure 12.11
Infrared spectrum of acetonitrile ($CH_3C\equiv N$). Arrow shows band due to the $C\equiv N$ group.

Samples can be liquids, solids, and gases. They can be organic or inorganic, although inorganic materials sometimes do not give very definitive spectra. The only molecules transparent to IR radiation under ordinary conditions are monoatomic and nonpolar molecules, such as Ne, He, O_2, N_2, and H_2. Liquid samples may be analyzed without dilution or dissolved in a solvent. It is a very desirable to run a sample without a solvent (*pure* or *"neat" sample*).

Analyzing solids without dissolving them is possible. There are two methods available for this analysis. The first method is the *potassium bromide (KBr) pellet technique*. In this technique, a small portion of the dry solid sample is mixed with potassium bromide. A small amount of this mixture is then transferred to a "pellet die" in which it is pressed into a potassium chloride pellet. It is a transparent half-inch disc that can be placed directly into the radiation path. The other method for analyzing undissolved solid sample is the *mull method*. In this method, the dry solid sample is mixed with mineral oil so that the substance becomes toothpaste-like. This mixture is then placed between two salt crystals and the spectrum recorded.

12.3 CALIBRATION OF SPECTROPHOTOMETERS

Calibrations are performed at the beginning of the analysis to make sure that the instrument is working properly. This *initial calibration* is determined for each parameter tested and is based on instrument response for different concentrations of standards (calibration standards) against a calibration blank. The optimum concentration range and the number of these standards are directed by the analytical method. The concentration of the calibration standards should be bracketed in this optimum range. The concentration of standards and the measured response (absorbance, transmittance, etc.) of the instruments should be plotted on the *calibration curve* and approved by calculating the corresponding *correlation coefficient*. Computerized, modern instruments display the curve and the value of the correlation coefficient. Its value should be >0.9950, and serves as a basis for acceptance or rejection of the calibration curve.

In UV/Vis spectrophotometers, the initial calibration is based on a 4 to 6 point standard curve in the optimum linear range as stated in each particular parameter. After the calibration curve is established, once every analytical batch (samples which are analyzed together with the same method and with the same lots of reagents) or at the 5% frequency, the curve should be validated with a *continuing calibration standard*. This validation includes the analysis of the *continuing calibration standard* (CCS) and *calibration verification standard* (CVS), and must be performed before samples are measured. The concentration of a CCS is equivalent to the midpoint on the initial calibration standard. The deviation from the original value should be within ±5%. The CVS should be a certified standard or independently prepared from a source other than the calibration standards. Its analyzed value is accepted within ±10% deviation from the 100% recovery.

When samples go through any pretreatment (digestion, distillation, extraction, filtration, etc.) prior to the actual analysis, it should be verified and the effects from the sample preparation should be monitored. To support these data, a blank, called a *preparation blank*, and one standard, called a *laboratory control standard* (LCS) should be prepared and analyzed together with the samples. A preparation blank or "prep blank" consists of analyte-free water treated as a sample. An LCS is a sample taken from the CVS, except that it is carried through the preparation. The accepted values are ±15% deviation from the 100% recovery.

The *frequency of calibration curve preparation* depends on instrumentation. For a UV/Vis spectrophotometer, use the available calibration curve until the correct calibration is approved. Recalibration is done at least once every 6 months or on the failure of any CCS.

Daily calibration is made by zeroing the instrument with a calibration blank and analyzing one CCS with a ±15% recovery, and with the ±10% recovery of a CVS. Once per analytical batch or with 10% frequency, this check should be repeated. If the CCS and CVS fail the calibration criteria, the analysis must be stopped, and a new initial calibration must be performed. Samples measured before the failed standard must be analyzed again.

12.3.1 General Rules in Preparation of Calibration Curves

The calibration curve is prepared by analyzing standards of known concentrations. For the preparation of a calibration curve, ordinary rectangular-coordinate paper is satisfactory. For some graphs (for example, measurement of millivolt response by using pH meter with ion selective electrodes), semilogarithmic paper is preferred. Plot the independent and dependent variables on abscissa and ordinate in a manner that can be comprehended easily, and cover as much of the graph paper as possible. Choose the scales so that the slope of the curve approaches unity as nearly as possible, and choose the variables to give a plot that will be close to a straight line. Present legends on the graph to give complete information about the conditions under which the data were obtained: parameter determination, method and reference, volume of the standards, wavelength used, time between addition of reagents and the reading, data obtained from the linear regression calculation, date of preparation, and name and signature of the preparer.

12.3.2 Linear Regression Calculation

When a calibration curve is prepared, the sample concentration is obtained by measuring the instrument's response under the same conditions that were used for the standards, and the concentration of the sample can be read from the horizontal axis of the plot. Although unknown concentrations can be read directly from the graphical plot, better accuracy is possible by using the linear regression calculation, also called the least of square calculation. In this calculation, of the possible straight lines that can be drawn through or near the data points, the one chosen minimizes the sum of the squared deviations. The deviation for each point is the difference between the actual data points with the same x-axis value that lies exactly on the straight line. It gives information about the best straight line through the points entered, including its correlation coefficient, intercept, slope and the predicted x- and y-values. The *correlation coefficient* gives the correlation between the x- and y-values in a set of data points. A result near 1 indicates that the values have a strong linear relationship. A relation near 0 indicates that the values are only slightly related. A value near to −1 indicates that the values are very closly related, but in a negative way; that is, an increase in one is related to a decrease in the other. The value of >0.9950 is accepted.

Intercept (*b*) tells whether there is a significant blank measurement even when the concentration of the blank is 0. The intercept value gives the y-intercept of the best straight line through the points. The calculated value of the slope (*m*) gives the slope of the best straight line. The formulas for calculating these values are as follow:

$$m = [nExy - ExEy]/[nEy^2 - (Ey)^2] \qquad \textbf{(12.7)}$$

$$b = [Ey^2Ex - EyExy]/[nEy^2 - (Ey)^2] \qquad \textbf{(12.8)}$$

where *m* = slope, *E* = sum, *y* = absorbance or other response, *b* = intercept, and *x* = concentration.

By knowing the absorbance, slope and intercept values, the concentration of the sample can be calculated according to the formula:

$$x = my + b \qquad\qquad (12.9)$$

For example, in a spectrophotometric analysis, the initial calibration gives the following data:

number of the standards (n)	= 6
concentration of the standards (x)	= 0.2, 0.4, 0.5, 0.6, 0.8, 1.00
measured absorbances (y)	= 0.206, 0.392, 0.503, 0.598, 0.789, 0.992
calculated correlation coefficient	= 0.99985

Using the linear regression calculation, predict the slope and intercept values for the above data:

x	y	y^2	xy
0.2	0.206	0.042	0.041
0.4	0.392	0.154	0.157
0.5	0.503	0.253	0.252
0.6	0.598	0.358	0.359
0.8	0.789	0.623	0.631
1.0	0.991	0.982	0.991
3.5	3.479	2.412	2.431

$$m = [(6 \times 2.431) - (3.5 \times 3.479)] / [(6 \times 2.412) - (3.479)^2]$$

$$= [14.586 - 12.177] / [14.472 - 12.103] = 2.409 / 2.369$$

$$= 1.017$$

$$b = [(2.412 \times 3.5) - (3.479 \times 2.431)] / [(6 \times 2.412) - (3.479)^2]$$

$$= [8.442 - 8.457] / [14.472 - 12.103] = -0.015 / 2.369$$

$$= -0.006$$

One unknown concentration sample absorbance was measured as 0.246. By using formula (12.9), the sample concentration will be

$$x = 1.017 \times 0.246 + (-0.006) = 0.244 \text{ ppm}$$

12.3.3 Performance Check of UV/Vis and IR Spectrophotometers

The design and model of the spectrophotometers are different. Correct operation and maintenance of the instruments are given in the manufacturer's manual and in the standard operation procedure (SOP) specified for each laboratory. In addition to the calibration, the correct performance of the instruments must be checked periodically.

Model						
	Absorbance Read at Wavelength					
Date	**500 nm**	**505 nm**	**510 nm**	**515 nm**	**520 nm**	**Remarks**	**Sign.**

nm = nanometer, unit of the wavelength. The calibration check is satisfied when maximum absorbance (or minimum transmittance) occurs between 505 and 515 nm wavelengths..

Figure 12.12
Documentation of spectrophotometer wavelength calibration check.

For a UV/Vis spectrophotometer, wavelength calibration and a linearity check are recommended. Wavelength accuracy may be checked by a commercially available "didymium" calibration filter, or with the very simple cobalt chloride test. In this test, measure the absorbance of a cobalt chloride solution (22 g $CoCl_2$ dissolved and diluted to 1 L with 1% of HCl solution) on 500, 505, 510, 515, and 520 nm wavelengths. The wavelength calibration check is satisfied when maximum absorbance or minimum transmittance occurs between 505 and 515 nm wavelengths. A linearity check of the instrument is given by the measurement of the absorbance at 510 nm of the cobalt chloride solution used for the wavelength calibration and at the same wavelength the absorbance of the 1:1 dilution of this solution. The absorbance of the 1:1 diluted solution should be half of the original reading in correct operation. Documentation of these checks is shown in Figures 12.12 and 12.13.

The correct operation of the IR spectrophotometer is determined with a commercially available 0.05-mm-thick polystyrene film. By following the manufacturer's instruction, record the spectrum of this film and compare with the commercially supplied one. If the test spectrum is not within the tolerance indicated, adjustment is necessary, probably by a service representative.

12.3.4 Maintenance of UV/Vis and IR Spectrophotometers

Good care and proper maintenance of the instruments are basic requirements for accurate and sufficient laboratory results.

Spectrophotometer model				
Date	**Checking Solution**	**Absorbance at 510 nm**	**Remark**	**Sign.**
	Stock Cobalt Soln.			
	1:1 Cobalt Soln.			
	Stock Cobalt Soln.			
	1:1 Cobalt Soln.			
	Stock Cobalt Soln.			
	1:1 Cobalt Soln.			
	Stock Cobalt Soln.			
	1:1 Cobalt Soln.			

The absorbance of the 1:1 diluted cobalt solution should be half of reading produced by the stock cobalt solution (22 g $CoCl_2$ in 1 L 1% HCl solution).

Figure 12.13
Documentation of spectrophotometer linearity check.

The recommended daily maintenance of the UV/Vis spectrophotometers is to keep the sample compartment and cuvettes sparkling clean. Weekly checking of the lamp alignment is advised, and the service contract's instrumentation specialist must clean the windows quarterly.

For IR spectrophotometers, the daily care includes sample cell cleaning and gas leakage checks. Monthly window cleaning and quarterly changing of the desiccant are criteria of the maintenance.

Chapter **13**

POTENTIOMETRY AND ION-SELECTIVE ELECTRODES

13.1 PRINCIPLES OF ELECTROCHEMISTRY

Chemical reactions may be classified as precipitation, acid–base, and oxidation–reduction reactions. A special class of oxidation–reduction reaction is combustion reactions. As discussed previously in Chapter 11, reactions in which one or more electrons are transferred are called oxidation–reduction reactions, or redox reactions. Oxidation is defined as loss of electrons, and reduction is defined as gain of electrons. For example,

$$2 \text{ Na(s)} + \text{Cl}_2(\text{g}) \rightarrow 2 \text{ NaCl(s)}$$

In the reaction of elemental sodium (Na) and chlorine (Cl_2), each sodium atom loses one electron, forming a 1+ ion (Na^+). Therefore, sodium is oxidized. Each chlorine atom gains one electron, forming a chloride ion (Cl^-), and is thus reduced. In other words, oxidation is an increase in oxidation number (loss of electrons) and reduction is a decrease in oxidation number (gain of electrons). *An oxidizing agent is the electron acceptor, and a reducing agent is the electron donor.* During oxidation–reduction reactions, an electric current is produced, which is used by devices called batteries. We are familiar with batteries from our everyday life (car, watches, calculators, radios, etc.).

Electrochemistry is the study of the interchange of chemical and electrical energy.

13.1.1 Galvanic Cell

An electrochemical cell, also called a *galvanic cell*, is a device powered by an oxidation–reduction reaction where the oxidizing agent is separated from the reducing agent so that the electrons must travel through a wire from the reducing agent to the oxidizing agent. The name "galvanic cell" honors Luigi Galvani (1737–1798), an Italian professor of anatomy, generally credited with the discovery of electricity. He discovered that electricity caused the muscles in frog legs to contract. He also noticed that different metals, when in contact, produced the same effect in frog muscles. These cells are often called *voltaic cells* after Alessandro Volta (1745–1827), an Italian scientist, who first constructed cells of this type around 1800. A typical galvanic cell is shown in Figure 13.1. Galvanic cells are commonly called batteries.

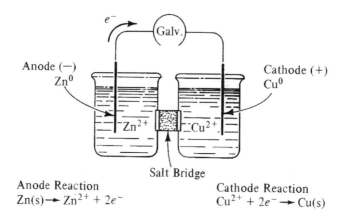

Figure 13.1
A galvanic cell.

A *battery* is nothing more than a device which consists of two separate containers (known as "half cells"), in which two redox half-reactions occur: the oxidation half-reaction, which is the source of the electrons, and the reduction half-reaction, which is the ultimate destination of the electrons. The pathway for the electrons to travel from one half-cell to the other is provided by connecting two metal strips (electrodes) with an electrical conductor metal wire. The electrode where the oxidation occurs is called the *anode*. At the other electrode, the oxidizing agent gains electrons and is thus reduced. The electrode where the reduction occurs is called the *cathode*. In addition, there must be some contact between the two solutions to allow the diffusion of ions between the two half-cells. Such diffusion is required for the current to flow and represents the final step in forming a completed circuit. The contact can made by a "salt bridge," which is a glass tube filled with a solution of some inert ions, such as K^+ and Cl^-. The flowing electrons represent the electrical current which is used to operate battery-driven devices. The use of batteries in everyday life is important. Automobiles have 12 V batteries, ordinary flashlights require 1.5 V batteries, many calculators need 9 V batteries, etc.

Cell potentials — The galvanic cell has an ability to push electrons through the external circuit. The magnitude of this ability is expressed as a *potential* or *electromotive force* (EMF), which can be conveniently thought of as a force with which an electric current is pushed through a wire. Potential is measured in terms of an electrical unit called the *volt* (V).

The voltage measured for a galvanic cell varies according to the amount of current that is flowing through the circuit. The maximum potential of a given cell is called its *cell potential*, E_{cell}. The value of E_{cell} depends on the composition of the electrodes, the concentration of the ions in each of the half-cells, and the temperature.

Standard cell potential (E°_{cell}) — This is the potential of the cell when all of the ion concentrations are 1 M, the temperature is 25°C, and any gases that are involved in the cell reaction are at a pressure of 1 atm.

The measured overall cell potential arises from a competition between the two half-cells for electrons, and is expressed as *reduction potential* or *standard reduction potential* (at 25°C, concentration 1 M, 1 atm pressure).

The standard cell potential is expressed as

$$E_{cell}^{\circ} = \frac{\text{standard reduction potential}}{\text{of substance reduced}} - \frac{\text{standard reduction potentia}}{\text{of substance oxidized}} \qquad \text{(13.1)}$$

$$E_{cell}^{\circ} = E^{\circ} \text{ cathode} - E^{\circ} \text{ anode} \qquad \text{(13.2)}$$

13.1.2 Dependence of EMF on Concentration

When all of the ion concentrations in a cell are 1 M and when the partial pressure of gases involved in the cell reaction is 1 atm, the cell potential is equal to the standard potential. When the concentrations or pressures change, however, so does the potential. For example, in an operating cell or battery, the potential gradually drops as the reactants are consumed. The cell approaches equilibrium, and when it gets there the potential has dropped to zero — the battery is dead.

Because the EMF of a cell depends on the concentration of the ions and on the gas pressure, cell EMFs provide a way to measure ion concentrations. The cell EMFs for various concentrations of ions and various gas pressures can be related to standard electrode potentials by means of an equation first derived by the German chemist Walther Nernst (1864–1941).

13.1.3 The Nernst Equation

The Nernst equation is an equation relating the cell EMF to its standard EMF and the reaction quotient.

$$E_{cell} - E_{cell}^{\circ} - (2.303 \; RT / nF) \log Q \qquad \text{(13.3)}$$

where R (constant) = 8.3145, T = 298 K, F (Faraday) = 96,500 coulombs (C). Substituting values for R and F at 25°C gives the equation

$$(2.303 \times 8.3145 \times 298)/96,500 = 0.0592$$

$$E_{cell} = E_{cell}^{\circ} - (0.0592 / n) \log Q \qquad \text{(13.4)}$$

where n is the number of electrons transferred in a reaction and Q is the reaction quotient. Reaction quotient is an expression that has the same form as the equilibrium constant expression but whose concentration values are not necessarily those at equilibrium. The equilibrium constant (K_c) is an expression obtained for a reaction by multiplying the concentrations of products together, dividing by the concentrations of reactants, and raising each concentration term to a power equal to the coefficient in the chemical equation.

Example — Measure Cu^{2+} concentration in water sample in which the copper ion concentration is expected to be quite small. The apparatus that was used consisted of a silver electrode, dipped into a 1 M solution of $AgNO_3$, connected by a salt bridge to a second half-cell containing a Cu electrode dipped into the analyzed water samples. In the analysis of one of the samples, the cell potential was measured to be 0.62 V, with the copper electrode serving as the anode. What was the concentration of the copper ion in the sample?

Since copper is the anode, it is being oxidized. This also means that Ag^+ is being reduced. Therefore, the correct equation for the cell reaction is

$$Cu(s) + 2Ag^+(aq) \rightarrow Cu^+(aq) + 2Ag(s)$$

Two electrons are transferred, so the Nernst equation is

$$E_{cell} = E°_{cell} - (0.0592/2) \log [Cu^{2+}]/[Ag^+]^2$$

$$E_{cell} = 0.62 \text{ V}$$

$$E°_{cell} = E°_{Ag^+} - E°_{Cu^{2+}}$$

$$= (0.80) - (0.34)$$

$$= 0.46 \text{ V}$$

Substituting these values,

$$0.62 = 0.46 - 0.0296 \log [Cu^{2+}]/(1)^2$$

$$\log [Cu^{2+}]/1 = -5.4$$

Taking the antilog and solving for $[Cu^{2+}]$,

$$[Cu^{2+}] = 3.98 \times 10^{-6} M$$

One of the most useful results of using the Nernst equation is that cell potential measurements provide a way indirectly to measure and monitor ion concentrations in aqueous solutions. A number of specialized electrodes have been developed that can be dipped into one solution after another and whose potential is affected in a reproducible way by the concentration of only one species in the solution.

13.2 DETERMINING CONCENTRATIONS FROM CELL POTENTIAL

One of the principal uses of the relationship between concentration and cell potential is the measurement of concentrations of some dissolved species. The measurement of the potential with an ordinary voltmeter (or equivalent) in a chemical analysis is called *potentiometry.*

The three components required for a potentiometric measurement are the *reference electrode,* the *indicating electrode,* and some type of *potentiometer* or *electronic voltmeter.* The indicating electrode senses the activity and concentration of an ion. The reference electrode has a fixed potential that is not affected by the test solution.

13.2.1 The Saturated Calomel Reference Electrode (SCE)

A voltage measurement is a relative measurement — that is, the measurement must be made relative to some point in the circuit which has a constant voltage. The reference electrode has a potential that never changes; it is always constant. Most commonly, a *saturated calomel reference electrode* (SCE) is used. The calomel electrode is a type of half-

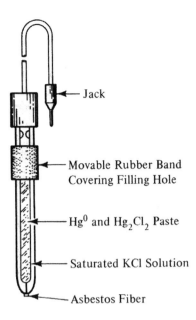

— Jack

— Movable Rubber Band
Covering Filling Hole

— Hg^0 and Hg_2Cl_2 Paste

— Saturated KCl Solution

— Asbestos Fiber

Figure 13.2
A saturated calomel reference electrode.

cell in which the electrode is mercury coated with calomel (Hg_2Cl_2) and the electrolyte solution is saturated potassium chloride (KCl). A typical SCE is shown in Figure 13.2. Actually, this is an electrode and a salt bridge encased in a single unit. It consists of two glass tubes, one inside the other. Within the internal tube there is a material known as calomel. It is a paste made by mixing mercury metal and mercurous chloride (Hg_2Cl_2). Between the external and internal tubes is a saturated potassium chloride (KCl) solution with the presence of some undissolved KCl. The KCl solution serves as a salt bridge to the test solution via a porous fiber plug in the tip of the external tube. The saturated calomel electrode has a potential of +0.246 V.

For measurements where chloride ion might leak into the sample solution and affect the indicator electrode potential, a *double-junction reference electrode* is used in which a second electrolyte is interspersed between the potassium chloride salt bridge and the sample solution.

13.2.2 Potentiometer and Electronic Voltmeter

If an indicator electrode and a reference electrode are connected through a voltmeter and inserted into a test solution, a galvanic cell is set up. The voltage (EMF) of the cell is determined by the concentration of the ions to be measured. Usually in analytical chemistry it is desirable to measure the voltage of a galvanic cell under conditions where the cell actually does no electrical work. A conventional voltmeter draws enough current from the cell to cause a decrease in the concentration of the ion to be measured, so the EMF in this type cell cannot be measured with a voltmeter. A simple device called a *potentiometer* is employed to measure the voltages of galvanic cells. Potentiometers worked very well in many cases but are unsuitable for cells involving very high internal resistances, such as in pH measurements with glass electrodes. *Electronic voltmeters* with very high input resistances were developed. Not only did these instruments work with cells of high resistance, but they also became so inexpensive and so accurate that they have replaced the potentiometer even for applications where the latter served well. Electronic voltmeters with direct digital readout are now very common.

The modern pH/mV meter measures pH or mV (millivolts) and is an electronic voltmeter. The voltage readout can be calibrated to read in pH units, mV, or concentration of a specific ion.

13.2.3 Ion-Selective Electrodes (ISE)

Ion-selective electrodes (ISE) are practical uses of electrochemistry that rely on the unique role that electrochemistry plays as an interface between chemical systems and electronic devices that display, record, and manipulate data. The key to using this interface has been the development of electrodes that are selectively sensitive to the kinds of chemicals of interest. These electrodes are called ion-selective electrodes because of their ability to respond to certain specific ions while ignoring others.

Ion-selective electrode systems are composed of a sensing electrode (half-cell), a reference electrode (half-cell), and a meter (pH meter read in mV, or ion analyzer measures mV and direct concentration mode).

When a sensing electrode is exposed to a sample solution of ions for which it is selective, a potential develops across the sensing membrane surface. This membrane potential varies with the concentration of the ion being measured. To make a measurement, a second unvarying potential against which the membrane potential may be compared is required. The reference electrode provides this function. A filling solution completes the electrical circuit between the sample and the internal cell of the reference electrode. The point of contact between the sample and filling solutions is the liquid junction. Reference electrodes used for ISEs are the sleeve-type electrodes. In the "single junction" reference electrode, the inner filling solution is silver chloride (AgCl), and in the "double junction" reference electrode the filling solution has an ion react with the ion analyzed.

The four types of ion-selective indicating electrodes are

1. glass electrodes: measure only +1 cations
2. liquid membrane electrodes: measure mainly +2 cations
3. solid state electrodes: measure mainly –1 anions
4. gas sensing electrodes: measure the concentration of certain gases dissolved in water or ions in solution converted to dissolved gases.

Glass electrode, measurement of pH — The glass electrode is a solid-state-membrane electrode because of the nature of the ion-sensitive material used in its construction. Commercial glass electrodes respond strongly only to +1 ions, including H^+. The glass electrodes using glasses with favorable physical properties and a specific composition such as 65% SiO_2, 28% Li_2O, 4% La_2O, and 3% Cs_2O. The +1 ions that can be measured potentiometrically are H^+, Li^+, Na^+, K^+, Ag^+, and NH_4^+.

The essential elements of a glass electrode are schematically shown in Figure 13.3. The electrode is constructed from a hollow glass tube sealed with a special thin-walled glass membrane at the bottom. The tube is filled partway with a dilute solution of HCl, and dipping into this HCl solution is a silver wire coated with a layer of silver chloride (AgCl). The potential of this electrode is controlled by the difference between the hydrogen ion concentrations inside and outside the thin glass membrane at the bottom. Since the H^+ concentration inside the electrode is constant, the electrode's potential varies only with the concentration of H^+ in the solution outside. In fact, this potential is proportional to the logarithm of the H^+ concentration and therefore to the pH of the outside solution.

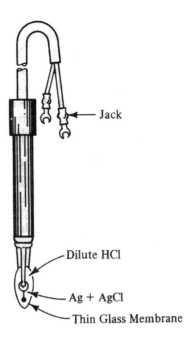

Figure 13.3
A glass electrode.

A glass electrode is always used with another reference electrode whose potential is constant, and this gives a galvanic cell whose potential depends on the pH of the solution into which the electrodes are immersed. The measured potential (measured against a reference electrode) is directly proportional to pH. The measurement of the potential and the translation of the potential into pH are the job of a pH meter. The glass electrode is relatively free from interferences caused by color, turbidity, colloidal matter oxidants, reductant, or high salt content of the sample. High sodium-containing samples give higher pH readings. It can be reduced or eliminated by using a "low-sodium error electrode." Coating of oily material or particulate matter can impair electrode response. The coating usually can be removed by gentle wiping or detergent washing followed by distilled water rinsing. Additional treatment with 1+9 HCl may be applied to remove any remaining film. Temperature effects may be eliminated by using a pH meter equipped with a temperature compensator.

More detailed information of pH measurement is given in Chapter 20.

Liquid membrane electrodes — Commercial liquid membrane electrodes are, like glass electrodes, designed to measure a single specific ion. A liquid membrane electrode contains a water-insoluble organic liquid that is capable of transporting some specific ion across the membrane boundary. This ability makes the electrode respond selectively to the particular ion being transported. Thus liquid membrane electrodes are made to measure Ca^{2+}, Mg^{2+}, Cu^{2+}, Pb^{2+}, Cl^-, NO_3^-, and ClO_4^-.

The key difference between liquid membrane electrodes and glass membrane electrodes is that the liquid membrane reactive sites are mobile and able to travel to the ion, whereas such sites are fixed in the glass membrane. The essential elements in a liquid ion exchange electrode are shown schematically in Figure 13.4. It is similar to the glass electrode in that it contains an internal reference electrode and an internal reference solution of fixed composition. Instead of glass, the membrane is a thin, porous organic polymer saturated with a liquid ion exchanger dissolved in a water-immiscible organic

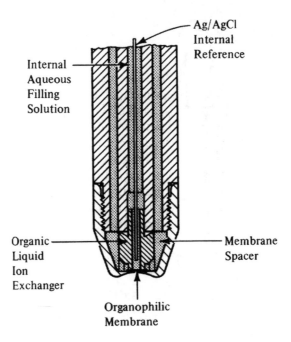

Figure 13.4
Liquid membrane ion-selective electrode.

solvent. Detailed information of the measurement of NO_3^- ion by ion-selective electrodes is given in Section 22.4.2.

Solid-state membrane electrodes — Like the glass membrane electrodes, the solid-state membrane electrodes have fixed reactive sites that cannot travel to the ions measured. Thus, to measure negative anions, a +1 reactive site like the Ag^+ ion must be imbedded in the membrane. To measure +2 cation, a –2 anion like S^{2-} must be arrayed in the membrane as a reactive site.

The essential elements in a solid-state membrane electrode are shown in Figure 13.5. The sensing element is a conducting solid, either a single crystal or a pellet pressed from

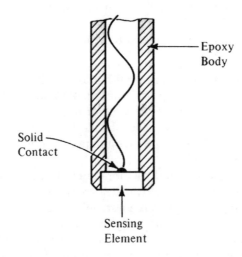

Figure 13.5
Solid-state ion-selective electrode.

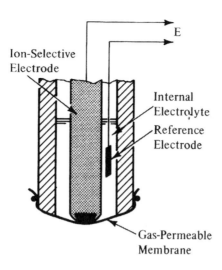

Figure 13.6
Gas-sensing electrode.

crystalline material. For example, the chloride-sensing electrode is made from silver sulfide (Ag_2S), the fluoride-indicating electrode is made from a singe crystal of lanthanum fluoride. Analysis of fluoride ions with ion-selective electrodes is discussed in Section 1.1.3.

Gas-sensing electrodes — These electrodes measure the concentration of certain gases dissolved in aqueous solution, or the concentration of ions in solution that can be converted to a dissolved gas by a simple chemical reaction. A diagram of a typical gas-sensitive electrode is presented in Figure 13.6. The membrane is made of a hydrophobic porous plastic that prevents water from entering the pores or passing through the membrane. The gas in the sample solution diffuses through the membrane and comes to equilibrium with a liquid film inside the electrode, where it chemically reacts with some substance to form ions. These are detected by an ion-selective electrode. An internal reference electrode is measured in millivolts by using a potentiometric instrument similar to a pH meter. Gas-sensing electrodes are commercially available for measuring carbon dioxide (CO_2), ammonia (NH_3), hydrogen sulfide (H_2S), and nitrite (NO_2^-) salts. A detailed method of ammonia determination by ISE is discussed in the methodology section.

According to the Nernst equation (13.4), the measured potential, measured relative to the SCE, is proportional to the logarithm of the concentration of the ion.

$$E = E° - \log [ION] \qquad (13.5)$$

Logarithmic regression calculations — When ions are measured by ion-selective electrode (ISE) methods, a specially designed meter called an *ion analyzer* may be used, which expresses the results in direct concentration. If the ion analyzer is not available, a pH meter may be used by changing the pH and its reference electrode to one of the ion-selective electrode and the accompanied reference electrode, and the reading position of pH turns to a millivolt (mV) readings on the instrument. Different concentrations of standards will produce different mV readings on the instrument. With the increase in the concentration, mV readings will decrease. The curve is prepared by plotting the concentrations against the mV readings. Therefore, the plot will be different from the usual calibration curves where the increased concentration increases the absorbance reading. For plotting the curve, use semilog graphpaper by using the x-axis for mV and the y-axis for the logarithm of the concentration. The result must be a straight line expressed with the equation of

$$y = mx + b \qquad\qquad (13.6)$$

where $y = E$ (potential), m = slope of the curve, x = log [ION], and b = intercept = $E°$. Instead of using linear regression calculations (see Section 12.3.2) use *logarithmic regression calculation*. It is similar to the linear regression calculation, except the concentration values should be logarithmic values.

The final prediction of the x- and y-values are calculated the same way as the linear regression, except the final result will be converted to reporting form by calculating it as an "antilog."

Example — Determine the fluoride concentration of a water sample according to the following data using logarithmic regression calculation:

Values of calibration standards (x):   0.2, 0.5, 0.8, 1.0, 2.0 ppm
Values of corresponding mV readings: 262, 248, 234, 224, 210 mV

y	x (log)	x^2	xy
262	−0.699	0.488	−183.138
248	−0.301	0.090	−74.648
234	−0.097	0.009	−22.698
224	0.000	0.000	0.000
210	0.301	0.090	63.210
1178	−0.795	0.677	−217.274

Values are substituted in formula (12.6):

$$m = 5(-217.274) - (0.795)(1178) / (5(0.677) - (0.795)^2 = -54.4$$

$$b = (0.677)(1178) - (-0.795)(-217.274) / 2.753 = 228.85$$

If one sample with an unknown concentration gives a mV reading of 250, the ppm value for the sample is calculated by using formula (13.6):

$$250 = -54.4x + 229$$

$$x = -0.386$$

$$x = \text{antilog}\ (-0.386)$$

$$x = 0.411\ \text{ppm}$$

The reported value for this particular sample will be 0.411 ppm.

Commonly used ISEs — Commonly used ion-selective electrodes in environmental analytical work are ammonia, fluoride, and nitrate. Detailed methods for using these electrodes are in the methodology section.

TABLE 13.1

Proper Storage of Ion-Selective Electrodes

Electrode	Short-Term Storage	Long-Term Storage
pH	In pH buffer 7.00 or 0.1 M KCl solution or "electrode storage solution." Level of reference electrode filling solution should be at least 1 in. above storage solution level.	Reference chamber filled. Sensing element covered with protective cap containing a few drops of storage solution. Before use, rinse well with DI water, refill, and soak in pH 7.00 buffer for 2 hours.
Ammonia (NH_3)	1000 ppm NH_3-N standard.	Disassemble completely. No filling solution, no membrane. Store dry.
Bromide (Br^-)	Store dry; protective cap must cover sensing element.	Same as short-term storage.
Fluoride (F^-)	Store dry; protective cap must cover sensing element.	Same as short-term storage.
Nitrate (NO_3^-)	1000 ppm NO_3-N standard (overnight only).	Store module in its vial. Cover tip with protective cap.
Oxygen (O_2)	Meter "off" position. Electrode in BOD bottle (used for calibration), with enough DI water to just cover the bottom of the bottle.	Same as short-term storage.

Ion-selective electrodes are also used widely in the clinical laboratory field and have made biochemical reactions potentially easier to study. In 1991, the Nobel Prize in physiology or medicine was awarded to two German scientists, Erwin Neher and Bert Sackmann, for their use of microelectrodes in the study of ion transportation through cell membranes. Their work has contributed to a better understanding of the cellular mechanisms of several diseases, including diabetes and cystic fibrosis.

Practically, however, while such ion-selective electrodes exist, and are very important analytically, many suffer from disadvantages mostly due to interference from other ions. Still, they are commonly used and provide very rapid, easy analysis for many chemical species.

Proper storage of selected ISEs is presented in Table 13.1.

Chapter **14**

ENVIRONMENTAL TESTING LABORATORIES

14.1 ENVIRONMENTAL REGULATIONS

The complicated operations of environmental testing laboratories today grow directly out of laws and regulations that have been passed in the U.S. over the past quarter century. Previously, environmental testing laboratories existed primarily for the purpose of monitoring public drinking water for bacteriological contamination. But today, large environmental testing facilities are available to analyze all media that can carry contamination: groundwater, surface water, potable water, industrial and domestic wastewater, agricultural discharges, air, soil, hazardous material, etc. Federal and state laws requiring new and more demanding analytical performance dictate the increase of analyzed parameters, necessitating sensitive and complex analytical instrumentation, and establish sets of procedures and protocols to document and control each step of the analytical process under various regulatory programs.

In 1970, the U.S. Environmental Protection Agency (EPA) was established. Now the EPA manages many of the environmental protection laws. Current responsibilities of the EPA include research on the health and environmental impact of a wide range of pollutants as well as development and enforcement of environmental standards for pollutants outside of the workplaces.

Individual states began drafting legislation to regulate environmental problems in their own states. The final decision between the federal government and the individual states regarding environmental issues is to develop federal standards while allowing the states to manage their own programs as long as they are at least as stringent as the ones set up by federal laws.

14.1.1 Overview of Federal Regulations

Clean Water Act (CWA) — The CWA is the current program designed to control water pollution of our surface waterways. The major objective of the CWA is to restore and maintain the "chemical, physical, and biological integrity of the nation's waters." The CWA

- controls the discharge of toxic materials into surface streams. Each state should establish water quality standards for all surface waters and specify their usage.

- controls the operation of Wastewater Treatment Plants.
- mandates the National Pollutant Discharge Elimination System (NPDES), requiring special discharge permits both for domestic and industrial wastewaters under the NPDES. These permits are based on specified effluent standards and limitations. EPA has 34 industrial categories covering about 130 toxic pollutants regulated for discharge into surface waters.
- mandates the Water Quality Management Program, providing funding to the states for planning their control strategies on surface water quality.
- mandates the Dredge and Fill Permit System, for wetland protection. The EPA and the U.S. Army Corps of Engineers are jointly responsible for protecting waters against degradation and destruction caused by disposal of dredged spoils and fills. Wetlands are vital elements of the natural water systems because they control floods, provide a natural filtration system for pollutants, and provide a habitat for fish and wildlife. Wetland protection is provided by the Office of Wetland Protection established by the EPA.
- the Marine Protection, Research, and Sanctuary Act protects the oceans from discriminate dumping of wastes.

Safe Drinking Water Act (SDWA) — The SDWA is designed to protect drinking water sources (groundwaters and surface waters). The SDWA provides for

- primary drinking water standards with maximum contaminant levels (MCLs) and maximum contaminant level goals (MCLGs). These standards are designed to protect human health.
- protection of aquifers against contamination from the disposal of wastes by injection into deep wells. States regulate their underground injection system.
- groundwater protection strategy.
  - Wellhead Protection Program is state-developed and administered by the states.
  - The Sole Source Aquifer Program prevents contamination of aquifers designated as a sole source of water supply for an area or community.

Toxic Substance Control Act (TSCA) — The TSCA provides for

- premanufacture notification process: manufacturers should notify the EPA at least 90 days before producing or importing a new chemical substance.
- chemical inventory: the TSCA requires the EPA to develop and keep a current comprehensive inventory of commercially used chemicals. It contains all the information, properties, health effects, and environmental risks of the substances. This inventory is called the Material Safety Data Sheets (MSDS) and should be available in each workplace where chemicals are used in daily jobs.
- good laboratory practices: the TSCA requires the EPA to establish approved standard methods for environmental testing laboratories, prepare guidelines for laboratories on how they can prove that their reported values are correct and defendable and how to document all of the activities involved in the generation of analytical data and the final reports (laboratory QA/QC).
- ban of further production of toxic substances: some toxic substances require special attention because they are so widespread in the environment or because they pose serious health threats even in extremely low levels of contamination. For example, polychlorinated biphenyls (PCBs), asbestos, dioxin, etc.

Resource Conservation and Recovery Act (RCRA) and its amendment, the Hazardous and Solid Waste Act (HSWA) — These acts deal with management of solid wastes with emphasis on hazardous wastes. RCRA regulations include:

1. Identifying and classifying hazardous waste.

2. Issuance of publications for generators about activities and developing standards for generators.

3. Developing standards for transporting hazardous wastes.

4. Developing standards for treatment, storage and disposal.

5. Making enforcement standards available within the program.

6. The few characteristic properties that qualify a waste as a RCRA waste material are ignitability, corrosivity, reactivity, and toxicity. *Ignitability* refers to the characteristics of being able to sustain combustion and includes the category of flammability (ability to start fires when heated to temperatures less than 60°C or 140°F). *Corrosive wastes* may destroy containers, soil, groundwater, or react with other materias to cause toxic gas emissions. Corrosive materials provide a very specific hazard to human tissue and aquatic life where the pH levels are extremes. *Reactive wastes* may be unstable or have a tendency to react, create explosions or toxic fumes, gases or vapors when mixed with water or acids. *Toxicity* shown by wastes which are harmful or fatal if ingested or absorbed. A critical test to determine toxicity is the Toxic Characteristic Leachate Procedure (TCLP).

7. Underground storage tank. Leaking underground storage tanks containing petroleum products and other hazardous substances have been identified as a major source of groundwater contamination. In 1986, amendments of RCRA provided funds to clean up petroleum residues originated from leaking underground storage tanks. The fund was financed by tax on motor fuels.

Comprehensive Environmental Response, Compensation and Liability Act (CERCLA) (Superfund) — The act is concerned primarily with funding cleanups of past activities and present spills. There are four key concepts of CERCLA: (1) removal action, (2) remedial action, (3) funding, and (4) emergency planning. Major changes to CERCLA under the Superfund Amendments and Reauthorization Act of 1986 (SARA) require the cleanup of Superfund sites to "assure protection of human health and the environment." Under Title III of SARA, which is the Emergency Planning and Community Right-to-Know Act, companies are used to report annually the amount of certain substances used and discharged into the environment. The Community Right-to-Know Act provides information to the public about the presence of hazardous chemicals in their communities.

Occupational Safety and Health Act (OSHA) — The act was organized to provide allowable levels of exposure to toxic chemicals for workers exposed to the hazard for 40 hours per week. OSHA issued *permissible exposure levels* (PEL) based on exposure for a healthy 70 kg male. It also stipulates that in every workplace, Material Safety Data Sheets (MSDSs) should be available for complete information on every chemical substance associated with the work.

Federal Insecticide, Fungicide, and Rodenticide Act (FIFRA) — Under this act, the EPA must control the manufacture and use of pesticides.

Clean Air Act (CAA) — Over the past two decades, the Clean Air Act has evolved from a set of principles designed to generally guide states in controlling sources of air pollution (1967), to a series of detailed control requirements (1970, 1977, and 1990 amendments of the act) that the federal government implemented and the states administered. The CAA regulatory programs have traditionally fallen into three categories.

First, all new and existing sources of air pollution are subject to ambient air quality regulation, through source-specific emission limits contained in the states' implementation plans (SIPs). Second, new sources are subject to more stringent control technology and permitting requirements. Third, the act addresses specific pollution problems, including hazardous air pollution and visibility impairment.

In 1990, Congress amended this three-part system of regulation and added a fourth program — a comprehensive operating permit program to focus in one place all of the Clean Air Act requirements that apply to a given source of air pollution. Over the coming years, EPA will continue to face implementation responsibilities that far surpass those that have been assigned to vitually any other administrative agency, due in large part to the activities required by the Clean Air Act.

As the 1990 amendments are implemented, clean air regulation will have a greater impact on day-to-day business decisions than perhaps any other piece of environmental legislation.

14.1.2 Overview of State Programs

- Groundwater program development: every state received a grant from the EPA to develop a groundwater program.

- Local land use control: control the locations of industrial facilities and incorporate land use for groundwater protection.

- State Superfund program: clean up hazardous waste sites.

- Regulations on solid waste disposal sites: states are the primary regulators of solid waste disposal sites.

- Regulations for commercial and industrial facilities: states control discharge permits, storage and management of hazardous materials, recycling programs, and cleanup plans.

- Waste disposal wells: each state is responsible for its underground injection control (UIC) program. The program includes waste characterization, monitoring wells, location of the wells, sampling, and inspection. Each state program is unique.

- Underground storage tank (UST) program: the state is responsible for permits, design of construction, installation, testing and monitoring of the quality, and for the selection of the out-of-service tanks.

- Agricultural sources: state laws regulate pesticides, fertilizers, and irrigation waters. States work closely with the U.S. Department of Agriculture. States regulate pest and weed control programs and the proper disposal and treatment of animal wastes.

- Mining regulations: each state has its own mining regulations according to local requirements.

- Septic systems: states are charged with the special design criteria, site selection, operation, and maintenance of residental septic systems, as well as septage disposal regulations, and regulations related to commercial and industrial septic systems.

- Transportation of hazardous materials: transportation and special packaging of hazardous materials are regulated by the U.S. Department of Transportation (DOT). According to the DOT requirements, the state has to identify and document: (1) what is the hazardous material and its quantity, (2) the method used for transportation, (3) special potential hazards of the material, and (4) the safety of the transportation of the specified material.

14.2 LABORATORY OPERATIONS FOR THE ANALYSIS OF REGULATED SUBSTANCES

Definitions of "pollution" and "toxic and hazardous materials" depend upon analytical chemical measurements. Each scientific advance in the methodology and instrumentation that measure chemicals in the environment has resulted in more stringent regulation. Laws required analytical measurements to verify compliance. The Safe Drinking Water Act (SDWA) was the federal legislation that had the most immediate impact on the expansion of analytical laboratories. State governments, through their health departments and environmental agencies, have primary enforcement responsibility for the water quality standard. They also have authority to certify independent testing laboratories as qualified to analyze samples under the programs and laws. The new analytical needs are enormous. Hazardous waste analysis is regulated by the Resource Conservation and Recovery Act (RCRA), and whether or not a waste falls under the stringent provisions of RCRA regulations is an issue with great financial, legal, and liability consequences. Analytical verification has therefore become a vital concern to property owners and potentially affected waste generators. The Comprehensive Environmental Response, Compensation and Liability Act (CERCLA), commonly known as Superfund, stipulates that all the complex mixtures of industrial hazardous wastes at individual sites (generated by industries, consisting mostly of organic chemicals) should be analyzed.

The Clean Water Act (CWA) sets goals to measure and monitor surface water quality according to their uses. The National Pollution Discharge Elimination System (NPDES) requires regular sampling and monitoring programs of industrial discharges for permitting and pretreatment programs.

The Clean Air Act was established in 1970 with air quality standards based on a few regulated pollutants. The 1990 amendment to the Clean Air Act increased the number of chemicals regulated, so analytical programs had to be developed and air quality monitoring requirements become a new important section in environmental testing laboratories.

14.2.1 Methodology

For each of the major regulatory programs described above, methods were developed to analyze the various media for different parameters. Each method approved by the EPA specifies the procedures, instrument calibration, sample preparation, the actual analytical procedures, and the quality control requirements for the specified analytical work. EPA methods are differentiated according to the media (matrix) of the sample analyzed.

Each laboratory has a written guidebook which details the specific procedures used, called Standard Operating Procedures (SOPs). It should be constantly revised to include new methodologies and procedural changes. The SOP is an important tool for the QA/QC operation of the laboratory.

14.2.2 EPA-Approved Methods and References for Analyzing Water Samples

Methods and references for analyzing drinking water

- *Methods for Chemical Analysis of Water and Wastes*, EPA 600/4-79-020, revised March 1983.
- *Methods for Determining Organic Compounds in Drinking Water*, EPA 600/4-88-039, December 1988.
- *Standard Methods for the Examination of Water and Wastewater*, APHA-AWWA-WPCF, 18th ed., 1995. (Updated, new edition issued every 5 years).
- *Manual for Certification of Laboratories Analyzing Drinking Water*, EPA 570/9-90/008, April 1990.
- *CFR* Part 141, Subpart C and Subpart E*, monitoring and analytical requirements.

---

* CFR = Code of Federal Regulations.

- *EPA 500 series*, should be used for organic analyses of drinking waters and raw source waters.

Methods and references for analyzing surface waters and wastewater effluents

- *Methods for Chemical Analysis of Water and Wastes*, EPA 600-4-79-020, revised March 1983.
- *Test Methods for Evaluating Solid Waste*, EPA SW-846, 3rd ed., 1986, and its Revision I, December 1987.
- *40 CFR Part 136*, Tables IA, IB, IC, ID, and IE, July 1989.

Method and references for analyzing water sources (surface and groundwaters) pursuant to 40 CFR Part 261 (RCRA) and the DEP Chapter 17-700 Series

- *Test Methods for Evaluating Solid Waste*, EPA SW-846, 3rd ed., 1986, and its Revision I, 1987.
- *Methods listed in 40 CFR Part 261*, Appendix III, 1989.
- *USEPA Contract Laboratory Program Statement of Work for Inorganic Analyses*, EPA SOW ILMO3.0, March 1990.
- *USEPA Contract Laboratory Program Statement of Work for Organic Analyses*, EPA SOW OLMO3.1, August 1994.

Methods and references for microbiological and biological tests of water samples

- *Microbiological Methods for Monitoring the Environment*, EPA 600/8-87-017, 1987.
- *40 CFR Part 141, Subpart C (July 1989)*, monitoring and analytical requirements.
- *40 CFR Part 136, Table IA (July 1989)*.
- *Methods for Measuring the Acute Toxicity of Effluents to Freshwater and Marine Organisms*, EPA 600/4-85-013, 1985, 3rd ed.
- *Short-Term Methods for Estimating the Chronic Toxicity of Effluents and Receiving Waters to Freshwater Organisms*, EPA 600/4-89-001, 1989.
- *Short-Term Methods for Estimating the Chronic Toxicity of Effluents and Receiving Waters to Marine and Estuarine Organisms*, EPA 600/4-87-028, 1988.

14.2.3 EPA-Approved Methods and References for Analyzing Soils, Sediments, and Residuals

Methods and references for analyzing soils, sediments, domestic and industrial sludges, solid and hazardous wastes

- *Test Methods for Evaluating Solid Waste*, EPA SW-846, 3rd ed., 1986, Revision December 1987.
- *40 CFR Part 261, Appendix III*, July 1989.
- *Procedures for Handling and Chemical Analysis of Sediments and Water Samples*, EPA/Corps of Engineers, CE-81-1, 1981.
- *USEPA Contract Laboratory Program Statement of Work for Inorganic Analysis*, EPA SOW ILMO3.0 March 1990.
- *USEPA Contract Laboratory Program Statement of Work for Organic Analysis*, EPA SOW OLMO3.1, August 1994.

- *POTW Sludge Sampling and Analysis Guidance Document*, EPA Permits Division, August 1989.

14.2.4 Approved Modifications of EPA Methods

- EPA Method 300.0: This method may be used for the analysis of specified ions in groundwater and surface water, except fluoride. Currently approved for drinking water analysis.

- EPA Methods 601, 602, 624, and 625: Capillary columns may be used instead of the specified packed columns if the laboratory meets the accuracy and precision criteria and detection limit with this modification.

- EPA Methods 601 and 602: The photoionization detector and electrolytic conductivity detector may be used in a series if the laboratory can meet the performance criteria.

- EPA Methods 602, 8020, 8021: May include the analysis of xyelene and methyl-tert-buthyl-ether (MTBT).

- EPA Methods 610, 625, 8100, 8310, 8250, 8270: May include the analysis of methylnaphtalenes.

- EPA Method 5030/8010: Must be modified to analyze EDB in soils. An electron capture detector instead of an electrolytic conductivity detector must be used.

14.2.5 EPA Contract Laboratory Protocol (CLP)

The EPA contract laboratory protocol was developed for the Superfund program. CLP specifies a set of methods which are based on the existing methodology for organic and inorganic parameters, but which are modified to incorporate certain quality control, calibration, and deliverable requirements. The data package includes a full reporting of quality control procedures and data, making it particularly useful if litigation is a possibility.

The results of the analyses are provided in many different formats, ranging from a sample report only to a full documentation data package. The CLP, as stated in the EPA Statement of Work (SOW), has a level of quality assurance requirement. The deliverable requirements include quality control summaries (method blank, initial calibration verification, duplicate analysis, matrix spike/matrix spike duplicates), quality control data, as well as data on a diskette.

CLP, therefore, has become a commonly requested methodology, and has the effect of separating larger laboratories that have the equipment, certifications, and trained personnel capable of producing data according to this protocol from the thousands of smaller environmental laboratories which do not.

Because EPA methods, as now written, are not interchangeable, it is very difficult for an analytical laboratory to accommodate all quality control criteria for all of the methods. Thus, EPA's current intent is to create a unified method to minimize the differences in requirements.

14.3 LABORATORY QUALITY ASSURANCE AND QUALITY CONTROL (QA/QC)

Detailed laboratory quality assurance/quality control (QA/QC) requirements are discussed in Chapter 3.

14.4 STANDARDS

14.4.1 Drinking Water Standards

The correct definition of drinking or potable water is water delivered to the customer that can be safely used for drinking, cooking, and washing. Potable water must meet the physical, chemical, bacteriological, and radiological quality standards criteria established by the regulatory agencies. The sources of water supply are being endangered by hundreds of new chemicals and pollutants every year. New, modern equipment make it possible to detect contaminants in a very small quantity.

The Safe Drinking Water Act (SDWA) was the federal legislation that had the most immediate impact on the expansion of analytical laboratories and development of instrumentation for environmental testing. The experience of the first 12 years of implementing the SDWA led to substantial amendments to the act, which were passed June 1986 and are still in the process of being implemented.

Drinking water regulations fall into two basic categories: primary and secondary. *Primary standards* determine how clean drinking water must be to protect public health. The parameters are health-related. *Secondary standards* are not health-related. Recommended levels of these substances are mainly to provide aesthetic and taste characteristics.

Primary and secondary drinking water standards with maximum contaminant levels (MCLs), established by the Amendments of the Safe Drinking Water Act (SDWA) in 1986, are shown in Table 14.1.

The most recent requirements for drinking water standards with MCLs, detection limits (DLs), and analytical methods are listed in Table 14.2.

The 1986 Amendments to SDWA strengthened the enforcement provisions to provide additional authority to the EPA Administrator to take enforcement action against public water systems in violation of any regulation. Despite the strengthening of the enforcement authorities, EPA and the primary states continue to be harshly criticized for a lack of legal action. The Natural Resource Council (NRDC) in "Think Before You Drink", 1992–94 Update, (July, 1994, p. 4) alleged that "nearly 18 million people were served by drinking water systems that violated EPA regulations and that 24 states failed to report violations of chemical and radiological MCLs and testing requirements."

14.4.2 Waste Characterization

The characteristic properties that qualify a waste as a RCRA waste material are ignitability, corrosivity, reactivity, and toxicity. *Ignitability* refers to the characteristics of being able to sustain combustion and includes the category of flammability (ability to start fires when heated to temperatures less than 60°C or 140°F). *Corrosive wastes* may destroy containers, soil, and groundwater, or react with other materials to cause toxic gas emissions. Corrosive materials provide a very specific hazard to human tissue and aquatic life where the pH levels are extreme. *Reactive wastes* may be unstable or have a tendency to react, explode, or generate pressure during handling. Also, pressure-sensitive or water-reactive materials are included in this category. *Toxicity* is a function of the effect of waste materials that may come into contact with water or air and be leached into the groundwater or dispersed in the environment. The toxic effects that may occur to humans, fish, or wildlife are the principal concerns.

Criteria for hazardous waste evaluation — The criteria for hazardous waste evaluation are as follows:

TABLE 14.1

Drinking Water Standards Established by the 1986 Amendments to the Safe Drinking Water Act

	MCLs
Primary Standards	
Inorganics	
Arsenic	0.05 mg/L
Barium	1.00 mg/L
Cadmium	0.01 mg/L
Chromium	0.05 mg/L
Lead	0.05 mg/L
Mercury	0.002 mg/L
Nitrate-Nitrogen	10.0 mg/L
Selenium	0.01 mg/L
Fluoride	4.00 mg/L
Silver	0.05 mg/L
Turbidity	1–5 NTU
Organics	
Endrine	0.0002 mg/L
Lindane	0.004 mg/L
Methoxychlor	0.10 mg/L
Toxaphene	0.005 mg/L
2,4-D	0.10 mg/L
2,4,5-T (Silvex)	0.01 mg/L
Volatile Organics (VOCs)	
Trichloroethylene	0.005 mg/L
Carbontetrachloride	0.005 mg/L
Vinyl chloride	0.002 mg/L
1,2-Dichloroethane	0.005 mg/L
Benzene	0.005 mg/L
para-Dichlorobenzene	0.075 mg/L
1,1-Dichloroethylene	0.007 mg/L
1,1,1-Trichloroethane	0.20 mg/L
Total trihalomethanes	0.10 mg/L
Microbiology	
Total Coliform bacteria	zero counts/100 mL
Radionuclides	
Radium 226 and 228 (total)	5 pCi/L
Gross alpha particle activity	15 pCi/L
Gross beta particle activity	50 pCi/L

Parameter	Recommended Level
Secondary Standard	
Chloride	250 mg/L
Color	15 C.U.
Copper	1.0 mg/L
Corrosivity	noncorrosive
Foaming agent	0.5 mg/L
Hardness, as CaCO$_3$	50 mg/L
Iron	0.3 mg/L
Manganese	0.05 mg/L
Odor	3 T.O.N.
pH	6.5–8.5 pH unit
Sulfate	250 mg/L
Total dissolved solids (TDS)	500 mg/L
Zinc	5.0 mg/L

Note: Recommended levels for these substituents are mainly to provide aesthetic and taste characteristics. Secondary drinking water regulations are not health-related. They are intended to protect "public welfare" by offering unenforceable guidelines on the taste, odor, or color of drinking water. Abbreviations: MCLs = maximum contaminant levels; N.T.U. = nephelometric turbidity unit; PCi/L = picoCurie per liter; C.U. = color unit; T.O.N. = threshold odor number; 2,4-D = 2,4-dichlorophenoxyacetic acid; 2,4,5-T = 2,4,5-Trichlorophenoxyacetic acid (Silvex).

TABLE 14.2

Recent Requirements for Drinking Water Standards

Parameters	MCL	Analytical Method	Detection Limit
Primary Standard			
Inorganic (PWSO30)	**(mg/L)**		**(mg/L)**
Arsenic	0.05	EPA 206.2	0.0020
Barium	2.00	EPA 200.7	0.0140
Cadmium	0.005	EPA 200.7	0.0010
Chromium	0.10	EPA 200.7	0.0090
Cyanide	0.20	EPA 335.2	0.0050
Fluoride	4.00	EPA 340.2	0.01
Lead	0.015	EPA 239.2	0.0010
Mercury	0.002	EPA 245.1	0.0002
Nickel	0.100	EPA 200.7	0.0110
Nitrate nitrogen	10.00	EPA 353.2	0.01
Nitrite nitrogen	1.00	EPA 354.2	0.01
Selenium	0.05	EPA 270.2	0.0010
Sodium	160	EPA 200.7	0.226
Antimony	0.006	EPA 204.2	0.0020
Beryllium	0.004	EPA 200.7	0.0020
Thallium	0.002	EPA 279.2	0.0010
Secondary Standard			
Aluminum	0.200	EPA 200.7	0.100
Chloride	250	EPA 300.0	0.5
Copper	1.00	EPA 200.7	0.0040
Iron	0.30	EPA 200.7	0.0500
Manganese	0.05	EPA 200.7	0.5000
Silver	0.10	EPA 200.7	0.0050
Sulfate	250	EPA 300.0	0.0100
Zinc	5.00	EPA 200.7	0.0140
Color	15 C.U.	SM 204A	5.00
Odor	3 T.O.N.	SM 207	1.00
pH	6.5–8.5	EPA 150.1	—
Total dissolved solids (TDS)	500	EPA 160.1	10.0
Foaming agents	0.5	SM 512B	0.0500
Trihalomethane (PWSO27)	**(µg/L)**		**(µg/L)**
Bromoform		EPA 502.2	0.00013
Chloroform		EPA 502.2	0.00005
Dibromochloromethane		EPA 502.2	0.00013
Dichlorobromomethane		EPA 502.2	0.00007
Total THMs	0.10	EPA 502.2	
Volatile Organics (PWSO28)			
1,2,4-Trichlorobenzene	70	EPA 502.2	0.310
cis-1,2-Dichloroethylene	70	EPA 502.2	0.0300
Xylenes (total)	10,000	EPA 502.2	0.170
Dichloromethane	5	EPA 502.2	1.40
o-Dichlorobenzene	600	EPA 502.2	0.140
p-Dichlorobenzene	75	EPA 502.2	0.190
Vinyl chloride	1	EPA 502.2	0.290
1,1-Dichloroethylene	7	EPA 502.2	0.170
trans-1,2-Dichloroethylene	100	EPA 502.2	0.180
1,2-Dichloroethane	3	EPA 502.2	0.0400
1,1,1-Trichloroethane	200	EPA 502.2	0.0300
Carbon tetrachloride	3	EPA 502.2	0.0400

TABLE 14.2 *(continued)*

Recent Requirements for Drinking Water Standards

Parameters	MCL	Analytical Method	Detection Limit
	(µg/L)		(µg/L)
1,2-Dichloropropene	3	EPA 502.2	0.0400
Trichloroethylene	3	EPA 502.2	0.0400
1,1,2-Trichloroethane	5	EPA 502.2	0.0400
Tetrachloroethylene	3	EPA 502.2	0.0800
Monochlorobenzene	100	EPA 502.2	0.0700
Benzene	1	EPA 502.2	0.0500
Toluene	1000	EPA 502.2	0.0800
Ethylene benzene	700	EPA 502.2	0.0600
Styrene	100	EPA 502.2	0.0700
Pesticides and PCBs (PWSO29)			
Endrin	2	EPA 508	0.01
Lindane	0.2	EPA 508	0.01
Methoxychlor	40	EPA 508	0.02
Toxaphene	3	EPA 508	0.2
Dalapon	200	EPA 515.1	1
Diquat	20	EPA 549	4
Endothal	100	EPA 548	10
Glyphosate	700	EPA 547	10
Di(2-Ethylhexyl)adipate	400	EPA 506	1
Oxamyl (Vydate)	200	EPA 531.1	0.5
Simazine	4	EPA 507	0.1
Picloram	500	EPA 515.1	0.2
Dinoseb	7	EPA 515.1	0.2
Hexachlorocyclopentadiene	15	EPA 512	0.1
Carbofuran	40	EPA 531.1	0.5
Atrazine	3	EPA 507	0.1
Alachlor	2	EPA 507	0.3
2,3,7,8-TCDD (Dioxin)	0.00003		
Heptachlor	0.4	EPA 508	0.01
Heptachlor epoxide	0.2	EPA 508	0.01
2,4-D	70	EPA 515.1	0.5
2,4,5-T (Silvex)	50	EPA 515.1	0.05
Hexachlorobenzene	1	EPA 508	0.01
Di(2-ethylene hexyl)-phthalate	6	EPA 506	1
Benzo(a)pyrene	0.2	EPA 550	0.01
Pentachlorophenol	1	EPA 515.1	0.05
PCB	0.5	EPA 508	0.05
Dibromochloropropane	0.2	EPA 504	0.005
Ethylene dibromide	0.02	EPA 504	0.005
Chlordane	2	EPA 508	0.05
Radiochemical Analysis	(pCi/L)		
Gross alpha	5	EPA 900.0	
Radium-226	15	EPA 900.0	
Radium-228	50	EPA 900.0	
Microbiology			
Total coliform	zero	count/100 mL	

Note: Secondary drinking water regulations are not health-related. They are intended to protect "public welfare" by offering unenforceable guidelines on the taste, odor, or color of drinking water. Recommended levels for these substituents are mainly to provide aesthetic and taste characteristics. Abbreviations: MCL = maximum contaminant level; C.U. = color unit; T.O.N. = threshold odor number; pCi/L = picoCurie per liter; 2,4-D = Dichlorophenoxyacetic acid; 2,4,5-T = Trichlorophenoxyacetic acid.

TABLE 14.3

Maximum Concentration of Contaminants in
Characterization of EP Toxicity

Contaminant	Maximum Concentration (mg/L)
Arsenic	5.0
Barium	100.00
Cadmium	1.0
Chromium	5.0
Lead	5.0
Mercury	0.2
Selenium	1.0
Silver	5.0
Endrin	0.02
Lindane	0.4
Methoxychlor	10.0
Toxaphene	0.5
2,4-D	10.0
2,4,5-TP Silvex	1.0

Note: 2,4-D = 2,4-Dichlorophenoxyacetic acid; 2,4,5-TP
= 2,4,5-Trichlorophenoxyacetic acid; EP toxicity =
extraction procedure toxicity. The extraction procedure toxicity test was developed to characterize
hazardous wastes based on the leaching ability of
toxic substances in significant concentrations. A
liquid extract (leachate) of the material is analyzed for the above 14 parameters. During the
migration of the leachate, attenuation and dilution occur with a ratio factor of 100. This is used
to establish the maximum concentration levels.
(U.S. Environmental Protection Agency, "Extraction Procedure Toxicity Characteristics," C.F.R.,
Vol. 40, No. 261.24, May 19, 1980.)

- Ignitability: Flashpoint <60°C or 140°F.
- Corrosivity: pH less than 2.00 or higher than 12.00.
- Reactivity: Reacts violently or generates pressure. The substance should be free from cyanide (CN) and sulfide (S).
- Toxicity: Leaching tests extraction procedure (EP Toxicity) and toxicity characteristics leachate procedures (TCLP) parameters should meet regulatory MCL criteria.

Lists of the analyzed parameters with MCLs are shown in Tables 14.3 and 14.4.

14.4.3 Irrigation Water Quality

Water quality problems in irrigation include salinity and toxicity. Salinity affects crop production, because crop roots have great difficulty extracting enough water and nutrients from saline solution. Consequently, crop production is limited because sufficient water cannot reach the root zone. Toxicity is also a problem in maintaining good yeild. Boron, chlorides, and sodium are common toxic substances.

Of particular consequence is the ratio of sodium to calcium and magnesium. When sodium-rich water is applied to soil, some of the sodium is taken up by clay and the clay gives up calcium and magnesium in exchange. Clay that takes up sodium becomes sticky and slick when wet and has low permeability. Then the dry clay shrinks into hard clots

TABLE 14.4

Toxicity Characteristic Leachate Contaminants and Regulatory Levels

Contaminant	Regulatory Level (mg/L)	Contaminant	Regulatory Level (mg/L)
Acrylonitrile	5.0	Isobutanol	36.0
Arsenic	5.0	Lead	5.0
Barium	100.0	Lindane	0.06
Benzene	0.07	Mercury	0.2
Bis(2-chloroethyl)ether	0.05	Methoxychlor	1.4
Cadmium	1.0	Methylene chloride	6.6
Carbon disulfide	14.4	Methyl ethyl ketone	7.2
Carbon tetrachloride	0.07	Nitrobenzene	0.13
Chlordane	0.03	Pentachlorophenol	3.6
Chlorobenzene	1.4	Phenol	14.4
Chloroform	0.07	Pyridine	5.0
Chromium	5.0	Selenium	1.0
o-Cresol[a]	10.0	Silver	5.0
m-Cresol[a]	10.0	1,1,1,2-Tetrachloroethane	10.0
p-Cresol[a]	10.0	1,1,2,2-Tetrachloroethane	1.3
2,4-D	1.4	Tetrachloroethylene	0.1
1,2-Dichlorobenzene	4.3	2,3,4,6-Tetrachlorophenol	1.5
1,4-Dichlorobenzene	10.8	Toluene	14.4
1,2-Dichloroethane	0.40	Toxaphene	0.07
1,1-Dichloroethylene	0.1	1,1,1-Trichloroethane	30.0
2,4-Dinitrotoluene	0.13	1,1,2-Trichloroethane	1.2
Endrin	0.003	Trichloroethylene	0.07
Heptachlor (and its hydroxide)	0.001	2,4,5-Trichlorophenol	5.8
Hexachlorobenzene	0.13	2,4,6-Trichlorophenol	0.30
Hexachlorobutadiene	0.72	2,4,5-TP (Silvex)	0.14
Hexachloroethane	4.3	Vinyl chloride	0.05

Note: In 1986, the EPA expanded the EP toxicity characteristic substances (containing 8 metals, 4 insecticides, and 2 herbicides) with the additional 38 organic substances. The new procedure is called toxicity characteristic leachate procedure (TCLP) test. By the application of the TCLP test, the extract or leachate of the waste containing any of these 52 substances at or above the regulatory level qualified as hazardous toxic waste. Also, the TCLP test used compound-specific dilution/attenuation factors instead of the 100 in EP toxicity characterization. The listed parameters with the regulatory levels are in the U.S. Environmental Protection Agency, "Hazardous Waste Management System," C.F.R., Vol. 51, No. 114, June 13, 1986. The TCLP procedure is outlined in C.F.R., Vol. 40, No. 261.24, May 19, 1980.

[a] o (ortho), m (meta), and p (para)-Cresol concentrations are added together and compared to a threshold of 10.0 mg/L.

that are difficult to cultivate. Even worse, high concentration of sodium salts can produce alkali soils in which little or no vegetation can grow. On the other hand, when the same clay carries excess calcium and magnesium ions, it tills easily and has good permeability. If an irrigation water contains calcium and magnesium ions sufficient to equal or exceed the sodium ion, enough calcium and magnesium is retained on clay particles to maintain good tilth and permeability. These waters serve well for irrigation.

The sodium effect can be calculated by the sodium absorption ratio (SAR) method:

$$SAR = Na / \sqrt{[(Ca + Mg) / 2]}$$

Na, Ca, and Mg values are expressed in milliequivalents per liter. Water with SAR values below 10 is sufficient for irrigation, and SAR values of 18 or higher are not recommended for irrigation. Table 14.5 contains the trace elements and their recommended

TABLE 14.5

Recommended Maximum Concentrations of Trace Elements in
Irrigation Water

Elements	For Waters Used Continuously on Soils (mg/L)	For Waters Used up to 20 Years on Fine-Textured Soils pH 6.0–8.5 (mg/L)
Aluminum (Al)	5.0	20.0
Arsenic (As)	0.1	2.0
Beryllium (Be)	0.1	0.5
Boron (B)	[a]	2.0
Cadmium (Cd)	0.01	0.05
Chromium (Cr)	0.1	1.0
Cobalt (Co)	0.05	5.0
Copper (Cu)	0.2	5.0
Fluoride (F)	1.0	15.0
Iron (Fe)	5.0	20.0
Lead (Pb)	5.0	10.0
Lithium (Li)	2.5	2.5
Manganese (Mn)	0.2	10.0
Molybdenum (Mo)	0.01	0.05[b]
Nickel (Ni)	0.2	2.0
Selenium (Se)	0.02	0.02
Vanadium (V)	0.1	1.0
Zinc (Zn)	2.0	10.0

[a] No problem when less than 0.75 mg/L. Increasing problem when between
    0.75 and 2.0 mg/L. Severe problem when greater than 2.0 mg/L.
[b] For only acid fine-textured soils with relatively high iron oxide content.

National Academy of Sciences and National Academy of Engineering, 1972;
Groundwater and Wells, Fletcher G. Driscoll, 2nd ed., 1987.

maximum concentrations in irrigation water, and Table 14.6 shows the analytical parameters needed to evaluate water used for irrigation.

14.5 ANALYTICAL DEPARTMENTS IN ENVIRONMENTAL TESTING LABORATORIES

The complicated operations of environmental testing laboratories today grow directly
out of the laws and regulations that have been passed in the U.S. over the past quarter
century. In earlier decades, independent environmental laboratories existed primarily for
the purpose of monitoring public drinking water for bacteriological contamination. This
is still an important function for many laboratories. But today, large, independent environmental testing facilities analyze all media (matrix) that can carry chemical contamination, such as potable water, air, soil, hazardous materials and waste, petroleum products,
and industrial and municipal wastewaters. Federal and state laws require the use of
expensive, sensitive, and complex analytical instruments, and dictate procedures and
protocols to document and control each step of the analytical process under regulatory
programs.

According to the nature of the analytical parameters, environmental testing laboratories are divided into four different departments: (1) wet chemistry (physical properties
and inorganic nonmetallic parameters), (2) metals, (3) organics, and (4) microbiology.
Experienced environmental chemists and other laboratory personnel represent a rich
source for further refining technology and its practical application in the laboratory.

TABLE 14.6

Analytical Values Needed to Evaluate
Irrigation Waters

Acidity — Alkalinity as $CaCO_3$
Ammonia-Nitrogen (NH_3-N)
Bicarbonate (HCO_3^-)
Boron (B)
Calcium (Ca)
Carbonate (CO_3^{2-})
Chloride (Cl)
Electrical conductivity
Iron (Fe)[a]
Lithium (Li)[a]
Magnesium (Mg)
Nitrate-Nitrogen (NO_3-N)
Phosphate-Phosphorus (PO_4-P)[a]
Potassium (K)[a]
Sodium (Na)
Sodium absorption ratio (SAR)
Sulfate (SO_4^{2-})

Note: Calculation of the sodium absorption ratio:

$$SAR = Na / \sqrt{(Ca + Mg)/2}$$

The concentration of Na, Ca, and Mg is in
milliequivalent per liter (meq/L).

[a] Marked parameters are determined only in special situations.

TABLE 14.7

Ambient Air Quality Standard

Contaminant	Primary (Federal) Standard
Ozone	0.12 ppm
Carbon monoxide	35 ppm
	9 ppm (8 h)
Nitrogen dioxide	0.05 ppm (annual)
Sulfur dioxide	0.14 ppm (24 h)
	0.03 ppm (annual)
Particulate matter	150 µg/m³ (24 h)
	50 µg/m³ (annual)
Sulfates	25 µg/m³ (24 h)[a]
Lead	1.5 µg/m³ (3 months)
Hydrogen sulfide	0.03 ppm[a]
Vinyl chloride	0.010 ppm (24 h)[a]

Note: Standard values are for 1-hour averages unless otherwise indicated.

[a] Standard values are California Ambient
Air Quality Standards.

Roger D. Griffen, *Principles of Hazardous Materials Management*, 1989, p. 44.

Chapter **15**

RAW DATA CONVERSION INTO REPORTABLE RESULTS

15.1 RAW DATA AND REPORTABLE DATA

Raw data are data generated by analytical performances, including quality control checks. *Reportable data* or reportable results are generated from raw data by mathematical or statistical calculations. Final results may be produced by direct reading from the instrument or calculating from readings or instrument outputs by using formulas. Before starting calculations, assure that all readings or instrument outputs are correct and the selected formula is appropriate. The formula and calculations used should be recorded in the laboratory notebook or on the worksheet. As for all records, calculations are entered in ink and mistakes are never erased, just crossed through with a single line, dated, and signed. Generation of raw data and all of the related calculations and complete record-keeping are the responsibility of the analyst.

15.2 CALCULATIONS FOR FINAL VALUES

Raw data produced by analytical performances should be converted to reportable, final results in the appropriate unit. By using the proper formula, the analyst calculates the final values and registers them on the laboratory "result" sheet or enters them into the laboratory computer system, to be ready for further supervision and checks. The most commonly used calculations for the final results of physical properties and inorganic nonmetallic parameters are as follows.

15.2.1 Temperature Correction for pH Values

pH meters are usually automatically adjusted to the desired temperature. If this correction is not available, the temperature of the sample should be associated with the pH value. When the temperature is not specified in the report, it should be 25°C (pH measurement is discussed in Section 20.2.1).

15.2.2 Temperature Compensation for Conductivity

The value of the electrical conductance changes with temperature at a rate of approximately 2% deviation per degree Celsius. Always report conductivity at 25°C. If the temperature of the sample is below 25°C, add 2% of the reading per degree difference. If the temperature of the sample is above 25°C, substract 2% of the reading per degree. The following formula for temperature correction may be used as an alternative:

$$\mu mhos/cm \text{ at } 25°C = [\text{measured conductivity}]/[1 + 0.019(t - 25)] \qquad \textbf{(15.1)}$$

Measurement of electrical conductance is detailed in Section 18.4.3.

Example — The measured conductivity of a sample is 350 μmhos/cm. The temperature of the sample at the time of measurement is 23°C. The reported value for the conductivity will be 2% of 350 = 7, difference in temperature is 25 − 23 = 2°C, 2 × 7 = 14. The reported conductivity will be 350 + 14 = *364 μmhos/cm*. Or,

$$350/[1 + (0.019)(-2)] = 350/(1 - 0.038) = 350/0.962 = 364 \text{ } \mu mhos/cm$$

15.2.3 Solids, Moisture, and Ash Calculation

Values for solids, moisture and ash results may be reported as milligrams per liter, milligrams per kilogram, or percentage. Usually total, suspended, and dissolved solids (TS, SS, TDS) are reported in milligrams per liter, moisture and ash results are expressed in percentage.

Solids

$$mg/L \text{ Solids (TS, TSS, TDS)} = [(A - B) \times 1000]/mL \text{ sample} \qquad \textbf{(15.2)}$$

$$\% \text{ Solids (TS, TSS, TDS)} = [(A - B) \times 100]/mL \text{ sample} \qquad \textbf{(15.3)}$$

$$\% \text{ Solids (TS, TSS, TDS)} = mg/L/10,1000 \qquad \textbf{(15.4)}$$

where A = for TS, weight of dish and residue dried at 105°C, in grams, for TDS, weight of dish and residue dried at 180°C, in grams, or for TSS, weight of filter and residue dried at 105°C, in grams, and B = weight of dish in grams.

Determination of total solids (TS), total suspended solids (TSS) and total dissolved solids (TDS) is discussed in Section 19.1.

Moisture — Moisture of any solid is determined by drying a known quantity aliquot of the well-mixed sample at 103–105°C in a laboratory oven. After the dried sample is cooled in a desiccator, weigh it and calculate its percent moisture by using the following formula:

$$\% \text{ moisture} = [(g \text{ of solid} - g \text{ of dried solid}) \times 100]/g \text{ of solid} \qquad \textbf{(15.5)}$$

Ash — The ash content of any solid is determined by igniting a known quantity aliquot of the well-mixed sample at 1000°C in a muffle furnace. By knowing the weight of the original sample and the weight of the remaining ash (ignited residue), and using the following formula, the percentage of ash content of the sample is determined and calculated:

$$\% \text{ ash} = \text{g of ash} \times 100/\text{g of original sample} \qquad \textbf{(15.6)}$$

Percentage composition of a solid sample

- Determine the percentage of moisture at 105°C as described above, using formula (15.5).
- The remaining residue from the above determination will be ignited at 1000°C. The diference between the weight of the ignited residue (at 1000°C) and the dried solid (at 105°C) will give the volatile and organic compounds in the sample. It expressed also as a percentage.
- The remaning residue is considered as ash (inorganics) and calculated as a percentage.
- The sum of the percentage of moisture, percentage of volatiles, and percentage of ash values should be 100%. If the sum of the calculated fractions is not 100% and the checked calculations are correct, the whole process should be repeated.

15.2.4 Calculations Related to Titrimetric Analysis

The general formula for calculating results for titrimetric determination expressed as mg/L is as follows:

$$\text{mg/L} = (\text{mL} \times N \times \text{Eqw} \times 1000)/\text{mL sample} \qquad \textbf{(15.7)}$$

where mL = volume of the used titrant solution, N = normality of the used titrant solution, and Eqw = equivalent weight of the determined substance. The formula used for expressing the result as a percentage is as follows:

$$\% \text{ w/v} = (\text{mL} \times N \times \text{mEqw} \times 100)/\text{mL sample}$$

$$\% \text{ w/w} = (\text{mL} \times N \times \text{mEqw} \times 100)/\text{g sample}$$

where % w/v = weight per volume percentage (g/100 mL), % w/w = weight per weight percentage (g/100 g), mL = volume of the used titrant solution, N = normality of the used titrant solution, and mEqw = millequivalent weight of the determined substance (mEqw is calculated by dividing the equivalent weight with 1000).

Calculating titrimetric analysis results is discussed in Section 9.2 and equivalent weights in Section 5.3.2.

15.2.5 Calculations Related to Spectrophotometric and Potentiometric Analytical Performances

Linear regression calculations and predicting the concentrations of the unknown samples are given along with examples in Section 12.4.2. Logarithmic regression calculations and predicting results of unknown samples are given along with examples in Section 13.2.3.

15.2.6 Calculating the Results for Solid Matrices

For solid matrices, the report is expressed as ppm, milligram per kilogram (mg/kg) or ppb, microgram per kilogram (µg/kg). The report should state that the reported value is calculated on the "*wet base*," sometimes called the "as is base," or on the "*dry base*." Wet base means that the original solid sample (soil, sediment, sludge, etc.) contains various quantities of moisture in it; therefore, the original weight of the sample incorporates the weight of the moisture too. Consequently, the weight of the sample is incorrect. With knowledge of the moisture content of the sample, the analyst can make the necessary corrections. For this calculation, use the following formula:

$$\text{mg/kg on wet base} = [\text{mg/L} \times \text{final volume of the sample after treatment}]/\text{g sample} \qquad (15.8)$$

Knowing the percentage of moisture of the sample, correct the error caused by the moisture content by using the following formula:

$$\text{mg/kg on dry base} = \text{mg/kg on wet base}/\text{decimal fraction of dry solid} \qquad (15.9)$$

Example — A 5 g soil sample was weighed and digested for lead (Pb) analysis. After the preparatory process (digestion), the final volume of the "ready for analysis sample" was 100 mL. The Pb content of the digestate was found to be 0.56 mg/L (0.56 mg per 1000 mL), which is 0.056 mg/100 mL. The 100 mL digestate corresponds to the 5 g soil sample; therefore, the 5 g soil sample contains 0.056 mg Pb, and the question is how much will be in a 1000 g (1 kg) sample? A 1 kg sample will contain 200 times more Pb; $200 \times 0.056 = 11.2$ mg. So, the result is 11.2 mg/kg (ppm) Pb is on wet base. To avoid the long calculations, use formula (15.8):

$$\text{mg/kg Pb on wet base} = (0.56 \times 100)/5 = 11.2 \text{ mg/kg}$$

The moisture content of the soil sample was 12%. The dry soil is $100 - 12 = 88\%$. Using formula (15.9):

$$\text{mg/kg Pb on the dry base} = 11.2/0.88 = 12.72 \text{ mg/kg}$$

15.2.7 Calculation of Total Hardness from Calcium and Magnesium Values

With available calcium (Ca) and magnesium (Mg) values, the total hardness expressed as calcium carbonate ($CaCO_3$) may be calculated. Total hardness of a water is principally caused by Ca and Mg salts. Therefore, hardness in terms of $CaCO_3$ refers to the sum of Ca as $CaCO_3$ plus Mg as $CaCO_3$.

$$\text{Total Hardness, as } CaCO_3 \text{ mg/L} = \text{Ca as } CaCO_3 + \text{Mg as } CaCO_3 \qquad (15.10)$$

Convert Ca values to CaCO₃ — Convert Ca values to $CaCO_3$ by dividing the Ca value with a factor of 0.4.

$$\text{Ca, } CaCO_3 \text{ mg/L} = \text{Ca mg/L}/0.4 \qquad (15.11)$$

The factor of 0.4 is derived from the ratio of the equivalent weight of Ca and the equivalent weight of $CaCO_3$ ($20/50 = 0.4$). The equivalent weight of Ca is its atomic weight divided by its valency, $40/2 = 20$. The equivalent weight of $CaCO_3$ is the molecular weight divided by the valency of the Ca, $100/2 = 50$. (Equivalent weight calculations are discussed in Section 5.3.1).

Convert Mg values to CaCO₃ — Convert Mg values to $CaCO_3$ by dividing the Mg values with a factor of 0.24.

$$Mg \text{ as } CaCO_3 \text{ mg/L} = \text{mg/L}/0.24 \qquad (15.12)$$

The factor of 0.24 is derived from the ratio of the equivalent weight of Mg and the equivalent weight of $CaCO_3$ ($12/50 = 0.24$).

Example —

Calcium, as Ca	= 24 mg/L
Magnesium, as Mg	= 8 mg/L
Calcium, as $CaCO_3$	= 24/0.4 = 60 mg/L
Magnesium, as $CaCO_3$	= 8/0.24 = 33 mg/L
Total Hardness as $CaCO_3$	= 60 + 33 = 93 mg/L

15.2.8 Calculation of Magnesium Values from Total Hardness as CaCO₃ and Calcium as CaCO₃

With knowledge of the total hardness as $CaCO_3$ and the calcium as $CaCO_3$, the value of magnesium can be calculated.

$$\text{Total Hardness as } CaCO_3 - Ca \text{ as } CaCO_3 = Mg \text{ as } CaCO_3 \qquad (15.14)$$

Ca and Mg may be calculated from their $CaCO_3$ form by multiplying it with the factors of 0.4 and 0.24, respectively (these factors are discussed above).

Example —

Total Hardness as $CaCO_3$	= 120 mg/L
Calcium, as $CaCO_3$	= 70 mg/L
Magnesium as $CaCO_3$	= 120 − 70 = 50 mg/L
Calcium as Ca	= 70 × 0.4 = 28 mg/L
Magnesium as Mg	= 50 × 0.24 = 12 mg/L

15.2.9 Calculation of Carbonate, Bicarbonate, and Hydroxide from Alkalinity

There are three kinds of alkalinity: hydroxide (OH^-), carbonate (CO_3^{2-}), and bicarbonate (HCO_3^-). In order to distinguish between the kinds of alkalinity present in a sample and to determine the quantities of each, a titration is made with a standard acid using two indicators successively. Alkalinity is expressed in terms of milligrams per liter as $CaCO_3$, and titrated in the presence of phenolphthalein and methylorange indicators. Detailed titration procedures and calculations are discussed in Chapter 20. "P" (phenolphthalein) alkalinity is a result of the titration of the sample with a standard acid solution in the presence of phenolphthalein indicator. "M" (methylorange) alkalinity is obtained by titration in the presence of methylorange indicator. "T" (total) alkalinity is the sum of the results of the previously mentioned "P" and "M" alkalinities. The updated method for the determination of "M" alkalinity uses the mixed bromcresol green–methyl red indicator instead of the methylorange indicator. Alternatively, a potentiometric titration may be applied for alkalinity analysis by using preselected pH end points.

There are five alkalinity conditions possible in the sample:

1. Hydroxide alone
2. Hydroxide and carbonate
3. Carbonate alone
4. Carbonate and bicarbonate
5. Bicarbonate alone

The results obtained from the "P" and "T" alkalinity determinations offer the calculation of these alkalinity conditions:

1. $P = T$ Hydroxide alkalinity, as $CaCO_3$ $= T$
2. $P > \frac{1}{2}T$ Hydroxide alkalinity as $CaCO_3$ $= 2P - T$
                 Carbonate alkalinity as $CaCO_3$ $= 2(T - P)$
3. $P = \frac{1}{2}T$ Carbonate alkalinity as $CaCO_3$ $= 2P$
4. $P < \frac{1}{2}T$ Carbonate alkalinity as $CaCO_3$ $= 2P$
                 Bicarbonate alkalinity as $CaCO_3$ $= T - 2P$
5. $P = 0$ Bicarbonate alkalinity as $CaCO_3$ $= T$

Converting hydroxide, carbonate, and bicarbonate as $CaCO_3$ to the corresponding anions by using conversion factors — Conversion factors are obtained from the ratio of the equivalent weight of the anion and the equivalent weight of the $CaCO_3$.

Anion	Ratio	Factor
Hydroxide, OH^-	17/50	0.34
Carbonate, CO_{32-}	30/50	0.60
Bicarbonate, HCO^-	61/50	1.22

Example — The result of one alkalinity determination is "P" alkalinity as $CaCO_3 = 12$ mg/L, "T" alkalinity as $CaCO_3 = 126$ mg/L. The situation corresponds to alkalinity condition 4 above, $P < \frac{1}{2}T$. Therefore,

$$\text{Carbonate alkalinity as } CaCO_3 = 2 \times 12 = 24 \text{ mg/L}$$

$$\text{Bicarbonate alkalinity as } CaCO_3 = 126 - 24 = 102 \text{ mg/L}$$

The values for the participating anions are calculated by using the conversion factors:

$$\text{Carbonate as } CO_3^{2-} = 24 \times 0.6 = 14.4 \text{ mg/L}$$

$$\text{Bicarbonate as } HCO_3^- = 102 \times 1.22 = 124 \text{ mg/L}$$

TABLE 15.1

Constant A as a Function of
Water Temperature

Water Temperature (°C)	Constant A
0	2.60
4	2.50
8	2.40
12	2.30
16	2.20
20	2.10
25	2.00
30	1.90
40	1.70
50	1.55
60	1.40
70	1.25
80	1.15

Analytical procedures for the determination of "P" and "T" alkalinities and the corresponding calculations are given in Chapter 20.

15.3 CALCULATION OF CALCIUM CARBONATE SATURATION, CORROSIVITY

Corrosivity expressed as Langelier Saturation Index (SI) is defined as the actual pH minus pH of saturation, pH_s. Saturation pH is the pH at which a water is just saturated with calcium carbonate, $CaCO_3$. A negative index indicates a tendency to dissolve $CaCO_3$, and a positive index indicates the tendency to deposit $CaCO_3$.

To calculate the saturation index (SI), determine alkalinity, calcium ion concentration, pH, temperature and total filtrable residue.

$$pH_s = [(A + B) - \log Ca] - [\log \text{Alkalinity}] \qquad (15.13)$$

where A and B = constants, Ca = calcium hardness, as $CaCO_3$ mg/L, and Alkalinity = total alkalinity, as $CaCO_3$ mg/L. Values of constants and logarithms in equation (15.13) are given in Tables 15.1, 15.2, and 15.3.

TABLE 15.2

Constant B as a Function of Total
Dissolved Solids

Total Dissolved Solids, TDS (mg/L)	Constant B
0	9.70
100	9.77
200	9.83
400	9.86
800	9.89
1000	9.90

TABLE 15.3

Logarithms of Calcium and
Alkalinity Concentrations

Ca^{2+} and Alkalinity $CaCO_3$ (mg/L)	log
10	1.00
20	1.30
30	1.48
40	1.60
50	1.70
60	1.78
70	1.84
80	1.90
100	2.00
200	2.30
300	2.48
400	2.60
500	2.70
600	2.78
700	2.84
800	2.90
900	2.95
1000	3.00

Example — A water sample has the following analytical results:

Calcium, as $CaCO_3$ .... 200 mg/L

Total Alkalinity as $CaCO_3$ .... 60 mg/L

TDS .... 650 mg/L

Temperature .... 16°C

pH 9.0

What is the saturation index (SI) of this water? Is the water stable, corrosive, or scale-forming? According to equation (15.13):

$$pH_s = 2.20 + 9.88 - 2.30 - 1.78 = 8.0$$

$$SI = pH - pH_s = 9.0 - 8.0 = 1.0$$

The water sample is scale-forming.

15.4 DILUTIONS AND CONCENTRATIONS

Any *dilution* applied during the test should be recorded, and the final result will be obtained by multiplying the measured value by the dilution factor. Any *concentration* technique used during sample preparation should be documented, and the analytical value divided by the concentration factor will produce the result.

Examples

1. A water sample analyzed for phosphorus and for the appropriate reading should be diluted five times. If the analyitical result of the diluted sample was 0.8 mg/L, the final result will be $0.8 \times 5 = 4.0$ mg/L phosphorus.

2. During the digestion technique, a water sample analyzed for total chromium content was concentrated 10 times (original 100 mL sample cooked down to 10 mL final volume), because of the low concentration of the metal in the sample. The reading of this concentrated sample was 0.06 mg/L Cr; therefore, the reported result will be 0.06/10 = 0.006 ppm (mg/L) or 6 ppb (μg/L).

15.5 SIGNIFICANT FIGURES

The number of significant figures refers to the number of digits reported for the value of a measured or calculated quantity indicating the accuracy and precision of the value. The general rule is to report only such figures that are justified by the accuracy of the method.

The number of significant figures is said to be the number of digits remaining when the data are rounded. For example, when a measured value is 10.6 mg/L, the analyst should be quite certain of the 10, but may be uncertain as to whether the 0.6 should be 0.5 or 0.7, or even 0.4 or 0.8, because of the unavoidable uncertainty in the analytical procedure. The 10.6 mg/L reported value has three significant figures. To count the number of significant figures, the following rules exist:

- Nonzero integers always count as significant figures.
- Leading zeros are zeros that precede all of the zero digits. They do not count as significant figures. The zeros simply indicate the position of the decimal point. For example 0.00065 has two significant figures.
- Captive zeros are zeros between nonzero digits, and always count as significant figures. For example, 1.034 has four significant figures.
- Trailing zeros are zeros at the right end of the number. They are significant only if the number contains a decimal point. For example, when reporting 4600 ppb, it is better to report 4.6 ppm if the two significant figures are trusted.
- Significant figures in multiplication and division: the number of significant figures in the result is the same as the number in the least precise measurement used in the calculation. For example, $1.6 \times 5.24 = 8.38$, by rounding 8.4. The correct result will be 8.4, because 1.6 has two significant figures.
- Significant figures in addition and subtraction: the result has the same number of decimal places as the least precise measurement used in the calculation. For example, 15.62 + 12.5 + 20.4 = 48.52. The correct result is 48.5, because 12.5 has only one decimal place.

15.5.1 Exponential Notation

Exponential notation, for example 1.15×10^3 or 11.5×10^2, is an acceptable way to express both the number and the significant figures. This form is not used in analytical reports, because it would not be consistent with the normal expression of the results and might be confusing.

Chapter **16**

EVALUATION AND APPROVAL OF ANALYTICAL DATA

To evaluate and approve reportable results, checks should be performed on sampling, sample handling and storage, methodology, calibration, and quality control. If any problems arise through doubtful data, initiation of corrective actions and providing the final approved results is the responsibility of the analyst, the laboratory supervisor, and the QC department.

16.1 VALIDATION OF ANALYTICAL QC CHECKS

As discussed previously, QC is the check of the result of an analytical measurement. When any error occurs, it should be identified in detail, and should be corrected into a valid, reportable value. The reviewer should be concerned about the acceptance criteria of the QC check, the probable source of the out-of-control item, and the steps taken in the specified corrective action. The QC check must include QC measures, such as blanks, calibration processes, QC check standards, QC check samples, certified reference standards, matrix and reagent water spikes, duplicates, split samples, standardization of titrants, instrument performances, proper storage of the reagents and standards, and quality check of the laboratory pure water, as outlined in Chapter 3. No data should be released until statistically supported limits are reported for them. Data should be technically sound and defensible before they can be reported. The first requirement for reporting is that all pertinent documentation is available and referenced so that it can be found at any time it might be needed.

The checklist of documents needed for the evaluation of correct analytical results are as follows:

- Sample documentation
    - Sampling plan
    - Sampling methodology
    - Sample identification
    - Chain of custody
- Quality assurance plan

    Control charts

    Statistical controls

- Measurement documentation

    Methodology

    Calibration

    Instrumentation

    Analyst

- Data documentation

    Notebook, worksheets

    Data reduction procedure

Blanks — Analytical blanks measure the degree of contamination within a test and seriously affect the accuracy of low-level determinations. Blanks can be a method blank, reagent blank, calibration blank, instrument blank, preparation blank, equipment blank, and trip blank, as described in Section 3.2.9. *Blank values should be less than the method detection limit (MDL).*

Possible causes of unaccepted blank values are contaminated reagents, contaminated analyte-free water, contaminated glassware or sample containers, contaminated apparatus, unclean sample collection and sample handling, dirty garments of the analyst, and filthy laboratory environment.

Calibration — Calibration in chemical measurements refers to the process by which the response of a measurement system is related to the different concentrations of the analyte of interest. Generally, a measurement is a comparison process in which the unknown is compared with a known standard (calibration procedures can be found in Section 3.2.5, Section 12.3, and Section 13.2.3).

The first criteria for correct calibration is the purity and accuracy of the standards. Standards may be prepared by the analyst or suppliers, and should have an assigned expiration date indicating the stable life expectancy. Standards should never be used beyond such dates. Storage and life expectancies of standard solutions are given in Table 5.1.

Acceptance criteria of calibration — As discussed in Section 12.3, the correlation coefficient should be at least 0.9950, but it can be variable according to the method and instrumentation. The deviation of the continuing calibration standard (CCS) from the original calibration standard should be ±10% in inorganic analysis, and ±15% in organic analysis. The accepted criteria for calibration verification standard (CVS) or QC check sample (reference standard) is in-house generated, which cannot exceed method range.

Probable sources of unacceptable calibration — These can include calculation error, incorrectly prepared stock, intermediate or calibration standards, outdated stocks and standards, faulty or expired QC check standard (CVS), improperly stored stocks and standards, improperly selected and improperly used volumetric glassware, incorrect responses of the instrument, incorrectly cleaned glassware or containers, or dirty environment. Reanalyze all samples measured between the acceptable and unacceptable calibration check with the corrected calibration.

Duplicates and split samples — Duplicate and split samples are used to measure the precision of the analytical system. Duplicates may be field duplicates, laboratory duplicates, or matrix spike duplicates (see Section 3.2.9). The accepted limit or target limit for the calculated precision is different according to laboratories, parameters, and methods. When the measured and calculated precision values are out of this limit, corrective action is needed to find the problem.

Probable sources of unacceptable precision values — These include sampling error for field duplicates, preparation error for laboratory duplicates, use of unidentical samples, contamination, or calculation error. Other samples analyzed with the same analytical run are questionable, and have to be reanalyzed after the problem is corrected.

Spikes — All daily spiking data should agree with the spike accuracy limit established for each parameter and method.

Probable sources of unacceptable spikes values — These include calculation error, error in spiking, improperly prepared and stored spike stock solution, expired spike stock solution, contamination during spiking, and malfunction of the instrument. All associated samples analyzed with the unacceptable spiked sample must be reprocessed and analyzed again after correction.

Standard titrant solutions — Titrant solutions must be prepared with extreme care and standardized to exact normality each time before use. Titrant solutions are normal (N) solutions with standardized exact normality (see Chapter 6). The normality of the titrant is given by each particular method, or chosen according to the expected concentration of the analyte in the samples.

The acceptance criteria of the good titrant solution is the production of a ±5% deviation from the true value of the QC check standard, and all QC checks (blanks duplicates, spiked samples) should fall in the acceptable range.

Probable sources of unacceptable titrant solutions — These include bad titrant solution caused by preparation error, mistakes in dilution, overdated stock solution, standardization or calculation errors, contaminated or expired reagents or indicators, improperly used volumetric glassware, mistake in end point reading, or contaminated glassware or laboratory facility.

Analytical performance checks — To validate the analytical system or measurement process, laboratories used performance evaluation samples. *Internal reference standards* are prepared by the laboratory. *External reference materials* provided by an outside source, and according to the origin of the material, can be a certified reference material originated from a technically accepted and certified organization. Analytical values should be within the certified limits.

Causes of failing this criteria — These include calculation error, improper sample preparation, improper analysis. If samples have to be concentrated or diluted to the proper range of the analysis, carefully check these techniques and the accompanying calculations. Contamination error is also a possible source of unacceptable results. Reprocess and reanalyze all samples measured together with the failed reference standard after correction.

Quality of laboratory pure water — The acceptance criteria for the quality of laboratory pure water is discussed in Section 5.2. If the measured values show alteration from this

Lab Section:			
Analyst:	Date:	Supervisor:	Date:

Description:

Samples Affected:

Resolution:

Follow Up:

Review:

Approval:

Signature	Title	Date

DISTRIBUTION: Project file,

Figure 16.1
Deficient incident report.

standard, corrective action is needed immediately. Change the cartridges and call the contracted company for regular maintenance of the system.

16.1.1 Documentation of the Out-of-Control Conditions

All out-of-control conditions must be documented immediately by all levels of responsibility. A typical Deficient Incident Report form is given in Figure 16.1.

16.2 CHECKING THE CORRECTNESS OF THE ANALYSIS

Analytical data for each parameter, produced by the analyst, are checked by QC acceptance criteria, validated, corrected if necessary, and then converted to a reportable value. The approved and checked analytical results for each analyzed parameter will be transferred to the sample report form. The sample report contains the results for each requested parameter analyzed for the particular sample. By comparing selected analytical results, there is a procedure to check the correctness of the analyses in one sample.

16.2.1 Calculation of Total Dissolved Solids (TDS)

The correctness of the measured value of the TDS may be verified by using the following formula:

$$TDS = 0.6 \text{ (Alkalinity)} + Na + K + Ca + Mg + Cl + SO_4 + SiO_2 + NO_3\text{-}N + F \qquad \textbf{(16.1)}$$

The concentration of these constituents is expressed in milligrams per liter. The measured TDS concentration should be higher than the calculated one, because significant contributors may not be included in the calculation. If the measured value is less than the calculated one, the sample should be reanalyzed. If the measured TDS concentration is more than 20% higher than the calculated one, the selected constituents for the calculation should be reanalyzed.

The accepted ratio of the measured and the calculated TDS value is

$$1.0 < \text{measured TDS/calculated TDS} < 1.2$$

16.2.2 TDS and Electrical Conductance Ratio

The ratio of TDS, expressed as milligrams per liter and conductivity, expressed as micromhos per centimeter should be between 0.55 and 0.7. If the ratio is outside this limit, the measured TDS and conductivity are suspect and should be reanalyzed. As the opposite, when conductivity is multiplied by a factor between 0.55 and 0.7, the value should give the measured TDS. If the value is outside of this limit, reanalyze both TDS and conductivity.

16.2.3 Anion–Cation Balance

The sum of the cations, expressed in milliequivalents per liter (meq/L), and the sum of the anions, expressed in milliequivalents per liter (meq/L), must be balanced, because all potable waters are electrochemically neutral. The test is based on the percentage of difference between the sum of the cations and sum of the anion. The percentage of difference is calculated as follows:

% Difference = [(sum of cations – sum of anions)/(sum of cations + sum of anions)] × 100

The acceptance criteria depends on the sum of the anions:

Sum of Anions, meq/L	Acceptance Criteria
0–3.0	±0.2 meq/L
3.0–10.0	±2%
20.0–800	±2–5%

Table 16.1 contains the conversion factors to convert mg/L to meq/L and vice versa. The factors are derived from the valency and the atomic weight or the formula weight of the ions.

TABLE 16.1

Conversion Factors Milligrams per Liter to Milliequivalent per Liter

Cations	Factor mg/L x = meq/L	Factor meq/L x = mg/L	Anions	Factor mg/L x = meq/L	Factor meq/L x = mg/L
Al^{3+}	0.1112	8.994	BO_2^-	0.02336	42.81
B^{3+}	0.2775	3.603	Br^-	0.01257	79.90
Ba^{2+}	0.01456	68.67	Cl^-	0.02821	35.45
Ca^{2+}	0.04990	20.04	CO_3^{2-}	0.03333	30.00
Cr^{3+}	0.05770	17.33	CrO_4^{2-}	0.01724	58.00
Cu^{2+}	0.03147	31.77	F^-	0.05264	19.0
Fe^{2+}	0.03581	27.92	HCO_3^-	0.01639	61.02
Fe^{3+}	0.05372	18.62	HPO_4^{3+}	0.02084	47.99
H^+	0.9922	1.008	$H_2PO_4^{-}$	0.01031	96.99
K^+	0.02558	39.10	HS^-	0.03024	33.07
Li^+	0.1441	6.941	HSO_3^-	0.01234	81.07
Mg^{2+}	0.08229	12.15	HSO_4^-	0.01030	97.07
Mn^{2+}	0.03640	27.47	I^-	0.00788	126.9
Mn^{4+}	0.07281	13.73	NO_2^-	0.02174	46.01
Na^+	0.04350	22.29	NO_3^-	0.01613	62.0
NH_4^+	0.05544	18.04	OH^-	0.05880	17.01
Pb_{2+}	0.009653	103.6	PO_4^{3-}	0.03159	31.66
Sr^{2+}	0.02283	43.81	S^{2-}	0.06238	16.03
Zn^{2+}	0.03059	32.69	SO_4^{2-}	0.02082	48.03

Note: mg/L = milligrams/liter; meq/L = milliequivalent/liter; meq/L = mg/L × factor; Factor = ionic charge/atomic or formula weight (Cl_f = 1/35.45 = 0.02821); mg/L = meq/L × factor; Factor = atomic or formula weight/ionic charge (Cl_f = 35.45/1 = 35.45).

When the analyzed mg/L value of the ion is converted to meq/L, the mg/L value is multiplied with the factor derived from the ratio of the valency and the atomic or formula weight.

$$meq/L = mg/L \times (V/Aw) \qquad (16.2)$$

For example, Na has a valency of 1 and the atomic weight is 22.9897; the factor will be 1/22.9897 = 0.04340. For a sulfate, SO_4^{2-} the valency is 2 and the formula weight is 96.0636; therefore, the factor is 2/96.0636 = 0.0282.

When the meq/L value is converted to mg/L, the meq/L value is multiplied by a factor derived from the ratio of the atomic or formula weight divided by the valency.

$$mg/L = meq/L \times (Aw/V) \qquad (16.3)$$

For example, for meq/L Na converted to mg/L, the factor will be 22.9897/1 = 22.9897. For sulfate meq/L converted to mg/L, the factor will be 96.0636/2 = 48.03.

16.2.4 Relation Between Conductivity and Total Cations and Anions

The calculated total cations or total anions expressed as meq/L multiplied by 100 should be close to the measured conductivity, as µmhos/cm. Otherwise, check conductivity, and if it gives the same value, reanalyze cations and anions.

16.2.5 Relation Between Total Hardness and Calcium and Magnesium Values

The analytical value of the total hardness, as $CaCO_3$, should be equal to the sum of the calcium and magnesium values expressed as $CaCO_3$. The conversion of calcium and magnesium values into their $CaCO_3$ forms is discussed in Section 15.2.7. If the criteria is failed, reanalyzing the two metals and the total hardness is necessary.

16.2.6 Relation of the Different Forms of Nitrogens

The analytical groups of nutrients includes the forms of nitrogen of greatest interest in environmental samples, namely ammonia-nitrogen (NH_3-N), nitrite-nitrogen (NO_2-N), nitrate-nitrogen (NO_3-N) and organic nitrogen. To determine organic nitrogen, analyze total Kjeldahl nitrogen (TKN), which is the sum of the ammonia nitrogen and organic nitrogen in the sample. When the analytical result of NH_3-N is greater than the TKN, the two tests should be repeated to rectify the problem. The total-nitrogen of the sample should be the sum of the concentrations of the individually analyzed nitrogen compounds. Otherwise, check the analytical and QC data, or the analytical tests should be repeated until the correct result is achieved.

16.2.7 Relation of the Different Forms of Phosphorus

Phosphates in environmental samples are classified as ortho (water-soluble), condensed, and organically bound phosphates, reported as PO_4-P. When the analyzed total phosphate value is different from the sum of the individual forms of phosphates, the tests should be repeated to verify the problem.

Chapter 17

REPORTING ANALYTICAL DATA

Reports of analyses are the written records of analytical work. The content of the report should include all of the necessary information and should be clear and understandable for the end-user. The designer of the final report always has to remember that the reader frequently does not understand the issued results and needs more specification. Brief interpretation of the reported values is essential.

Chapters 15 and 16 have given an account of how analytical data is calculated and checked. Never issue any report until all the checks have been completed and approved.

17.1 REQUIRED DOCUMENTATION

All documentation used to approve and defend reported data must be collected and should be available and referenced so it can be found at any time it may be needed. The content of these documents should be detailed and clear enough to explain to the interested party how the final reported values have been generated.

17.1.1 Documentation Required to Approve and Defend Reported Data

Documentation for sample collection and identification

- Sampling plan, sample collection methods
- Field QC
- Sample identification, chain-of-custody
- Field notebook, documentation of field tests

Documentation of the analytical performance

- Analytical method used and method detection limit (MDL)
- Instrumentation (manufacturer, model, performance checks, maintenance log)
- Calibration data (initial and continuing)
- Detailed analytical work (working papers, standards and reagent preparation, calculations)

QA/QC documentation and data

- Analysis of blanks
- Precision and accuracy data
- QC charts
- Acceptable ranges for precision and accuracy per parameter
- Source of QC check standards
- Preparation of spike stock solution, how spikes were prepared

Checks and validation of analytical data

- Documentation, how analytical data are checked and validated
- Corrective actions (when applicable)
- Date and signature for approval of the reportable data of each parameter tested
- Date and signature for approval of the final analytical report

17.2 SIGNIFICANT FIGURES IN ANALYTICAL REPORTS

Numerical data are often obtained with more digits than are justified by their accuracy and precision. Report only those figures that are justified by the accuracy of the analytical method. Do not use the comon practice that all numbers in one column have the same number of significant figures to the right of the decimal point. The reported numbers belong to different parameters and they are analyzed by different methods, so the significant figures will also be different. Significant figures are discussed in Section 15.5.

For example, if an analytical result is 16.6 mg/L, and the analyst is certain about 16 but not certain of 0.6, the result should be rounded off and reported as 17. When an analytical result is 16.61 and generated by a certain method which justifies these significant figures, the reported number will be 16.61. If a calculated value is 2346 mg/L, but the analyst is not certain about the last two numbers, the number is rounded off and reported as 2350. If a number is written as 5.000, it is understood that the zeros are significant, or the number would have to be rounded as 5.00, 5.0, or 5, whichever is appropriate. If a result is 360 mg/L, it should be certain that the zero is significant, and cannot be deleted.

17.3 UNITS USED TO EXPRESS ANALYTICAL RESULTS

Units used to express analytical results depend on the analytical method used, the concentration of the analytes, and the matrices of the sample analyzed.

The most common units used to express analytical results in environmental samples is parts per million (ppm). It is equal to milligrams per liter (mg/L) or milligrams per kilogram (mg/kg). When the concentration is less than 0.1 ppm, it is more convenient to express the result as parts per billion (ppb), micrograms per liter (µg/L), or micrograms per kilogram (µg/kg). If the concentration is greater than 10,000 ppm, the result is expressed as a percentage. For example, 1% is equal 10,000 ppm when the specific gravity is 1.0.

If the result is issued in parts per million or percentage by weight for *solid samples or liquids with high specific gravity*, a correction is necessary as follows:

$$mg/kg = (mg/L)/(\text{specific gravity})$$

$$\% \text{ by weight} = (mg/L)/(10,000 \times \text{specific gravity})$$

Analytical results for solid samples are reported in milligrams per kilogram or micrograms per kilogram or, as stated above, in percentage by weight according to the concentration value. The report must also include the statement that the result is on "as is base," also called "wet base," of the solid, if the solid is not corrected to dry, moisture-free solid. Results reported as on the "dry base" means that the value is corrected to dry, moisture-free solid. Calculations of these units are presented in Section 15.2.6.

Calculations related to water treatment require analytical results in milliequivalents per liter (meq/L). Conversion of milligrams per liter values to milliequivalents per liter is discussed in Section 16.2.3.

Color is reported in color units (C.U.). A color unit is the color produced by 1 mg platinum per liter in the form of the chloroplatinate ion.

Turbidity values are reported as the nephelometric turbidity unit (NTU). This unit is related to the method that uses nephelometric measurement.

The method used to determine odor in water samples is the threshold odor test, and the unit used to report it is the threshold odor number (T.O.N.).

The pH at which water is just saturated with calcium carbonate ($CaCO_3$) is known as the pH of saturation, or pH_s. The Langelier Saturation Index (LSI) is defined as the actual pH minus pH_s. The LSI or shortly, saturation index, SI, is used as the unit for expression of calcium carbonate saturation, or corrosivity. A negative value indicates a tendency to dissolve $CaCO_3$ (the water is corrosive), and a positive value indicates a tendency to deposit $CaCO_3$ (the water is scale-forming). When the value is 0 ± 0.2, the water is stable. The index is not directly related to corrosion, but deposition of a thin, coherent carbonate scale may be protective; therefore, the positive index is associated with the noncorrosive condition and the negative value may be a sign of corrosivity.

Conductivity is the numerical expression of the ability of an aqueous solution to carry the electrical current. The measurement of conductivity is usually the resistance, expressed in ohms or megaohms. The reciprocal of the resistance is conductance, and expressed in the reciprocal of the ohms or mhos. A more convenient unit in water analysis is millimhos (mmhos/cm) or micromhos (μmhos/cm). In the SI (International System of Units) the reciprocal of the ohm is siemans (S), and conductivity is reported as millisiemens per meter (mS/m), with 1 mS/m = 10 μmhos/cm. To convert umhos/cm to mS/m, divide the μmhos/cm value by 10.

17.4 REPORT WITH CONFIDENCE INTERVAL

Reporting the calculated confidence interval (CI) of the measurement helps to estimate the reliability of the result. Calculated confidence interval is rounded to two significant figures as reported. Usually, the 95% CI is used in the calculation. Calculation of the CI and an example for the report format was discussed previously in Section 3.2.7.

17.5 REPORT FORMAT

A simple table form is accepted for regular, routine reporting. It should be clear, easy to follow, and should contain all of the necessary information to evaluate the analytical report. For special, nonroutine purposes, reports are more detailed, and a comprehensive QA/QC report should be attached to the analytical report. A report should include

- A title, which should be brief but descriptive to identify the goal of the analytical work.
- Identification of the organization or person for whom the work was done. Name, organization, address, phone number, work order, etc.
- A Laboratory Identification Number (ID number). The numbering system: calendar year/sequence number.
- The date report is completed.
- A short statement of the objective of the analysis.
- Sample identification, including a physical description of the sample, sampling area, all information related to the sample that may impact the data. If possible, include a photograph of the sampling area.
- Sampling details, including sampling procedures, sample type, sample preservation, name of the sample collector, date and time of the collection, chain of custody, sample field custody, and transportation.
- Analyzed parameters with method numbers and method references. References should be specific, and provide all of the necessary information (reference number, revision date, etc.). Any modification from the original method should be stated.
- Method detection limit (MDL) or practical quantitation limit (PQL).
- Numerical values reported with correct significant figures and with the corresponding unit.
- References, including any information related to previous analytical work, and references to other reports on the same sample location.
- Precision and accuracy data with the acceptance limit should be stated for each analytical performance.
- A discussion, including interpretation of the results, recommendation of additional work or corrections, and any special observations related to the sample or the analytical report that serves the objective.
- Signatures and titles of all persons responsible for the data and for the report.
- Full distribution list of the report.
- The QC report attached to the analytical report. It contains the basis of calibration, standardization, and statement where the documentation is found. Precision and accuracy data should be stated for each analytical performance in the analytical report with the acceptance limit.

The pages of the report should be properly numbered. The reader of the report must be certain that he or she has the full report for review. (For example, page 1 of 5, or page 5 of 5).

Chapter **18**

DETERMINATION OF COLOR, ODOR, AND CONDUCTIVITY

18.1 INTRODUCTION TO THE PHYSICAL PROPERTIES

This chapter deals with the introduction and general discussion of the physical properties of a sample. The typical, meaningful, and traditional parameters described here are color, conductance, odor, solids (dissolved, suspended, total, and volatile), and turbidity. However, physical properties cannot be separated entirely from chemical composition. For example, calcium carbonate saturation is listed under physical properties, but is dependent on chemical tests.

In the following chapters, each physical parameter will be discussed individually and a methodology will be given for its determination. Each method is written in an easy-to-follow, step-by-step format, and includes a discussion of the property, the theoretical background of the test and the measurement, along with quality control requirements and safety precautions.

18.2 COLOR

18.2.1 General Discussion

Color in water may result from the presence of natural metallic elements (iron and manganese), humus and peat materials, plankton, weeds, and industrial parameters.

The term "color" is used here to mean the *true color*, that is, the color of the water from which turbidity has been removed. The term *apparent color* includes not only color due to substances in solution, but also that due to suspended matter. Apparent color is determined on the original sample without filtration or centrifugation. In some highly colored industrial wastewaters, color is caused mostly by colloidal and suspended material. In this case, both true color and apparent color should be determined.

18.2.2 Sample Collection and Storage

Collect representative samples in clean glassware. Make the color determination within a reasonable period to avoid physical or biological changes occurring in storage.

18.2.3 Pretreatment

To determine color, turbidity must be removed by filtration or centrifugation prior to analysis. Some filtration procedures also may remove some true color. Centrifugation avoids interaction of color with filter materials, but results vary with the sample nature and size and speed of centrifugation. State the pretreatment method when reporting results.

18.2.4 Color Determination by Platinum–Cobalt Method

Method No. 2120 B (*Standard Methods*, APHA-AWWA-WEF, 18th ed., 1992).

Principle of the method — Color is determined by visual comparison of the sample with known concentrations of colored standards. Comparison also may be made with special, properly calibrated glass colored discs that give results in substantial agreement with the platinum–cobalt color standards.

Interferences — Even a slight turbidity causes the apparent color to be noticeably higher than the true color; therefore, turbid samples should be filtered or centrifuged before the actual measurement (see Section 18.2.3 above).

Apparatus — Volumetric flasks (1000 mL and 50 mL), Nessler tubes, matched (50 mL, tall form), and a comparator with glass colored discs are needed.

Reagents

Stock color solution — Dissolve 1.246 g potassium chloroplatinate (K_2PtCl_6) (equivalent to 500 mg metallic Pt) and 1.00 g cobaltous chloride hexahydrate ($CoCl_2 \cdot 6H_2O$) (equivalent to 250 mg metallic Co) in distilled water with 100 mL concentrated HCl and dilute to 1000 mL with distilled water. This stock solution contains 500 color units (C.U.).

Color standard preparation — From this stock solution, prepare standards with 5, 10, 15, 20, 25, 30, 35, 40, 45, 50, 55, 60, 65, and 70 C.U. by diluting 0.5, 1.0, 1.5, 2.0, 2.5, 3.0, 3.5, 4.0, 4.5, 5.0, 5.5, 6.0, 6.5, and 7.0 mL stock solution to 50 mL with distilled water.

Measurement by platinum–cobalt standards — Observe sample color by filling a matched Nessler tube to the 50 mL mark with sample and compare the color with the standards. Look vertically downward through tubes toward a white surface placed at such an angle that light is reflected upward through the columns of the liquid. If the color exceeds 70 units, dilute sample with distilled water in known proportions until the color is within the range of the standards. Calculate the color units (C.U.) by the following equation:

$$C.U. = (\text{estimated color} \times 50)/\text{mL sample} \qquad (17.1)$$

Measurement by comparator containing glass discs — The tubes in the comparator are filled with distilled water and the sample. Match sample color with the color of the tube with distilled water plus the calibrated colored glass when viewed by looking toward a white surface expressed in color units (C.U.).

The glass disc gives results in substantial agreement with those obtained by the platinum–cobalt method. Using glass discs is permissible only if these have been individually calibrated against platinum–cobalt standards.

The color value of water is extremely pH-dependent and invariably increases as the pH of the water is raised. *When reporting a color value, specify the pH at which color is determined.*

Reporting the result — A color unit (C.U.) is the color produced by 1 mg platinum (Pt) in the form of chloroplatinate ion. Report color units in whole numbers as follows:

Color Units	Record to the Nearest
1–50	1
51–100	5
101–250	10
251–500	20

The reported color units have to be accompanied with the actual pH of the sample at the time of the color measurement.

Quality control (QC) requirement — With a duplicate analysis, approve the precision of the measurement expressed by units as discussed in Section 3.2.7.

In the case of using the comparator with the colored glass discs, calibrate each disc to correspond with the colors on the platinum–cobalt scale.

18.3 ODOR

18.3.1 General Discussion

In its pure form, water does not produce any taste or odor. Many potentially toxic waters and foods can be avoided by smell and taste senses. These senses often provide the first warning of potential hazards in the environment. Most organic and inorganic chemicals contribute to taste or odor. These chemicals may originate from municipal or industrial waste discharges, from natural sources such as decomposition of vegetable matter, from microbial activities, and from disinfectants and their byproducts. The odor test is a qualitative description and approximate quantitative measurement of odor intensity.

Sensory tests are useful as a check of the quality of raw and finished water and for control of odor through a treatment process.

18.3.2 Sample Collection and Handling

Collect samples in glass bottles with glass or Teflon (TFE)-lined closures. Do not use plastic containers! Collect about 500 mL of sample and fill the bottle to the top. Analyze as soon as possible. If storage is necessary, cool to 4°C. Maximum holding time for the samples is 24 hours.

18.3.3 Sample Pretreatment

Most tapwaters and some wastewaters are chlorinated. Often it is desirable to determine the odor of the chlorinated sample as well as that of the same sample after dechlorination. A chlorinated sample needs to be dechlorinated before odor measurement with sodium thiosulfate ($Na_2S_2O_3$). Odor will vary with temperature. For most water samples, a sample temperature of 60°C permits the optimum detection of odor. If necessary, 40°C may be used as a standard temperature. Report odor results with the temperature at which the observations were made.

18.3.4 Measurement of Odor With Threshold Odor Test

Method No. 2150 B (*Standard Methods*, APHA-AWWA-WEF, 18th ed., 1992).

Principle of the method — Determine the threshold odor by diluting the sample with odor-free water until a definite perceptible odor is achieved. There is no absolute threshold odor concentration, because sensing any odor varies with individuals. A given person may vary in sensitivity over time. For precise work, use a panel of five or more testers, but measurement of odor by one person is often necessary at water treatment plants.

No smoking or eating is permitted before the test. Testers do not use scented soaps, perfumes, or shaving lotions, and they should be free from colds or allergies that affect the odor response. Keep the room free from disturbing odor, drafts, or distractions.

This threshold method is applicable to samples from nearly odorless natural waters to industrial wastes with threshold numbers in the thousands.

Apparatus — Odor-free glassware should be used. For the best result, reserve glassware for use only for this particular test. Do not use rubber, cork, or plastic stoppers. Do not use narrow-mouth containers. Apparatus for the threshold odor test include a constant-temperature water bath or electric hot plate, odor flasks (glass-stoppered 500 mL Erlenmeyer flask), graduated cylinders (200, 100, 50, and 25 mL), measuring pipets (10 mL, graduated in tenths), thermometer (0 to 110°C), and odor-free water (prepared by passing distilled or deionized water through activated carbon).

Threshold measurement — The threshold odor number (T.O.N.), is the greatest dilution of a sample with odor-free water yielding a definitely perceptible odor.

1. Determine the approximate range of the threshold number by adding 200 mL, 50 mL, 12 mL, and 2.8 mL samples to separate 500 mL glass-stoppered Erlenmeyer flasks and dilute to a total volume of 200 mL with odor-free water. Use a flask filled with 200 mL of sample as a reference. Heat dilutions and the reference sample to the desired temperature.

2. Note the dilution where odor was observed.

3. Prepare a dilution in the observed range according to Table 18.1 in the same way as described in step 1 above. Insert two or more blanks in the series. Heat to the desired temperature, read the corresponding threshold odor numbers from Table 18.2, and calculate according to the dilution. Use the formula

$$T.O.N. = (mL\ sample + mL\ odor\text{-}free\ water)/mL\ sample$$

18.4 CONDUCTIVITY

18.4.1 General Discussion

Conductivity is a numerical expression of the ability of a solution to carry an electric current. This ability depends on the presence of ions, on their total concentration, and on the temperature of the measurement.

TABLE 18.1

Conductivity of Potassium Chloride Solution at 25°C

Concentration (M)	Equivalent Conductivity (mho/cm/equiv.)	Conductivity (μmhos/cm)
0	149.85	
0.0001	149.43	14.94[a]
0.0005	147.81	73.90
0.001	146.95	147.0
0.005	143.55	717.8
0.01	141.27	1413
0.02	138.34	2767
0.05	133.37	6668
0.1	128.96	12,900
0.2	124.08	24,820
0.5	117.27	58,640
1	111.87	111.90

[a] Computed from equation given in Lind et al.

Data drawn from Robinson and Stokes.

In the laboratory the measurements are used to establish or monitor the degree of mineralization. Conductivity values are useful to estimate the total dissolved solids in a water sample in milligrams per liter by multiplying conductivity in micromhos per centimeter by an empirical factor. The factor may vary from 0.55 to 0.9, depending on the soluble components of the water. Conductivity measurement is also helpful in estimating sample size to be used for different analytical procedures. Multiplying conductivity values in micromhos per centimeter gives an approximate number of the total anions or cations expressed in milliequivalents per liter.

Solutions of most inorganic compounds are very good conductors; however, organic compounds do not dissociate in aqueous solution and therefore are poor conductors.

TABLE 18.2

Threshold Odor Numbers Corresponding to Various Dilutions

Sample Volume Diluted to 200 mL	Threshold Odor Number (T.O.N.)
200	1
140	1.4
100	2
70	3
50	4
35	6
25	8
17	12
12	17
8.3	24
5.7	35
4.0	50
2.8	70
2.0	100
1.4	140
1.0	200

Note: Threshold odor number is the greatest dilution of a sample with odor-free water yielding a definitely perceptible odor.

Distilled or deionized water has a conductivity of 0.5 to 2 µmhos/cm. Potable water ranges from 50 to 1500 µmhos/cm, and industrial waters can go above 10,000 µmhos/cm.

18.4.2 Sample Collection and Handling

Collect samples in glass or plastic containers and cool to 4°C until analysis. Holding time for samples is 28 days.

18.4.3 Measurement

Method No. 9050 (EPA SW-846, Vol. IC, 3rd ed., 1986); Method No. 2510 (*Standard Methods*, AWWA-APHA-WEF, 18th ed., 1992).

Principle of the method — Conductivity is measured directly using a self-contained conductivity meter. The meter is calibrated at the factory to read in submultiples of micromhos per centimeter when used with any conductivity cell constant of 1.0/cm. Conductance of a solution is measured between two spatially fixed and chemically inert electrodes. The conductance of the solution is directly proportional to the electrode surface area and inversely proportional to the distance between the electrodes.

Conductivity is customarily reported in micromhos per centimeter (µmhos/cm). In the International System of Units (SI), the unit used to report conductivity is millisiemens per meter (mS/m). 1 mS/m = 10 µmhos/cm and 1 µS/cm = 1 µmhos/cm.

Electrolytic conductance increases with temperature at a rate of approximately 2% deviation per degree Celsius. Whenever possible, samples are analyzed at 25°C. If samples are analyzed at a different temperature, temperature corrections must be made and the results reported at 25°C. For temperature corrections, see Section 15.2.1.

For example, a conductivity reading of a sample is 650 µmhos/cm and the temperature of the sample is 22°C. The 2% of the reading is 13 µmhos/cm. The temperature difference is 3°C, therefore 3 × 13 = 39 µmhos/cm should be added to the original reading. So, the conductivity value at 25°C is 650 + 39 = 689 µmhos/cm. If the 650 µmhos/cm reading is taken at a temperature of 28°C, the conductivity value will be 650 − 39 = 611 µmhos/cm.

Interferences — Platinum electrodes degrade over time. This is characterized by the flaking of the platinum black and/or erratic readings. If these problems are observed, the electrode needs to be replaced.

The cell becomes coated if the matrix of the sample is oily. Plastic cells should be cleaned with a foaming acid tile cleaner, rinsed with distilled water, and air-dried. The plastic will be damaged by concentrated sulfuric or nitric acids, and organic solvents. The probe must not be cleaned in aqua regia (1:3 HNO_3-HCl) or used in any solution known to etch gold or platinum. The cell must be air dried before storage.

Apparatus — Apparatus for determining conductivity include

- Self-contained conductometer, which should be capable of measuring conductivity with an error not exceeding 1% or 1 µmhos/cm.
- Conductivity cell, platinum electrode type, which should be cleaned and replatinized or changed, whenever the reading is erratic or when inspection shows that any platinum black has flaked off.

- Nonplatinum electrode type constructed from durable common metals (stainless steel). These are used for continuous monitoring and field studies.

Reagents

- Distilled water with conductivity value of 0.5 to 1 µmhos/cm.
- Stock 0.1 M potassium chloride (KCl) solution: dissolve 7.456 g of KCl (dried 2 hours at 105°C and stored in a desiccator) in distilled water and dilute to 1000 mL. Working KCl standards are diluted from this stock solution.
- Working KCl standards:
  - 0.001 M KCl solution: transfer 10 mL of 0.1 M stock KCl solution into a 1000 mL volumetric flask and dilute to the mark. Mix well.
  - 0.01 M KCL solution: transfer 100 mL of 0.1 M stock KCl solution into 1000 mL volumetric flask and dilute to the mark. Mix well.
  - 0.05 M KCl solution: dilute 500 mL 0.1 M stock KCl solution to 1000 mL.
- KCl reference check (independent) standard, such as Chempure Brand RC-306 with a true value of 1413 µmhos/cm. Alternatively, may prepare in the laboratory by using other source of KCl than have been used to prepare working standards.

Store KCl standard solutions in glass-stoppered borosilicate glass bottles. Conductivity values of different concentration KCl standards are given in Table 18.1.

Measurement — Conductometers have different ranges for reading conductivity. The most common YSI Model 32 Conductivity Meter has six ranges for readings. The conductivity must be read on the lowest possible range.

If a given solution is outside of the range, the meter will read "1...". The analyst needs to turn the knob up to the next range. If that reading is still too high, the analyst must go to the next range. The ranges are 2, 20, and 200 µmhos and 2, 20, and 200 mmhos. If the reading is taken on a mmhos range, the reading must be multiplied by 1000 to get the correct reporting unit, µmhos/cm.

Enough sample must be used to cover the hole in the side of the probe. Put a small magnetic stirrer into the sample and stir gently while taking the reading. If the temperature of the sample is not 25°C, adjust the result as discussed under the "Principle of the method" section.

Steps during the measurement

1. Electrode is stored in distilled water.
2. After instrument is in "on" position, wait about 10 to 15 minutes to warm up the instrument.
3. Rinse electrode with distilled water, wipe gently.
4. Rinse electrode three times with 0.001 M KCl standard solution.
5. Pour 0.001 M KCl standard into a small beaker. Add magnetic stirrer. Read and record conductivity value with the temperature of the sample.
6. Repeat steps 3 to 5 with 0.01 M KCl.
7. Repeat steps 3 to 5 with 0.05 M KCl.
8. Repeat steps 3 to 5 with reference (independent) standard.

9. If the reading of the working and reference standards are satisfactory, repeat steps 3 to 5 with each sample.

10. At the end of the run, rinse electrode with DI water and store.

Calculation — Apply temperature correction as discussed in Section 18.4.3 and Section 15.2.2. If any dilution have been applied, multiply the reading by the dilution factor.

Quality control

Duplicate analysis — One duplicate sample is to be analyzed of every ten samples in a sample batch. Duplicate samples are used to determine precision. The precision value should be within the in-house accepted limits.

Reference (independent) standard — Analyzed after calibration checks and before reading of the samples. Analyzed every ten samples in a sample batch, and at the end of the analytical run. The accepted accuracy limits are established by individual laboratories.

Chapter 19

ANALYSIS OF SOLIDS AND TURBIDITY

19.1 SOLIDS

19.1.1 General Discussion

Solids refers to matter suspended or dissolved in water and wastewater. Solids may affect water or effluent quality adversely in a number of ways. Waters with high dissolved solids generally are not desirable for drinking water purposes, and unsuitable for many industrial applications. Waters with high suspended solids are aesthetically unsatisfactory, for example, for bathing. Solids analyses are important in the control of biological and physical wastewater treatment processes and for assessing compliance with regulatory agency wastewater effluent limitations.

Total solids (total residue) is the residue left in a container after evaporation and drying at 103 to 105°C. Total residue is defined as the sum of the homogenous suspended and dissolved materials in a sample.

Suspended solids (nonfilterable residue) is a portion of the nitrate salts that may be lost.

19.1.2 Sample Collection and Handling

Collect samples in resistant-glass or plastic bottles, provided that the material in suspension does not adhere to container walls. Refrigerate samples at 4°C up to the time of analysis to minimize microbiological decomposition of solids. Do not hold samples more than 24 hours. Nonrepresentative particulates, such as leaves, twigs, insects, etc. must be excluded from the sample.

19.1.3 Sample Pretreatment

Bring samples to room temperature before analysis.

19.1.4 Determination of Total Solids (TS) Dried at 103–105°C

Method No. 160.3 (EPA 600/4-79-020).

Principle of the method — A well-mixed aliquot of the sample is quantitatively transferred to a preweighed dish, evaporated to dryness, and dried in an oven at 103–105°C to constant weight. The increase in weight over that of the empty dish represents the total solids.

Interferences — Highly mineralized water with a significant concentration of calcium, magnesium, chloride, and sulfate may be hygroscopic and require prolonged drying, proper desiccation, and rapid weighing.

Results for residues high in oil and grease may be questionable because of the difficulty of drying to constant weight in a reasonable time. Disperse visible floating oil and grease with a blender before withdrawing the sample portion to analysis.

Because excessive residue in the dish may form a water-trapping crust, limit sample to no more than 200 mg residue.

Apparatus and Material

- Evaporating dishes: porcelain, 90 mm, 100 mL capacity. High silica glass (Vycor, product of Corning Glassworks, Corning, N.Y.) or platinum dishes may be substituted and smaller size dishes may be used if required.
- Steam bath.
- Drying oven for operation at 103 to 105°C.
- Analytical balance capable to weighing to 0.1 mg.
- Desiccator provided with indicating desiccant.
- Muffle furnace for operation at 500 ± 50°C.

Procedure

1. Ignite evaporating dish at 550°C for 1 hour in muffle furnace. Cool and store in desiccator until use. Weigh immediately before use.
2. Transfer a measured volume of a well-mixed sample into preweighed dish.
3. Evaporate to dryness in a steam bath or in a drying oven. When evaporating in a drying oven, lower temperature to approximately 2°C below boiling point to prevent splattering.
4. Dry evaporated sample for at least 1 hour at 103 to 105°C.
5. Cool in desiccator and weigh.

Calculation

$$\text{mg/L Total Solids} = [(A - B) \times 1000]/\text{sample mL}$$

where A = weight of dried residue + dish in milligrams, B = weight of dish in milligrams.

Quality control — A method blank is analyzed once at the beginning of every analytical batch and once every 20 samples.

19.1.5 Determination of Total Dissolved Solids (TDS) Dried at 180°C

Method No. 160.1 (EPA 600/4-79-020).

Principle of the method — The method defines residue, filterable as those solids which are not retained by a glass fiber filter and dried at constant weight at 180°C. A well-mixed sample is filtered through a prepared glass fiber filter into a clean filter flask. The filtrate is transferred to a weighed evaporating dish and is evaporated on a steam bath to dryness. Afterwards, it is moved to a 180°C oven and dried until the weight is constant. The increase in dish weight represents the total dissolved solids.

Interferences — Highly mineralized waters with a considerable amount of calcium, magnesium, chloride, and sulfate content may be hygroscopic and require prolonged drying, proper desiccation, and rapid weighing. Samples high in bicarbonate require careful and prolonged drying to insure complete conversion of bicarbonate to carbonate.

   Because excessive residue in the dish may form a water-trapping crust, select sample size so that no more than 200 mg of residue remain after drying.

   Filter apparatus, filter material, filter preparation, and drying temperature are specified, because variations in these parameters have been shown to affect the results.

Apparatus and materials

- Glass fiber filter discs, 47 mm, without organic binder, 1 µm retention (Whatman grade 934 AH or Gelman type A/E or Millipore type AP40 or equivalent).
- Filtration apparatus: filter holder and membrane filter funnel (Gelman CMS Cat. No. 192-146 or equivalent) or Gooch crucible 25 to 40 mL capacity with Gooch crucible adapter.
- Filter flask of 500 mL capacity and vacuum with rubber vacuum tubing.
- Drying oven capable of achieving 180°C.
- Porcelain evaporating dishes 100 mL capacity (CMS Cat. No. 077-818 or equivalent. Vycor, product of Corning Glass Work, Corning N.Y. or equivalent may be substituted).
- Analytical balance capable of weighing to the nearest 0.0001 g.
- Desiccator with color-coded rechargeable desiccant.
- Calibrated thermometers, ambient to 200°C.
- Steam bath.

Reagents — Reference standard, 500 mg/L: dissolve 0.5000 g sodium chloride (NaCl) in DI water in a 1-L volumetric flask and fill up to the mark. The true TDS value of this standard is 500 mg/L.

Procedure

Preparation of evaporation dishes

1. Rinse a clean evaporating dish with DI water and place in an oven at 180°C for at least 1 hour. If volatile solids are to be measured, ignite clean evaporating dishes at 550°C for 1 hour in a muffle furnace.
2. Transfer to desiccator and cool for 90 minutes.
3. Number the dishes with a "sharpie" pen or other heat-resistant marker.
4. Weigh the dish on analytical balance. Record the weight to four decimal places.
5. Once the dishes have been weighed, they can be placed on a tray if there is going to be a delay before they are used. They must be covered with a lint-free towel to keep airborne

contaminants from falling on them. Alternatively, dishes may be stored in desiccator until they are needed, and weighed immediately before use.

Sample analysis

1. Place the filter disc on the filter holder. The wrinkled side must be up.
2. Turn on the vacuum and flush the filter with 3 aliquots of 20 mL DI water. Keep the vacuum on for a few seconds to dry the excess water out of the filter.
3. Remove the filter funnel with the rinsed filter and place on a 250-mL filter flask.
4. Assemble the filter apparatus.
5. Shake the sample well, being sure to suspend any sediment or particulate.
6. Pour 100 mL of sample into a graduated cylinder. Pour through the filter.
7. Rinse the cylinder with three aliquots of DI water, at least 5 mL each. Pour the rinses through the filter.
8. With a wash bottle, rinse down the sides of the filter holder with DI water.
9. Carefully rinse the filter and any visible residue with water from the wash bottle. Let the suction continue for a minute to dry the excess liquid.
10. Remove the filter with forceps and return to its marked weighing pan.
11. Transfer the filter from the filter flask to the tared evaporating dish. Carefully rinse the filter flask with 3 aliquots of DI water and combine the rinses with the filtrate in the dish.
12. If the sample is high in solids, the filter can clog and the time to process can become long. Therefore, if the sample takes longer than 10 minutes to filter, discard the filter and sample. Start over with a fresh filter and a clean flask. Shake the sample again and pour out a smaller aliquot. The size of the sample filtered would be based on how quickly the filter clogged. If clogging occurred almost immediately, process only 10 or 20 mL. If the sample begins to filter at a normal speed and then subsequently began to slow, 50 mL would be appropriate. Any samples that exceed 2000 mg/L must be returned with a lower sample volume.
13. The dishes are put on the steam bath to evaporate to dryness.
14. When dry, place them in the oven at 180°C. Check temperature of the oven with a calibrated thermometer. The samples must remain in the oven at least for 1 hour.
15. Place the dishes in a desiccator and cool for 90 minutes.
16. Weight the dishes with residues on the analytical balance. Record the weight.
17. After the dishes have been weighed, return them to the 180°C oven for 1 hour.
18. At the end of the drying time, transfer the dishes to the desiccator, cool for 90 minutes, and weigh. The result for this second weighing must be within ±5 mg of the first weighing for each sample. If the weights do not agree within the range, return the samples to the oven and continue the drying/weighing cycle until the weight is constant.
19. If the weights meet the criteria, calculate the concentration.
20. If volatile residue analysis is requested, save the dishes and residues for further analysis.

Calculation

$$\text{mg/L TDS} = [(A - B) \times 1000]/\text{sample mL used}$$

where A = weight of dried residue + dish, in milligrams, and B = weight of dish, in milligrams.

Quality control

- Method blank is analyzed once at the beginning of every analytical batch and once every 20 samples.
- Reference standard analyzed once in a sample batch and once every 20 samples. The acceptable range of the calculated accuracy is in-house generated.
- Duplicates analyzed once with each sample batch and in every 20 samples. The acceptable limit for the calculated precision is established by individual laboratories.

19.1.6 Determination of Total Suspended Solids (TSS) Dried at 103–105°C

Method No. 160.2 (EPA-600/4-79-020).

Principle of the method — The method defines residue, nonfilterable as those solids which are retained by a glass fiber filter and dried at constant weight at 103 to 105°C. A well-mixed sample is filtered through a prepared glass fiber filter, and the residue retained on the filter is dried to constant weight at 103 to 105°C. The filtrate from this test can be used for determination of filterable residue (TDS), EPA 160.3.

Interferences — Same as listed in total solids and total dissolved solids determinations. Samples high in filterable residue are subject to a positive interference. Careful washing of the filter and the residue will minimize this problem.

Apparatus and materials — The same apparatus and materials as those listed in Section 19.1.5 are required, except for evaporating dishes, steam bath, and 180°C drying oven. In addition, aluminum weighing dishes (CMS Cat. No. 079-053 or equivalent) are needed.

Procedure

Preparation of filter

1. Place the filter on the filter holder with the wrinkled side up. Turn on the vacuum and flush the filter with three aliquots of 20 mL each of DI water. Keep the vacuum on for a few seconds to dry the excess water out of the filter. Remove the filter with forceps and place in an aluminum weighing dish. Dry the filter in the oven for at least 1 hour at 103 to 105°C. Place the filters in the aluminum dishes with a "sharpie" pen or other heat-resistant marker.
2. Weigh the aluminum dish and filter paper on the analytical balance. Record the weight.
3. Once the dishes with the filter paper have been weighed, they can be placed on a tray. If there is going to be a delay before they are used, they must be covered with a lint-free towel to keep airborne contaminants from falling on them.

Sample analysis

1. Assemble the filter apparatus.
2. Using forceps, place the filter on the filter holder. The wrinkled side must be up. Attach the top of the magnetic filter funnel.

3. Shake the sample well, being sure to suspend any sediment or particulate.

4. Pour 100 mL of sample into a graduated cylinder. Pour through the filter.

5. Rinse the cylinder with three aliquots of DI water, at least 5 mL each. Pour the rinses through the filter.

6. With a wash bottle, rinse down the sides of the filter holder with DI water.

7. Carefully rinse the filter and any visible residue with water from the wash bottle. Let the suction continue for a minute to dry the excess liquid.

8. Remove the filter with forceps and return to its marked weighing pan.

9. If the sample is high in solids, the filter can clog and the time to process can become long. Therefore, if the sample takes longer than 10 minutes to filter, discard the filter and sample. Use a new filter that has been prepared and tared. Shake the sample again and pour out a smaller aliquot. The size of the sample filtered would be based on how quickly the filter clogged. If clogging occurred almost immediately, process only 10 or 20 mL. If the sample began to filter at a normal speed and then subsequently began to slow, 50 mL would be appropriate.

10. When all samples have been filtered, place the aluminum weighing dishes in an oven at 103 to 105°C for at least 1 hour.

11. The dishes must be cooled in a desiccator for 30 minutes.

12. Weigh the dishes and filters on the analytical balance and record the weights.

13. Afterwards, return them to the 103 to 105°C oven and dry for at least 1 hour.

14. After this drying time, transfer the dishes and filters to the desiccator and cool for 30 minutes.

15. Weigh the samples. The results of the second weighing must be within ±5 mg of the first weighing for each sample. If the weights meet the criteria, calculate the concentration. If the weights do not agree within the range, return the samples to the oven and continue the drying/weighing cycle until the weight is constant.

Calculation

$$\text{mg/L TSS} = [(A - B) \times 1000]/\text{sample mL}$$

where A = weight of filter + residue in milligrams, and B = weight of filter in milligrams.

19.1.7 Fixed and Volatile Solids Ignited at 500°C

Method No. 160.4 (EPA 600/4-79-020).

Principle of the method — The residue from the method 160.1, 160.2, and 160.3 is ignited to constant weight at $500 \pm 50°C$. The remaining solids represent the fixed total, dissolved, and suspended solids, while the weight lost on ignition is the volatile solids.

Interferences — Negative errors may be produced by loss of volatile matter during drying. Determination of low concentrations of volatile solids in the presence of high fixed solids concentrations may be subject to considerable error.

Apparatus and material — See methods 160.1, 160.2, and 160.3.

Procedures

1. Have muffle furnace up to temperature of $500 \pm 50°C$.
2. Place residues produced by methods 160.1, 160.2, and 160.3 in the furnace and ignite for 15 to 20 minutes. However, heavier residues may overtax the furnace and necessitate longer ignition time.
3. Let dish or filter paper cool partially in air until most of the heat has been dissipated.
4. Transfer sample to the desiccator for final cooling.
5. Weigh dishes and filters.
6. Repeat cycle of igniting, cooling, desiccating, and weighing until a weight change is less than 4% or 0.5 mg.

Calculation

$$\text{mg/L Volatile Solids} = [(A - B) \times 1000]/\text{mL sample}$$

$$\text{mg/L Fixed Solids} = [(B - C) \times 1000]/\text{mL sample}$$

where A = weight of residue + dish before ignition (mg), B = weight of residue + dish or filter after ignition (mg), and C = weight of dish or filter (mg).

19.1.8 Total, Fixed, and Volatile Solids in Solid and Semisolid Samples

Method No. 2540 G (*Standard Methods*, APHA-AWWA-WPF, 18th ed., 1992).

Principle of the method — The method is applicable to the determination of total solids and its fixed and volatile fractions in such solid and semisolid samples as river and lake sediments, sludges separated from water and treatment processes, and sludge cakes from vacuum filtration, centrifugation, or other dewatering processes.

Interferences — The determination of both total and volatile solids in these materials is subject to negative error due to loss of ammonium carbonate and volatile organic matter during drying. The mass of organic matter recovered from sludge and sediment requires a longer ignition time than that specified for waters. Accomplish all weighing quickly, because wet samples tend to lose weight by evaporation. After drying, residues are very hygroscopic and rapidly absorb moisture from air.

Apparatus and materials — These are the same as those listed in Section 19.1.4.

Procedure

Preparation of evaporation dishes — If volatile solids are to be measured, ignite a clean evaporating dish at $500 \pm 50°C$ for 1 hour in a muffle furnace. If only total solids are to be measured, heat dish at 103 to 105°C for 1 hour in an oven. Cool in desiccator and weigh as described in Section 19.1.4.

Sample analysis — Same as in the method for determining total solids (160.3.) or fixed and volatile solids (160.4.), with the exception of using the mass of the sample instead of volume. Preferable sample size is 25 to 50 g.

Calculation

$$\% \text{ Total Solids} = [(A - B) \times 100] / (C - B)$$

$$\% \text{ Volatile Solids} = [(A - D) \times 100] / (A - B)$$

$$\% \text{ Fixed Solids} = [(D - B) \times 100] / (A - B)$$

where A = weight of dried residue + dish (mg), B = weight of dish (mg), C = weight of wet sample + dish (mg), and D = weight of residue + dish after ignition (mg).

Quality control — Same as for methods 160.3 and 160.4.

19.1.8 Settleable Solids

Method No. 2540 F (*Standard Methods*, APHA-AWWA-WPCF, 17th ed., 1989).

Principle of the method — Measure the settleable solid volumetrically in Imhoff cone.

Apparatus — Imhoff cone.

Procedure

1. Fill the Imhoff cone to the 1 L mark with a well-mixed sample.
2. Settle for 45 minutes, gently stir sides of cone with a glassrod or by spinning, and settle 15 minutes longer.
3. Read the volume of settleable solids in the cone as milliliters per liter.
4. When a separation of settleable and floating materials occurs, do not estimate the floating material as settleable matter.

19.2 TURBIDITY

19.2.1 General Discussion

The clarity of a natural body of water is a major determinant of the condition and productivity of that system. Turbidity in water is caused by suspended matter, such as clay, silt, finely divided organic and inorganic matter, soluble colored organic compounds, and plankton and other microscopic organisms. Turbidity reporting units represent light-scattering and absorbing properties of suspended matter in water. The reportable unit is the nephelometric turbidity unit (NTU). Turbidity measurement is not limited to the determination of small particles, but is also related to the fact that these particles definitely tend to protect pathogens from disinfection treatment. Organic matter and microorganisms are presumed to protect pathogenic organisms from the bactericidal effect of disinfecting agents, while inorganic particles are not. The sample may also contain particles with toxic effect.

19.2.2 Sample Collection and Handling

Determine turbidity on the day sample is collected. Sample container may be glass or plastic. If storage is unavoidable, store samples in the dark and refrigerate. Maximum holding time is 48 hours.

19.2.3 Measurement of Turbidity by Nephelometric Method

Method No. 180.1 (EPA 600/4-79-020).

Principle of the method — The method is based on a comparison of the turbidity of light scattered by the sample under defined conditions with the intensity of light scattered by standard reference solution under the same conditions. The higher the intensity of the scattered light, the higher the turbidity. Formazine polymer is used as the reference turbidity standard suspension.

Interference — Dirty glassware, the presence of air bubbles, and the effects of vibrations all can contribute to interference. Color of the sample absorbs light, causes the turbidity value to be lower.

Apparatus and material

- Turbidimeter: the sensitivity of the instrument should permit detection of turbidity of less than 1 NTU with a range from 0 to 40 NTU. Differences in turbidimeter design will cause differences in measured values for turbidity, even though the same suspension is used to calibrate.
- Sample tubes in clear, colorless glass. Keep tubes scrupulously clean, both inside and outside. Discard them when they become scratched or etched. Never handle tubes where the light strikes them.

Reagents

- Stock turbidity suspensions:
  - Solution I. Dissolve 1.000 g hydrazine sulfate, $(NH_2)_2H_2SO_4$, in DI water and dilute to 100 mL in a volumetric flask. Prepare fresh monthly! *CAUTION:* Hydrazine sulfate is a carcinogen! Avoid inhalation, ingestion, and skin contact!
  - Solution II. Dissolve 10.000 g hexamethylenetetramine,$(CH_2)_6N_4$, in DI water and dilute to 100 mL in a volumetric flask. Prepare fresh monthly!
- Turbidity stock solution, 400 NTU: in a 100 mL volumetric flask, mix 5 mL of Solution I and 5 mL of Solution II. Let stand 24 hours at room temperature. After this period of time, dilute to 100 mL with DI water and mix well. The turbidity of this suspension is 400 NTU. Prepare fresh monthly!
- Turbidity standard solution, 40 NTU: dilute 10 mL of turbidity stock solution (400 NTU) to 100 mL with DI water. The value of the standard is 40 NTU. Prepare fresh daily!

- Dilute turbidity standards: prepare by dilution from the 40 NTU standard to check different turbidity ranges. Prepare daily!

Procedure

Calibration of turbidimeter — Follow the manufacturer's operating instruction. Check accuracy of any supplies' calibration scales on a precalibrated instrument by using appropriate standards. Run at least standard in each instrument range to be used.

Measurement

1. Thoroughly shake sample.
2. Wait until air bubbles disappear and pour sample into the turbidimeter tube. When possible, immerse tube with sample in an ultrasonic bath for 1 to 2 seconds (for complete bubble release).
3. Clean and dry tube outside with laboratory tissue paper (KIMWIPE). Be sure that tube is completely dry and clean. Never handle tubes where the light strikes them.
4. Read turbidity directly from the instrument scale.
5. When turbidity reading is above 40 NTU, sample should be diluted to the range that reads directly from the meter. The turbidity of the original sample will be given by the reading of the diluted sample multiplied with the dilution factor. For example, if a 10 mL sample is diluted to 100 mL, and the turbidity reading of this diluted sample is 25 NTU, the turbidity of the original sample will be $15 \times 10 = 150$ NTU.

Chapter **20**

ALKALINITY AND pH CALCIUM CARBONATE SATURATION

20.1 ALKALINITY

20.1.1 General Discussion

The alkalinity of a water is its quantitative capacity to react with a strong acid to a designated pH. The measured value may vary significantly with the end-point pH used. Alkalinity is a measure of an aggregate property of water and can be interpreted in terms of specific substances only when the chemical composition of the sample is known. Alkalinity is significant in many uses and treatments of natural and wastewaters. Because the alkalinity of many surface waters is primarily a function of *carbonate, bicarbonate, and hydroxide content*, it is taken as an indication of the concentration of these constituents. The measured values may include contributions from borates, phosphates, or silicates if these are present. Alkalinity in excess of alkaline earth metal concentrations is significant in determining the suitability of a water for irrigation. Alkalinity measurements are also used in the interpretation and control of water and wastewater treatment processes. Raw domestic wastewater has an alkalinity less than or only slightly greater than that of the water supply. Properly operating anaerobic digesters typically have supernatant alkalinities in the range of 2000 to 4000 mg/L $CaCO_3$.

20.1.2 Sample Collection and Storage

Collect samples in polyethylene or borosilicate glass bottles and store at a low temperature. Fill bottles completely and cap tightly. Because waste samples may be subject to microbial action and to loss or gain of carbon dioxide (CO_2) or other gases when exposed to air, analyze samples without delay, preferably within 1 day. If biological activity is evident, analyze within 6 hours. Avoid sample agitation and prolonged exposure to air. If the samples are not going to be analyzed immediately, they must be kept at 4°C. The holding time for alkalinity is 14 days from the time of sampling.

20.1.3 Potentiometric Titration Method

Method No. 310.1 (EPA-600/4-79-020), Method No. 2320 B (*Standard Methods*, 17th ed., 1989).

The method detection limit (MDL) is 1.0 mg/L total alkalinity as $CaCO_3$. The precision of the buret is usually the limiting factor.

Principle of the method — Alkalinity is the acid-neutralizing capacity of water. It is the sum of all the titratable bases in the water. The alkalinity is primarily a function of carbonate, bicarbonate, and hydroxide and is used as an indicator of the concentrations of these constituents. In this method, the levels of bicarbonate, carbonate, and hydroxide are calculated from the amount of titrant used to bring the pH from the beginning pH to pH 8.3 and then to pH 4.5.

The sample must not be filtered, diluted, or concentrated in any way. A sample is titrated to an electrometric end point of pH 8.3, known as *phenolphthalein alkalinity*. The volume of titrant is recorded, then the titration is continued to pH 4.5, known as the *total alkalinity*. The various species of interest are calculated from these two numbers.

Interferences — Anything that interferes with the response of the pH electrode will cause a slow response from the titration. Things such as soaps, oils, and greases or suspended material will coat the glass electrode and decrease response.

Substances such as the salts of inorganic acids present in large amounts will interfere with the pH measurements.

Samples with high concentration of mineral acids, such as mine wastes, need to be run by an alternate method. This would require previous knowledge of the sample.

Apparatus and materials — Use any commercial pH meter or electrically operated titrator that uses a glass electrode and can be read to 0.05 pH unit. Standardize and calibrate according to the manufacturer's instructions. Pay special attention to temperature compensation and electrode care. If automatic temperature compensation is not provided, titrate samples with a temperature of 25°C.

Glassware — Use a buret, volumetric flask (1 L).

The size and form of the titration vessels will depend on the electrodes and the sample size. Keep the free space above the sample as small as practicable, but allow room for titrant and full immersion of the indicating portions of electrodes. For conventional-sized electrodes, use a 200 mL, tall-form Berzelius beaker without a spout. Also, use a magnetic stirrer and magnetic stirring bars.

Reagents

- Sodium carbonate solution, 0.05 *N*. Dry 3 to 5 g primary standard Na_2CO_3 at 250°C for 4 hours and cool in desiccator. Weigh 2.5 ± 0.2 g (to the nearest mg), transfer to a 1-L volumetric flask. Dissolve in a small quantity of DI water and fill flask to the mark with DI water. Mix well. *Do not keep longer than 1 week!*
- Standard sulfuric acid (H_2SO_4), 0.10 *N*. Preparation and standardization is described in Section 6.3.1.
- Standard sulfuric acid (H_2SO_4), 0.02 *N*. In a 1-L volumetric flask, add 200 mL of 0.1 *N* Standard H_2SO_4 solution and bring to volume with DI water. Determine exact normality by standardization each time before use as described later in this section (titrant check).

- Reference standard stock, 10,000 mg/L. Dry primary standard-grade sodium carbonate, Na_2CO_3, at 250°C for 4 hours. Cool in desiccator. Weigh 10.6 g dried Na_2CO_3 and transfer into 1000 mL volumetric flask, dissolve, and dilute to the volume with DI water. The value is 10,000 mg/L alkalinity as $CaCO_3$. *Keep refrigerated!*
- Reference standard, 100 mg/L. Measure volumetrically 10 mL of stock QC check standard into a 1000 mL volumetric flask and dilute to the volume with DI water. *Prepare weekly! Keep refrigerated!*

Titrant check — The titrant must be verified at the beginning of each day of analysis. Standardize as outlined for 0.10 N H_2SO_4 in Section 6.3.1, except use approximately 0.02 g of primary standard sodium carbonate, Na_2CO_3, and titrate with 0.02 N H_2SO_4. Make three parallel titrations of the standard by titrating to pH 4.5 or to the end point of the mixed bromcresol green–methyl red indicator (see titration procedure below) with the acid that will be used for the samples. Calculate the exact normality of the titrant according to the formula in Section 6.3.1.

Titration procedure

1. Add 50 mL sample or smaller aliquot to a beaker.
2. Calibrate the pH meter according to the method of pH determination (Section 20.2.2).
3. Take the initial pH of the sample and record it. Note: A piece of styrofoam or thick cardboard must be placed under the beaker to protect it from temperature effects from the stirrer motor.
4. Add standardized 0.02 N H_2SO_4 drop-wise to the beaker until the pH is 8.3. Record milliliters used for "P" (phenolphthalein) alkalinity in the log form.
5. Continue adding the acid drop-wise until the pH reaches 4.5. Record this amount acid used for "T" (total) alkalinity in the log form.
6. If the total volume of the titrant is greater than 20 mL, the sample volume must be decreased to 25 mL. If the total volume is still more than 20 mL, repeat the process with 10 mL or 5 mL sample. If the total volume of the titrant is still more than 20 mL, the process must be repeated by using the 0.1 N H_2SO_4 titrant. This acid must be standardized (Section 6.2.3) before sample analysis begin. *The sample must not be diluted, concentrated, or altered in any way!*

Potentiometric titration of low alkalinity — For alkalinities less than 20 mg/L, titrate 100 to 200 mL sample according to the potentiometric procedure above, using a 10 mL microburette and 0.02 N standard acid solution. Stop the titration at a pH in the range 4.3 to 4.7 and record volume and exact pH. Carefully add additional titrant to reduce the pH exactly 0.3 pH units and again record volume.

Calculations

Alkalinity concentration for potentiometric titration to pH of 8.3

$$\text{Alkalinity as } CaCO_3 \text{ mg/L} = (\text{vol}_{acid} \times N \times 50 \times 1000)/\text{vol}_{sample} \qquad \textbf{(20.1)}$$

where vol_{acid} = volume of standard acid (mL) used from the beginning pH to pH 8.3, N = exact normality of the acid, vol_{sample} = volume of sample titrated, and 50 = equivalent weight of $CaCO_3$ (100/2).

Alkalinity concentration for potentiometric titration to pH 4.5 — Same as formula (20.1) with the exception that the vol_{acid} must be the volume of standard acid (mL) used from the beginning pH to pH 4.5.

Alkalinity Concentration for potentiometric titration using low-level method

$$\text{Total Alkalinity as } CaCO_3 \text{ mg/L} = [(2B - C) \times N \times 50 \times 1000] \, vol_{sample} \qquad \textbf{(20.2)}$$

where B = volume of titrant to first recorded pH (mL), C = volume of titrant required to lower pH by 0.30 units (mL), N = exact normality of the acid, and 50 = equivalent weight of $CaCO_3$.

20.1.4 Titrimetric Method by Color Change

Method No. 403 (SM, 15th ed.).

Principle of the method — The measured value of the alkalinity of water may vary significantly with the end point pH used, and is primarily the function of carbonate (CO_3^{2-}), bicarbonate (HCO_3^-), and hydroxide (OH^-) content. In order to distinguish between the kinds of alkalinity present in a sample and to determine the quantities of each, a titration is made with a standard acid using two indicators successively. The indicators used are phenolphthalein, which gives a pink color in the presence of alkalinity and the change from pink to colorless occurs at a pH value of 8.3, and methylorange, which changes from yellow to red color at a pH value of 4.5. Instead of methylorange, a mixed bromcresol green–methyl red indicator may be used to give a sharper end point. But the resulting titrations are identified traditionally as *"P" (phenolphthalein) alkalinity* (pH 8.3) and methylorange alkalinity (pH 4.5). The sum of the phenolphthalein and methylorange alkalinity is the *"T" (total) alkalinity.*

Regarding the range and color change of the indicators, phenolphthalein is pink in alkaline, and changes to colorless at the pH of 8.3; methylorange is yellow in alkaline, orange in neutral, and red in acidic media; and mixed bromcresol green–methyl red indicator is greenish-blue at pH 5.2, light blue at pH 5.0, light pink-gray at pH 4.8, and light pink at pH 4.5.

Apparatus and material — These are the same as those listed for potentiometric titration.

Reagents — These are also the same as listed for potentiometric titration, with the addition of

- Phenolphthalein indicator solution: dissolve 5 g phenolphthalein in 500 mL isopropyl alcohol and add 500 mL of DI water. Mix well.
- Mixed bromcresol green–methyl red indicator: mix 0.1000 g bromcresol green and 0.02 g methyl red in 100 mL isopropyl alcohol (IPA).
- Methylorange indicator: 0.5 g methylorange dissolved in DI water and diluted to 1 L.

Procedure

1. Obtain 100 mL or smaller aliquots of sample in a flask or beaker. Add a magnetic stirring bar.

2. Place the container on a magnetic stirrer and stir.

3. Add 3 drops of phenolphthalein indicator while stirring. If the sample becomes pink, add standardized 0.02 N H_2SO_4 titrant from a buret slowly until the pink color just disappears. Record the milliliters of acid used.

4. Add 3 drops of methylorange indicator. Sample becomes yellow. Add standardized 0.02 N H_2SO_4 until the color changes to orange. Record milliliters of acid used. Alternatively, instead of methylorange add 3 drops of mixed bromcresol green–methyl red indicator. Sample color becomes blue. Titrate with standardized 0.02 N H_2SO_4 until color changes to light pink. Record milliliters of acid used.

Calculations

$$\text{"P" Alkalinity, as } CaCO_3 \text{ mg/L} = (vol_{acid} \times N \times 50 \times 1000)/vol_{sample} \qquad \textbf{(20.3)}$$

where vol_{acid} = milliliters of acid used for phenolphthalein titration, N = exact normality of the acid, 50 = equivalent weight of $CaCO_3$, and vol_{sample} = milliliters of sample used.

$$\text{"T" Alkalinity as } CaCO_3 \text{ mg/L} = (vol_{acid}) \times N \times 50 \times 1000/vol_{sample} \qquad \textbf{(20.4)}$$

where vol_{acid} = total milliliters of acid used (phenolphthalein + methylorange titration), N = exact normality of the acid, 50 = equivalent weight of $CaCO_3$, and vol_{sample} = milliliters of sample used.

Quality control — For reference of QC items, calculations, and the accepted criteria, see Section 3.2.7, Section 3.2.9, and Section 16.1.

Method blank — Method blank is to be analyzed at the beginning of each analytical batch. The method blank demonstrates that interferences from the analytical system, glassware, and reagents are under control. The concentration of the blank must be lower than the detection limit of 1.0 mg/L.

Duplicate samples — One duplicate sample must be analyzed once with every sample batch or once every 20 samples whichever is most frequent. Duplicate samples are used to determine precision. The calculated value must be within the established limit by the laboratory.

Reference standard — The reference standard is titrated immediately after the blank, once every 10 samples, and once at the end of the sample batch. It is used to determine accuracy. The percentage of recovery of the value must be within the acceptance limit determined by the laboratory.

20.1.5 Calculation of Alkalinity Relationships

The results obtained from the phenolphthalein and total alkalinity determinations offer a means for stoichiometric classification of the three principal forms of alkalinity present in many waters. The classification ascribes the entire alkalinity to bicarbonate, carbonate, and hydroxide.

- Carbonate (CO_3^{2-}) alkalinity is present when phenolphthalein alkalinity is not zero but is less than the total alkalinity.

- Hydroxide (OH⁻) alkalinity is present if phenolphthalein is more than half the total alkalinity.
- Bicarbonate (HCO₃⁻) ions are present if phenolphthalein alkalinity is less than half the total alkalinity.

The mathematical conversion of the results is shown in Section 15.2.9.

20.1.6 Conversion of the Three Forms of Alkalinity (Carbonate, Bicarbonate, and Hydroxide Alkalinity) Expressed as CaCO₃ to Their Ionic Forms

Calculation is shown in Section 15.2.9.

20.2 DETERMINATION OF PH

20.2.1 General Discussion

Measurement of pH is one of the most important and frequently used tests in water chemistry. Practically every phase of the water supply and wastewater treatment, e.g., acid–base neutralization, coagulation, disinfection, and corrosion control, is pH-dependent. At a given temperature, the intensity of the acidic and basic character of the solution is indicated by pH.

pH is the negative logarithm of the hydrogen ion concentration,

$$pH = - \log [H^+] \tag{20.5}$$

The term "concentration of the hydrogen ion" is written $[H^+]$. The brackets mean "mole concentration" and H^+ is "hydrogen ion." The concentration of the H^+ is expressed in moles per liter.

Water dissociates by a very slight excess into H^+ and OH^- ions. It has been experimentally determined that in pure water $[H^+] = 1 \times 10^{-7}$, and $[OH^-] = 1 \times 10^{-7}$, therefore $[H^+] = [OH^-]$. The negative logarithm of 1×10^{-7} is 7.00; therefore, the pH of pure water is 7.00. Such a solution is neutral and has no excess of H^+ or OH^-. An excess of $[H^+]$ is indicated by a pH below 7.00 and the solution is said to be acidic. pH values above 7.00 indicate an excess in $[OH^-]$ and the solution is alkaline.

20.2.2 Sample Collection and Holding Time

Collect samples into glass or plastic bottles. No preservation required. Samples must be analyzed for pH within twenty four hours of sampling.

20.2.3 Potentiometric Determination of pH

Method No. 150.1 (EPA-600/4-79-020), Method No. 9040 and 9045 (EPA SW 846).

EPA 150.1 and SW 9040 are used to determine the pH of *water samples* and wastes that are at least 20% aqueous. SW 9045 is used to determine the pH of *solid samples. Nonaqueous liquids* are analyzed by SW 9041.

Principle of the method — EPA 150.1, SW 9040, and SW 9045 all use an electrometric method to determine the pH of a sample. Samples for 150.1 and 9040 are measured directly after being brought to ambient temperature. SW 9045 requires that the sample first be mixed with DI water or a calcium chloride solution, depending on whether the soil is considered calcareous or noncalcareous. Samples are considered noncalcareous unless specified otherwise.

A pH meter functions by measuring the electric potential between two electrodes that are immersed into the solution of interest. The basic principle is to determine the activity of the hydrogen ions by potentiometric measurement, using a glass electrode and a reference electrode. The most popular, called a combination electrode, incorporates the glass and the reference electrode into a single probe. The *glass electrode* is sensitive to the hydrogen ions, and changes its electrical potential with the change of the hydrogen ions. The *reference electrode* has a constant electric potential (see Section 12.2.3). The difference in potential of these electrodes, measured in millivolts (mV), is a linear function of the pH of the solution. The scale of the pH meters is designed so that the voltage can be read directly in terms of pH.

A glass electrode usually consists of a silver and a silver chloride electrode in contact with dilute aqueous HCl, surrounded by a glass bulb that acts as a conducting membrane. The hydrogen ion concentration of the HCl solution inside the electrode is constant; therefore, the potential of the glass electrode depends on the hydrogen ion concentration outside of the glass membrane.

The reference, also called "calomel" electrode, contains elemental mercury (Hg), calomel (Hg_2Cl_2) paste, and Hg metal. This paste is contacted with an aqueous solution of potassium chloride (KCl) solution. KCl serves as a salt bridge between the electrode and the measured solution. *The electrode must be visually inspected every month! The level of the solution should be checked every day and refilled as needed!*

Interferences — The glass electrode is relatively free from interferences caused by color, turbidity, colloidal matter, oxidants, reductants, or high salt content of the sample. High sodium content samples give higher pH reading interference. This can be eliminated by using a "low sodium error electrode."

The electrode can be coated with oily material. It can be cleaned with a detergent solution and rinsed with DI water, or it can be rinsed with dilute HCl (1+9), followed by DI water. Soaking the electrode with 5% pepsin solution may also be effective in removing organic deposits.

Temperature affects the response of the electrode in two ways. One is that the electrode response varies at different temperatures, and the second is that the pH changes at different temperatures. The best results are obtained when the temperature of the sample is within 5°C of the temperature of the buffers used to calibrate the instrument. Therefore, samples and buffers are allowed to equilibrate to room temperature prior to analysis. Temperature effects may be eliminated also by using pH meters equipped with a temperature compensator.

Apparatus and materials — These include a pH meter, pH and reference electrodes, or combination electrode, disposable cups (50 or 100 mL), analytical balance, capable of weighing to the nearest 0.01 g, magnetic stirrer, and magnetic stirring bars.

Reagents — Electrode filling solution, pH buffer solutions (pH 2.00, 4.00, 7.00, 10.00, and 13.00), potassium chloride (KCl) ACS grade, and pH electrode storage solution.

Meter calibration

1. Check the filling solution level in the electrode. If needed, add more filling solution until the appropriate volume is achieved.

2. Turn on the pH meter. The pH meter requires about 20 minutes to warm up. Electrode should be in the storage solution.

3. Set the temperature compensate knob to room temperature.

4. Add pH 7.00 buffer to a disposable cap. The depth of the buffer in the cap should be deep enough to cover the junction spot on the electrode when the electrode is emerged in the buffer.

5. Remove electrode from storage solution and rinse with DI water and dry by gently blotting with a kim-wipe.

6. Place the electrode into the buffer solution and stir gently either by using a magnetic stirrer and stirring bar or by hand-swirling. The rate of stirring should minimize the air transfer rate at the air–water interface of the sample. Be sure that the volume of the buffer is sufficient to cover the sensing elements of the electrode. Read the pH value. If the reading deviates from the actual value (7.00), adjust with the "calibration" knob.

7. Remove electrode from pH 7.00 buffer, rinse with DI water, wipe and place into pH 4.00 buffer. Read. If the reading deviates from 4.00, adjust to the exact reading by turning the "slope" knob.

8. Remove electrode from pH 4.00 buffer solution, rinse with DI water, wipe, and immerse into pH 10.00 buffer. Read. pH value should be 10.00 or very close. Do not adjust meter at this! The instrument is now calibrated.

9. Rinse the electrode with DI water and blot with a kim-wipe to remove excess water. Check the calibration with initial calibration verification, by measuring pH of the pH 4.00, 7.00, and 10.00 buffers. If the readings are correct, rinse electrode with DI water and wipe.

10. After calibration, but before sample measurement, immerse electrode in a reference/independent standard. If the reading is satisfactory, measurement of the samples may start.

Measurement of aqueous samples and wastes containing >20% water

1. Allow the temperature of the sample to equilibrate to room temperature.

2. Pour an aliquot of the sample into the disposable cap or beaker. The depth of the sample should be deep enough to cover the junction spot on the electrode when the electrode emerged in the sample.

3. Place the electrode into the sample and stir gently by using a magnetic stirrer and stirring bar or by hand-swirling.

4. Read the pH and record.

5. Rinse the electrode with DI water, and blot dry with a kim-wipe to remove excess water.

6. After the samples are measured, remove electrode, rinse, and blot to dry as in step 5.

7. Store the electrode in electrode storage solution until next use. Electrode storage solution is a commercially available "pH electrode storage solution" or may be a pH 7.00 buffer solution. *Keeping electrode in DI water is not recommended!*

Measurement of solid samples

1. Allow the temperature of the sample to equilibrate to room temperature.

2. Using a two decimal place balance, weigh about 20 g sample into the cap or beaker used for pH measurement.

3. Add 20 mL DI water to the cap or beaker.

4. Let stand for 30 minutes, mixing at 5-minute intervals.

5. Place the electrode into the aqueous portion of the sample and stir gently, either by using a magnetic stirrer and stirring bar or by hand-swirling.

6. Read the pH and record.

7. Repeat steps 5, 6, and 7 of the "Measurement of aqueous samples" procedure.

General rules — Samples are analyzed in a set referred to as a sample batch. A sample batch for pH consists of ten or fewer actual samples and a duplicate sample. Other QC samples are incorporated into a batch at the frequencies specified in the "Quality Control" section of this methodology.

If a sample reading shows a *high alkalinity character*, the meter must be recalibrated by using pH 7.00 and pH 10.00 buffer instead the pH 7.00 and pH 4.00 buffers as calibration standards. After recalibration, read the sample again.

If *field measurements* are being made, the electrode(s) may be immersed in the sample stream to an adequate depth and moved in a manner to insure sufficient sample movement across the electrode sensing element, as indicated by drift-free (0.1 pH) sample readings.

Calibration, measurement, and QC data are recorded on the pH working paper.

Because of the wide variety of pH meters, detailed operation procedures are given for each instrument. The analyst must be familiar with the operation of the system, and with the instrument functions. If field measurements are being made, the electrode(s) may be immersed in the sample stream to an adequate depth and moved.

Quality control — For reference of QC items, calculations, and the accepted criteria, see Section 3.2.7, Section 3.2.9, and Section 16.1.

Duplicate analysis — One duplicate sample is to be analyzed every 10 samples for each matrix in the sample batch. Duplicate samples are used to determine precision. The calculated RPD value should be documented, and its value should be within the acceptable range variate by individual laboratories.

Initial calibration verification (ICV) — This is performed immediately after the meter calibration reads pH 2.00, pH 10.00, and pH 13.00 buffers. If the buffer readings are acceptable, the meter is ready for sample measurement. The acceptable range for pH 2.00 is 1.90 to 2.10, for pH 10.00 is 9.95 to 10.05, and for pH 13.00 is 12.90 to 13.10.

Continuing calibration verification (CCS) — CCS is performed once after every 10 samples in a batch (if applicable) and once at the end of the sample batch. Read pH 2.00, pH 4.00, pH 7.00, pH 10.00, and pH 13.00 buffers. The acceptable range for each buffer is as follows: pH 2.00 is 1.90 to 2.10, pH 4.00 is 3.95 to 4.05, pH 7.00 is 6.95 to 7.05, pH 10.00 is 9.95 to 10.05, and pH 13.00 is 12.90 to 13.10.

20.3 CALCIUM CARBONATE SATURATION

20.3.1 General Discussion

The pH at which water is saturated with calcium carbonate ($CaCO_3$) is known as the *pH of saturation* (pH_s). The *Langelier Saturation Index* is defined as the actual pH minus pH_s. A negative index indicates the tendency to dissolve $CaCO_3$ (the water is corrosive).

In other words, water oversaturated with respect to $CaCO_3$ tends to precipitate $CaCO_3$, water undersaturated with $CaCO_3$ tends to dissolve $CaCO_3$, and water saturated or in equilibrium with $CaCO_3$ has neither precipitation nor dissolving tendencies; these waters are stable.

Calculation of the saturation index (SI) is based on the values of pH, total alkalinity as $CaCO_3$, total dissolved solids (TDS), temperature, and total hardness as $CaCO_3$. A positive index is associated with noncorrosive conditions, whereas a negative index indicates the possibility of corrosion.

Toxic metals such as lead and cadmium are corrosion products from lead and galvanized pipes. Similarly, high iron, copper, and zinc content of water is also proportional to the corrosive property of the water.

20.3.2 Calculation of the Saturation Index

Calculation of saturation index is outlined in Section 15.3.

Chapter 21

ANALYSIS OF FLUORIDE AND TOTAL HARDNESS

21.1 FLUORIDE

21.1.1 General Discussion

Fluoride is a pale, yellowish gas with a characteristic pungent odor. It belongs to the halogen family, and it is the most reactive nonmetallic element. Even though hydrogen fluoride, HF, and all its salts are poisonous, the fluoride ion (F^-) in trace concentration in drinking water helps growing bodies develop teeth that are particularly resistant to decay. Tooth enamel is mostly the mineral hydroxyapatite, $[Ca_3(PO_4)_2]_3 \cdot Ca(OH)_2$. When fluoride ion is available during the development of enamel, a much harder mineral tends to form instead, fluorapatite, $[Ca_3(PO_4)_2]_3.CaF_2$.

Hydroxyapatite contains hydroxide ions (OH_-), which are more avidly attacked by acids produced by mouth bacteria feeding on sugars than is the much weaker base fluoride ion (F^-) in fluorapatite. Fluoride has an influence on bone metabolism in all organisms and dental health in humans. At a concentration of 1.0 to 2.00 ppm in the drinking water, fluoride is beneficial. Over this limit, it is harmful, causing fluorosis and bone problems. Fluoride inhibits many enzymes. In most cases, the enzyme affected contains a metal ion with which the fluoride combines to form a metal–fluoride complex. This gives a general inhibiting effect on overall metabolism and the growth of cells.

The adjustment of the fluoride content of drinking water to 1 mg/L fluoride is in principle and in practice the most effective access to preventing dental cavities.

21.1.2 Sampling and Storage

Preferably, use polyethylene bottles for collecting samples, but glass bottles are also satisfactory. Rinse the bottles three times with the sample before actual sample collection. No special preservation is necessary. The holding time for the samples is 28 days.

21.1.3 Determination of Fluoride by Ion-Selective Electrode

Method No. 340.2 (EPA-600/4-79-020).

Principle of the method — This method is applicable to the measurement of fluoride in drinking, surface, and saline waters, domestic and industrial wastes, soils, and sediments. Different preparations include the following:

1. Aqueous samples can be read directly without any preparation. This will give the inorganic fluoride concentration.
2. Aqueous and solid samples can be distilled to get total fluoride, which includes both inorganic and organic fluoride.
3. Solid samples can be extracted to get extractable or soluble fluoride. This would be mostly the inorganic fluoride concentration.
4. For total and dissolved fluoride, the Bellack distillation is required for NPDES monitoring.
5. Distillation is not required for SDWA monitoring.

If distillation is required, the sample is distilled from boiling sulfuric acid to fluosilicic acid without etching the boiling flask. The sample size is relatively large to promote quantitative recoveries of the fluoride. The carryover of water and acid are minimized by controlling the temperature range of the distillation.

The fluoride is determined potentiometrically using a fluoride electrode in conjunction with a standard double junction reference electrode and a ion analyzer or a pH meter having an expanded millivolt scale.

The fluoride electrode consists of a lanthanum fluoride crystal across which a potential is developed by fluoride ions (see Section 13.2.3). The fluoride electrode is an ion-selective sensor. It measures the ion activity of fluoride in solution. Adding a buffer (TISAB) gives a nearly uniform ionic strength background, adjusts pH, and breaks up the complex so that, in effect, the electrode measures concentration.

Interferences — Common interferences are extremes of pH and polyvalent cations of Si^{+4}, Fe^{+3}, and Al^{+3}, which interfere by forming complexes with the fluoride. The degree of interference depends upon the concentration of the complexing cations, the concentration of fluoride, and the pH of the sample. The addition of the TISAB buffer controls most of the interferences, since it contains a chelating agent that preferentially complexes aluminum (the most common interference; it causes a negative bias at 3 ppm), silicon, and iron. It also controls the pH in the desired range of 5 to 9.

Apparatus and material

- Glassware
    - Boiling flask, 24/40 round bottom, 1000 mL
    - Volumetric flasks
    - Graduated cylinders
    - Clausen connectors, 24/40
    - Calibrated Teflon-coated thermometers that have a range from ambient to at least 200°C
    - Lurex water cooled condenser
- pH meter with full millivolt reading or one ion analyzer
- Fluoride-sensing electrode (ORION 94-09)
- Reference electrode single junction sleeve type (ORION 90-020)
- Fluoride electrode filling solution: equitransferent filling solution (ORION, Cat. No. 90-00-01)

- Plastic bottles with Teflon-lined lids, 500 mL
- Disposable beakers, 100 mL
- Magnetic stirrer
- TFE-coated stirring bars
- Styrofoam or cardboard
- Magnetic stirring bars
- Soft glass beads (soda lime)
- pH paper, full range
- Automatic shaker
- Top-loading balance
- Analytical balance
- Filter paper, Whatman 40 or equivalent

Reagents

Stock fluoride solution, 100 mg/L — Transfer 0.221 g anhydrous sodium fluoride (NaF) to 1 L volumetric flask, add DI water to dissolve and dilute to 1 L. Mix well.

$$1 \text{ mL} = 0.1 \text{ mg } (100 \text{ μg}) \text{ F}^-$$

Standard fluoride solution, 10 ppm — Measure 100 mL of stock fluoride solution and dilute to 1000 mL with DI water.

$$1 \text{ mL} = 0.1 \text{ mg } (10 \text{ μg}) \text{ F}^-$$

Total ionic strength adjustment buffer (TISAB) —

- Measure about 500 mL distilled water into one 1-L beakers.
- Add 57 mL of acetic acid glacial.
- Add 58 g of sodium chloride (NaCl) and 12 g sodium citrate dihydrate ($Na_3C_6H_8 \cdot 2H_2O$). Alternatively, use 4.0 g 1,2-cyclohexylenediamine-tetraacetic acid (CDTA) (Eastman 15411, J.T. Beaker 5-G083 or equivalent). Stir to dissolve.
- Place beaker in a cool water bath and add slowly 6 N NaOH solution (about 125 mL) with constant stirring until pH is between 5.0 and 5.5.
- Transfer solution carefully to a 1-L volumetric flask and add DI water to the mark.

Sodium hydroxide (NaOH) 6 N — Add about 500 mL DI water to a 1-L beaker, add a stirring bar, and place the beaker on a magnetic stirrer. Add 240 g NaOH and stir until dissolution is complete. Stirring helps disperse the heat generated. Cool and transfer into a 1-L volumetric flask and dilute the solution to the mark. Mix well. Prepare under laboratory hood! *CAUTION:* Dissolution of NaOH pellets in water generates heat!

Reference (independent) stock solution, 100 mg/L — Obtain anhydrous sodium fluoride from another source as used to prepare fluoride standards. Weigh 0.2210 g of this chemical and transfer into a 1-L volumetric flask, dissolve and fill up to the volume with DI water, and mix well.

Reference (independent) intermediate solution, 24 mg/L — Measure 240 mL of reference (independent) stock solution into a 1-L volumetric flask, fill up to the volume, and mix.

Working reference (independent) standard solution, 1.2 mg/L — Pipet 5 mL of the reference (independent) intermediate solution into a 100 mL volumetric flask. Dilute to volume with DI water. Mix by inversion. *Prepare fresh each time before use.*

Bellack distillation

1. Add 250 mL of DI water to a 1000 mL round-bottom flask. Under the hood, slowly add 200 mL concentrated H_2SO_4. Swirl the flask as the acid is added to aid in mixing.

2. Add a few boiling chips (Hengar granules) and soda lime beads. Mix well again and place on a heater. *NOTE:* Failure to completely mix the contents of the flask will result in hot spots that can erupt and spray boiling acid over the surrounding area! Heating a nonhomogenous acid–water mixture will result in bumping or possibly a violet explosion!

3. Assemble the distillation apparatus. Use standard taper joints for all connections in the direct vapor path. Position the thermometer so that the bulb is always immersed in the boiling mixture. Direct distillation apparatus for fluoride is shown in Figure 21.1.

4. Turn on the cold water and distill into a waste beaker until the temperature reaches 180°C.

5. Turn off heat and remove the distillation flask from the mantle. When the temperature has cooled to <120°C, the flask can be used again.

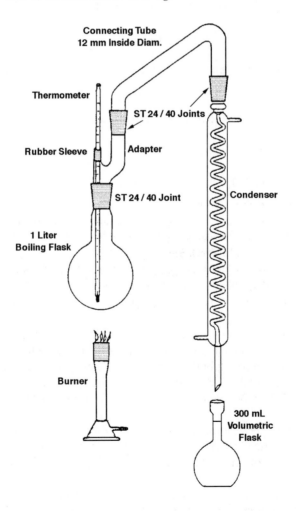

Figure 21.1
Direct distillation apparatus for fluoride.

6. **Distillation of water samples:** Into the flask with the acid/water mix, add 250 mL of sample. Mix well by swirling the flask. Assemble the distillation apparatus and distill the sample into a 250-mL volumetric flask. The distillation is complete when 250 mL are collected or the temperature reaches 180°C, whichever occurs first.

   Distillation of soil samples: Weigh 15 ± 0.5 g of sample on a top-loading balance and carefully add to the flask with the acid–water mixture. Add 250 mL DI water. Assemble the distillation apparatus and distill into a 250 mL volumetric flask. The distillation is complete when 250 mL are collected or the temperature reaches 180°C, whichever occurs first.

7. When the sample has been collected, the volume might have to be adjusted. If the volume is <250 mL, dilute to volume with DI water. If the volume is above the line, carefully measure the extra volume by pouring the sample into a 500-mL graduated cylinder. Record the exact volume and transfer the sample into a plastic container with a Teflon liner for storage until the readings are taken.

8. The boiling flask must be cleaned between each sample by adding 250 mL DI water to the acid–water mix and distilling until the temperature reaches 180°C. If the sample is dirty and leaves a residue on the flask or if the acid–water mixture turns yellow, dispose of the mix and clean the flask and begin again with fresh DI water and acid (step 1).

Extraction procedure

1. On the top-loading balance, weigh 15 g (±1.0 g) into a 250 mL wide-mouth plastic bottle with a Teflon-lined lid. Record the exact weight.
2. Add 100 mL DI water to the plastic bottle.
3. Close the cap securely and put the bottles on the automatic shaker for 30 minutes.
4. Let settle and filter through a 0.45 μm glass fiber filter.
5. Measure the volume of the filtrate and record it.

Measurement by ion analyzer — A variety of instruments are available for the test. Read carefully and follow manufacturer's instruction before starting analysis. The following procedure is suitable for the *ORION 901 Ion Analyzer.*

1. Prepare the electrodes by injecting the proper filling solution. Follow instruction of the electrode's manual. Refill the probe with the appropriate solution and soak in 100 mg/L standard solution at least 30 minutes.
2. Instrument setup: fluoride electrode plugs into "Sense A" connection, reference electrode plugs into "Ref A" connection, shorting strap between Sense B and Ref B.
3. Slope determination:
   - Put instrument into MV mode
   - Measure 50 mL of distilled water and 50 mL TISAB into 150 mL beaker. Add stirring bar and place the beaker on a magnetic stirrer. Stir at medium speed.
   - Immerse fluoride and reference electrode into the solution.
   - Add 1 mL of 100 ppm fluoride standard by volumetric pipet. After stable reading, record millivolts (mV).
   - Add 10 mL of 100 ppm fluoride standard by volumetric pipet. After stable reading, record millivolts (mV).
   - The difference between the two readings will give the slope. Good operation is indicated by a slope of –54 to –60 mV.
   - Adjust the slope switches of the instrument to the obtained slope value.

4. Standardization:

   - Prepare 1.0 and 10.0 ppm fluoride standards, 100 mL each, by pipeting 1.0 and 10 mL stock fluoride solution (100 ppm) into 100 mL volumetric flask and dilute to the mark. Mark the values on both volumetric flasks.
   - Put instrument into "CONC." mode.
   - Measure 50 mL 1.0 ppm standard into beaker and add 50 mL TISAB.
   - Turn STD value switches on the prepared concentration: 1.00.
   - Place electrodes into the standard and stir.
   - Press "CLEAR/READ MV." Allow time for reading to stabilize.
   - Press "SET CONC." Display will be the number set on the STD value switches: 1.00.
   - Place electrodes into the second standard, prepared by measuring 50 mL of 10-ppm standard and add 50 mL TISAB. Allow time for stabilized reading for the 10 ppm standard.
   - Slowly adjust the SLOPE switches until display reads the correct concentration of the second standard: 10.00.
   - If the slope value is correct, measure reference standard to verify correct calibration of the instrument. Measure 50 mL standard into a beaker and add 50 mL TISAB. Add stirring bar and by stirring immerse electrodes and read concentration.

5. Measurement of samples: if everything shows correct operation, measure samples by using 50 mL sample and 50 mL TISAB. Confirm calibration after every two to three samples and recalibrate instrument if necessary. Samples are analyzed in a set referred to as a sample batch, which consists of QC samples and 20 or fewer actual samples that are distilled or analyzed together. If dilution is necessary to get samples into the calibration range, the amount must be recorded, and the reading must be multiplied by the dilution factor.

Analysis using pH meter

1. Instrument and calibration curve preparation
   - An initial calibration curve is prepared and analyzed daily. The standards for the calibration curve are not distilled.
   - Prepare a series of standards by diluting with DI water 2, 5, 15, and 20 mL of the 100 mg/L fluoride stock solution to 100 mL in volumetric flasks. These standards are equivalent to 0.2, 0.5, 1.0, 1.5, and 2.0 mg/L fluoride.
   - Connect the fluoride and the reference electrodes to the meter.
   - Place electrodes into 100 mg/L fluoride standard for at least 30 minutes or until the reading stabilizes.
   - Label disposable beakers by the appropriate concentration of the standards and one blank.
   - Measure 50 mL DI water for the blank and 50 mL from each standard into each marked beaker.
   - Place blank on the magnetic stirrer and place the electrodes into the solution. When the reading has stabilized, read the result in millivolts. Record the millivolts reading. Repeat this step for each of the remaining standards.
   - Using a calculator with linear regression capabilities, enter the millivolts as the independent (x) variable and the log or natural log (ln) of the concentration as the dependent (y) variable. Perform the linear regression and obtain a value for the correlation coefficient (regression factor, R). This value should be >0.995. *NOTE:* If the calculator has exponential regression capabilities, the millivolts can be entered directly and the program will do the line fitting.

2. Slope check: immediately after the calibration curve is generated, a slope check must be performed at two different points of the curve to verify the electrode sensitivity. The ideal difference in millivolts for a tenfold millivolts difference in concentration is –56.4. The equation for the slope check is

$$SC_{0.2-2.0} = PR_{0.2} - PR_{2.0} \qquad\qquad (21.1)$$

where SC = slope check and PR = potentiometer reading (mV). The difference in millivolt readings must be between 50 and 60. If the slope does not fall within the range, change the filling solution of the electrode and soak again for 30 minutes in 100 mg/L fluoride standard. Check the slope again. If the result of the repeat slope check is still not satisfying, the electrode performance may be helped by coating the lanthanum fluoride crystal with a fluoride toothpaste and allowing it to sit for an hour. Rinse the toothpaste off thoroughly with DI water. Recheck the slope. If the result of the repeat slope check still do not fall between 50 and 60, the probe may need to be replaced.

3. Measurement of samples
   - If the results meet the calibration criteria, proceed with the analysis.
   - For undistilled samples, check sample pH using full-scale pH paper and record the actual pH of the sample.
   - Place the beaker on a magnetic stirrer and place the electrode into the solution. The electrode must remain in the solution until a stable reading is obtained (at least 3 minutes). Time spans of 5 minutes or longer might be required in solutions less than 0.5 mg/L. When the reading has stabilized, read the result in millivolts.

4. Calculations:

$$conc_{water} = curve_{inv} \times DF \qquad\qquad (21.2)$$

$$conc_{solids} = (curve_{inv} \times DF) \times (FV/Wt) \times (1000 \text{ g}/1000 \text{ kg}) \qquad\qquad (21.3)$$

where $conc_{water}$ = fluoride concentration in mg/L, $conc_{solids}$ = fluoride concentration in mg/kg, $curve_{inv}$ = mg/L value as obtained from calibration curve, DF = dilution factor, FV = final volume (mL) after distillation, and Wt = weight (g) of sample aliquot. Using the potentiometer reading, the calibration curve yields the log or ln of the concentration. To obtain the concentration, the inverse of the log or ln (10^x or e^x) must be calculated. Discussion of logarithmic regression calculation with example is presented in Section 12.4.7.

Quality control

Method blank — A method blank is analyzed once with every sample batch or once every 20 samples, whichever is more frequent. If distilled samples are being analyzed, the blank must also be distilled. Undistilled samples must be accompanied by an undistilled bank.

Matrix spike/matrix spike duplicates (MS/MSDs) — A MS/MSD is performed once for each sample batch or once every 20 samples, whichever is more frequent. If both distilled and undistilled samples are analyzed, a MS/MSD must be prepared for each method. For example:

1. For aqueous, undistilled samples, the matrix spike (MS) consists of 3.0 mL of fluoride stock solution (100 mg/L) pipetted into a 250-mL volumetric flask and diluted to the volume by the sample. Calculate the added spike value as mg/L:

$$A \times B = C \times D \hspace{5cm} \text{(21.4)}$$

where A = concentration of fluoride stock solution added, B = milliliters of fluoride stock solution added, C = concentration of the added spike value (unknown), and D = total milliliters of the spiked sample.

2. For aqueous distilled samples, the matrix spike consists of 3.0 mL of fluoride stock solution (100 mg/L) pipetted into a 250 mL volumetric flask, diluted to the volume with the sample, and poured into the distilling flask and distilled. The final volume of the distillate is 250 mL as in Section 21.1.3. (Bellack Distillation, step 6). Calculate the concentration of the added spike as the aqueous nondistilled samples above.

3. For solid extracted samples, matrix spike (MS) is prepared by adding 3 mL of 100 mg/L fluoride stock solution directly onto the solid sample in the plastic bottle. Add 100 mL DI water. Shake for the allotted time and filter (see Section 21.1.3, Extraction). Calculate the concentration of the spike as follows:

$$mg/kg = (Std) \times (100 \ \mu g/mL) \times (Ext/Wt) \hspace{3cm} \text{(21.5)}$$

where Std = volume of standard added (mL), Ext = volume of extract after filtration (mL), and Wt = dry weight of extracted sample (g).

4. For solid distilled samples, to prepare matrix spike (MS) add 3 mL of stock fluoride solution to a 250 mL volumetric flask and dilute to the volume with DI water. Add to the distillation boiling flask on top of the weighed sample (see Section 21.1.3, Bellack distillation, step 7). The concentration of the spike is calculated as follows:

$$mg/kg = (Std) \times (100 \ \mu g/mL) \times (Dist/Wt) \hspace{3cm} \text{(21.6)}$$

where Std = volume of stock fluoride solution added, Dist = volume of distillate collected (mL), and Wt = dry weight of the sample put into the distillation flask (g).

21.2 TOTAL HARDNESS AS CaCO$_3$

21.2.1 General Discussion

Total hardness is commonly reported as equivalent concentration of calcium carbonate, $CaCO_3$. Other reporting systems are also used, such as French, German, and British hardness degrees: 100 mg/L hardness as $CaCO_3$ is equal to 10 French hardness degrees, 5.6 German hardness degrees, and 7 British hardness degrees. Hardness in water is caused by calcium and magnesium salts; however, iron, manganese, and strontium also contribute to hardness if appreciable concentrations occur. Most of the calcium and magnesium present in natural water occurs as bicarbonates, sulfates, chlorides, and nitrates.

Temporary hardness is removed by boiling and caused principally by the presence of bicarbonates of calcium and magnesium. *Permanent hardness* is mostly due to calcium sulfate, which is precipitated at temperature above 150°C (300°F). *Carbonate hardness* is due to the presence of calcium and magnesium carbonates and bicarbonates. *Noncarbonate hardness* includes the calcium and magnesium sulfates, chlorides, and nitrates. Sulfates are often the only noncarbonate hardness compounds present.

The American Water Works Association indicates that "ideal" quality water should not contain more than 80 mg/L of hardness, as $CaCO_3$. Hardness in many instances exceeds this level, especially where waters have contacted limestone (calcium and magnesium carbonates) or gypsum (hydrated calcium sulfate, $CaSO_4 \cdot 2H_2O$). Large-scale studies in Japan, U.S., England, Sweden, and Finland have shown that there are statistical interrelationships between cardiovascular diseases and certain ingredients in drinking water, including hardness. Lower death rates from heart and circulatory diseases occurred in states where public water supplies were higher in hardness. The theorized protective agents include calcium, magnesium, vanadium, lithium, chromium, and manganese. The suspect harmful agents include the metals cadmium, lead, copper, and zinc, all of which tend to be found in higher concentrations in soft waters as a result of the relative corrosiveness of soft water.

21.2.2 Sampling and Storage

Plastic or glass bottles may be used. Preserve samples with the addition of 3 mL 1+1 nitric acid (HNO_3) or 1.5 mL concentrated HNO_3 per 1 L of sample. If samples are collected in nonpreserved bottles, rinse bottle with the sample three times before filling up bottle. In the case of using preserved sample bottles, obviously this step should be neglected. Check the pH of the preserved sample to be sure it is less than pH 2.0. Holding time is 6 months; store samples at room temperature.

21.2.3 Determination of Hardness by EDTA Titrimetric Method

Method No. 130.2 (EPA-600/4-79-020).

Principle of the method — Ethylenediaminotetraacetic acid disodium salt (EDTA) forms a chelated soluble complex when added to a solution of certain metal cations (see Section 11.1). If a small amount of dye such as Eriochrom Black T is added to an aqueous solution containing calcium and magnesium ions at a pH of 10, the solution becomes wine-red. If EDTA is added as a titrant, the calcium and magnesium will be complexed, and the solution turns from wine-red to blue, marking the end point of the titration. The sharpness of the end point increases with increasing pH. Magnesium ions must be present for a satisfactory end point. To insure this, a small amount of complexometrically neutral magnesium salt of EDTA is added to the buffer.

Interferences — Heavy metals are the main interference in this analysis. They interfere by causing fading or indistinct endpoints or by stoichiometric consumption of the EDTA. Inhibitors can be used when metals are interfering or else a non-EDTA method should be used to determine total hardness, such as calculation from Ca and Mg values.

Suspended or colloidal organic matter may interfere with the end point. The sample can be evaporated to dryness on a steam bath, then heated in a muffle furnace at 550°C until the organic matter is completely oxidized. The residue is dissolved in 20 mL 1 N HCl, neutralized to pH 7.0 with 1 N NaOH, and made to 50 mL with DI water. Cool to room temperature and titrate with EDTA.

Conduct titration at or near normal room temperature! The color change becomes very slow as the sample has lower temperature. In hot sample, indicator decomposition is the problem. A limit of 5 minutes set for the duration of the titration to minimize the tendency toward $CaCO_3$ precipitation.

Apparatus and materials

- Glassware
  - Buret 25 or 50 mL

- Volumetric pipets 5, 10, 25 mL
- Graduated cylinders 100 mL
- Mohr pipets 5 mL
- Erlenmeyer flasks 250 mL
- Volumetric flasks various sizes
- Disposable transfer pipets
- Magnetic stirrer
- Teflon magnetic stirring bars
- pH paper, full range
- Analytical balance, capable of weighing to the nearest 0.0001 g
- Top-loading balance, capable of weighing to the nearest 0.01 g

Reagents

Buffer solution

- Solution 1
  - Weigh 1.179 g disodium salt of EDTA (Na_2EDTA) and transfer into a 150 mL beaker.
  - Weigh 0.780 g magnesium sulfate heptahydrate ($MgSO_4 \cdot 7HO$) or 0.644 g magnesium chloride hexahydrate ($MgCl_2 \cdot 6H_2O$) and transfer into the same beaker.
  - Add DI water into the beaker until the volume is about 100 mL and mix until the solids are dissolved.
- Solution 2
  - Weigh 16.9 g of ammonium chloride (NH_4Cl), transfer into a 250 mL volumetric flask, and add 143 mL concentrated ammonia solution (NH_4OH).
  - Transfer solution 1 from the beaker into solution 2 in the 250 mL volumetric flask. Rinse the beaker well with DI water and add the rinsate to the volumetric flask. Fill to 250 mL with DI water. Stopper the volumetric flask and mix well. Store buffer solution in polyethylene bottle and put stopper tightly. The buffer solution is good about 1 month. Discard the buffer when 1 or 2 mL are added to the sample and it fails to produce a pH of 10.0 at the titration end point.

Eriochrom black T indicator — Weigh 0.5 g indicator and 100 g NaCl into a porcelain mortar, and mix well. Alternatively, a coffee grinder may be used for complete mixing.

0.02 N EDTA titrant — Dissolve 3.723 g of disodium EDTA in about 700 mL DI water in a 1 L volumetric flask and dilute to the mark with DI water. Standardize against standard 0.02 N $CaCO_3$ solution.

Calcium carbonate standard solution — Weigh 1.0000 g of anhydrous $CaCO_3$ (primary standard) and transfer into a 500 mL Erlenmeyer flask. Place a funnel in the flask neck and add dropwise 1+1 HCl solution until all $CaCO_3$ is completely dissolved. Add 200 mL distilled water and boil for a few minutes to expel CO_2. Cool at room temperature. Add a few drops of methyl red indicator, while stirring. Adjust the color, while stirring, to an intermediate orange color with 3 N NH_4OH or 1+1 HCl. Transfer quantitatively into a volumetric flask and dilute to 1 L with DI water. Store in a polyethylene bottle.

$$1 \text{ mL} = 1 \text{ mg } CaCO_3$$

Reference stock solution, 33,333 mg/L total hardness, as CaCO$_3$ —

- Transfer 12.4860 g anhydrous, primary standard calcium carbonate (CaCO$_3$) into a 1 L volumetric flask. Add about 200 mL of DI water and slowly add concentrated hydrochloric acid (HCl) until complete dissolution of the calcium carbonate.
- Transfer 19.5847 g anhydrous magnesium chloride (MgCl$_2$) into the 1 L volumetric flask containing the calcium carbonate solution. Mix well for complete dissolution and fill up to the 1 L volume. Mix well again. The solution contains 33,333 mg/L total hardness, as CaCO$_3$.

Reference (independent) standard, 166 mg/L total hardness as CaCO$_3$ — Pipet volumetrically 5 mL of the reference stock solution into a 1 L volumetric flask and dilute to the volume with DI water. It is the actual working reference solution with the value of 166 mg/L total hardness as CaCO$_3$.

Standardization of EDTA titrant with CaCO$_3$ standard solution

1. Pipet 10 mL CaCO$_3$ standard solution into 100 mL Erlenmeyer flask.
2. Using a Mohr pipet, add 5 mL buffer solution and one scoopful Eriochrom Black T indicator. Mix well. Solution should be wine-red.
3. Rinse the buret three times with the EDTA titrant.
4. Fill buret with the EDTA titrant.
5. Remove any air bubbles from the buret and bring level of titrant to 0.00 mL.
6. Titrate the content of the Erlenmeyer flask with EDTA solution until red tint disappears. The color will turn purple. Continue titration slowly, until the solution turns blue. This is the end point. Record the volume of EDTA used.
7. Perform this titrant check two more times.
8. Calculate the normality of the EDTA as follows:

$$\text{Normality}_{\text{EDTA}} = [(0.02)(S)]/V \qquad\qquad (21.7)$$

where 0.02 = prepared normality of EDTA, S = volume of CaCO$_3$ solution titrated (mL), and V = volume of EDTA used for titration. Determine the normality with three parallel titrations. The exact normality is calculated by averaging of the three results.

Titrimetric procedure

1. Measure 100 mL sample or a portion diluted to 100 mL into one 250 mL Erlenmeyer flask.
2. Add 5 mL buffer solution and a scoopful of Eriochrom Black T indicator and mix. Solution becomes wine-red.
3. Rinse the buret with standardized EDTA titrant three times.
4. Fill buret with standardized EDTA titrant.
5. Remove air bubbles from the buret and check 0.00 level.
6. Titrate sample until red tint disappears. The color turns pink. Continue titration slowly, until the solution turns blue.
7. If the volume of the titrant used is over 25 mL, repeat titration by using smaller sample size or appropriate dilution.

Calculation

$$\text{mg/L Hardness, as } CaCO_3 = (V - B) \times N \times 50 \times 1000/SV \qquad \textbf{(21.8)}$$

where V = volume of titrant used for sample (mL), B = volume of titrant used for blank (mL), N = the determined normality of EDTA, 50 = equivalent weight of $CaCO_3$ (100/2), and SV = sample volume (mL). Use appropriate dilution factor if necessary.

Quality control — For reference of QC items, calculations, and the accepted criteria, see Section 3.2.7 and Section 3.2.9.

- Method blank is analyzed once at the beginning of every analytical batch and once every 20 samples.
- Reference (independent) standard is analyzed once at the beginning of every analytical batch, once every 10 samples, and once at the end of the analytical batch.
- Duplicate is analyzed every 10 samples in the sample batch.

21.2.4 Total Hardness by Calculation

The preferred method for determining hardness is to compute it from the results of separate determinations of calcium and magnesium. See Section 15.2.7.

$$\text{Hardness as } CaCO_3 \text{ mg/L} = 2.497 \text{ (Ca mg/L)} + 4.118 \text{ (Mg mg/L)}$$

or

$$\text{Hardness as } CaCO_3 \text{ mg/L} = [(Ca, \text{ mg/L})/0.4] + [(Mg, \text{ mg/L})/0.24]$$

For example, calcium and magnesium have been determined by atomic absorption technique, with the following results: Ca = 16 mg/L and Mg = 9.6 mg/L. The calculated total hardness value as $CaCO_3$ =

$$(2.497 \times 16) + (4.118 \times 9.6) = 79.5 \text{ mg/L}$$

or

$$(16/0.4) + (9.6/0.24) = 80 \text{ mg/L}$$

21.2.5 Calculation of Carbonate and Noncarbonate Hardness

The relationship of alkalinity and total hardness expressed as $CaCO_3$ may give the values of carbonate and bicarbonate hardness.

If the normal carbonate and bicarbonate alkalinity expressed as $CaCO_3$ is greater than the total hardness, normal carbonates of sodium and potassium are present. These compounds are not the cause of hardness and, in this case, the carbonate hardness would be equal to the total hardness.

If the sum of the normal carbonate and bicarbonate alkalinities is equal to the total hardness, the carbonate hardness is also equal to the total hardness. If the sum of the normal carbonate and bicarbonate alkalinities is less the total hardness, the carbonate hardness is equal to the noncarbonate hardness. To illustrate,

ALKALINITY > TOTAL HARDNESS

Carbonate Hardness = Total Hardness
Noncarbonate Hardness = Not Present

ALKALINITY = TOTAL HARDNESS

Carbonate Hardness = Total Hardness
Noncarbonate Hardness = Not Present

ALKALINITY < TOTAL HARDNESS

Carbonate Hardness = Alkalinity
Noncarbonate Hardness = Total Hardness – Alkalinity

21.3 CALCIUM DETERMINATION BY EDTA TITRIMETRIC METHOD

Method No. 215.2 (EPA-600/4-79-020).

Principle of the method — When EDTA is added to water containing both calcium and magnesium, it combines first with calcium. Calcium can be determined directly with EDTA when the pH is made sufficiently high that the magnesium is largely precipitated as the hydroxide and an indicator is used that combines with calcium only.

Interferences — Orthophosphate precipitates calcium at the pH of the test. Strontium and barium give a positive interference and with alkalinity in excess of 300 mg/L may cause an indistinct end point in hard waters.

Apparatus and material — These are the same as those listed in Section 21.2.

Reagents

Sodium hydroxide, NaOH, 1 N — Place a 2 L beaker or Erlenmeyer flask on a magnetic stirrer under laboratory hood. Add about 500 mL DI water and a magnetic stirring bar and add 40 g of NaOH slowly to the water while stirring. *CAUTION:* The reaction liberates heat! After complete dissolution, transfer into a 1 L volumetric flask and fill up to the mark with DI water. Mix well. Store in polyethylene bottle.

Murexide (ammonium purpurate) indicator — A ground mixture of dye powder and sodium chloride provides a stables form of the indicator. Weigh 0.200 g murexide (ammonium purpurate) and 100 g NaCl, and grind the mixture to 40 to 50 mesh in a porcelain mortar, or in a coffee grinder used for this purpose.

Standard EDTA titrant, 0.02 N — Prepare as described in the total hardness determination, Section 21.2.

$$1 \text{ mL} = 400.8 \text{ μg Ca}$$

Reference stock solution, 12,500 mg/L Ca as CaCO₃ — Same as for total hardness in Section 21.2.3, with a value of 12,501 mg/L of Ca as $CaCO_3$.

Reference standard solution, 62.5 mg/L — Pipet 5 mL of reference stock solution into a 1 L volumetric flask and dilute to the volume with DI water.

Procedure

1. Measure 100 mL sample or smaller portion diluted to 100 mL.
2. Add 2 mL 1 N NaOH solution or a volume sufficient to produce a pH of 12 to 13. Stir.
3. Add a scoopful of indicator. The color of the sample becomes pink.
4. Titrate with standardized EDTA solution until the pink color changes to purple. This is the end point. Titrate immediately after adding indicator, because it is unstable under alkaline conditions.
5. Check end point by adding 1 to 2 drops of titrant in excess to make certain that no further color change occurs. Facilitate end-point recognition by preparing a color comparison blank containing 2 mL 1 N NaOH and a scoopful of indicator powder and sufficient EDTA titrant (0.05 to 0.10 mL) to produce an unchanging color.

Calculation

$$\text{Calcium, as } CaCO_3 = (\text{mL} \times N \times 50 \times 1000)/\text{mL sample} \qquad \textbf{(21.9)}$$

where mL = milliliters of EDTA standard used for titration, N = exact normality of the EDTA titrant, and 50 = equivalent weight of $CaCO_3$.

$$\text{Calcium as Ca, mg/L} = \text{Ca as } CaCO_3 \text{ mg/L} \times 0.4 \qquad \textbf{(21.10)}$$

Quality control — Same as listed for total hardness in Section 21.2.

21.4 DETERMINATION OF MAGNESIUM BY CALCULATION

Magnesium may be estimated based on the difference between total hardness and calcium as $CaCO_3$.

$$\begin{aligned}\text{Mg as } CaCO_3 \text{ mg/L} = \text{Total Hardness,} \\ \text{as } CaCO_3 \text{ mg/L} - \text{Calcium, as } CaCO_3 \text{ mg/L}\end{aligned} \qquad \textbf{(21.11)}$$

$$\text{Magnesium as Mg mg/L} = \text{Magnesium as } CaCO_3 \text{ mg/L} \times 0.24 \qquad \textbf{(21.12)}$$

A detailed calculation is presented in Section 15.2.8.

Chapter 22

DETERMINATION OF RESIDUAL CHLORINE, CHLORIDE, AND CYANIDE

22.1 CHLORINE RESIDUAL

22.1.1 General Discussion

Chlorine is a member of the halogen group. It is a greenish-yellow gas and its name originated from the Greek word *chloros*, meaning yellow-green. As an elemental gas, it exists as diatomic molecules as Cl_2. Chlorine has an irritating, characteristic odor, which is detectable at 2 to 3 mg/L concentration, and is fatal at 1000 mg/L in aqueous solution. Chlorine is used in the chemical industry and in the disinfection of drinking water and effluents.

Chlorine is soluble in water according to the following equation:

$$Cl_2 + H_2O \rightarrow HOCl + H^+ + Cl^-$$

Hypochlorous acid (HOCl) is very unstable, and dissociates to

$$HOCl \rightarrow H^+ + OCl^-$$

Chlorine with its strong disinfectant power is applied to water to destroy disease-producing microorganisms. Chlorine applied to water in its elemental or hypochlorite form initially undergoes hydrolysis to form *free available chlorine* consisting of the chlorine molecule (Cl_2), hypochlorous acid (HOCl), and hypochlorite ion (OCl^-). The relative proportion of these free chlorine forms depends on pH and temperature. Free chlorine reacts with ammonia, and forms monochloramine (NH_2Cl), dichloramine ($NHCl_2$), and nitrogen trichloride (NCl_3), called *combined available chlorine*. The presence and concentration of these compounds are also pH- and temperature-dependent. Chlorinated wastewater effluents, as well as certain chlorinated industrial effluents, normally contain only combined chlorine.

Besides the beneficial action of chlorination, it may produce adverse effects. When chlorine residuals combine with phenols, an objectionable taste and odor for drinking water may be produced. As disinfection by-products, cancer-causing trihalomethanes

(THMs) are formed by the reaction of natural organic compounds with chlorine. These substances are regulated by the EPA with the current standard for total THMs of 0.1 mg/L for systems serving more than 10,000 people.

22.1.2 Sample Collection and Handling

Chlorine in aqueous solution is unstable, and the chlorine content of samples or solutions, particularly weak solutions, will decrease rapidly. Exposure to sunlight or other strong light or agitation will accelerate the reduction of chlorine. Therefore, start chlorine determinations immediately after sampling, avoiding excessive light and agitation. Do not store samples!

22.1.3 Measurement

Comparator with orthotolidine reagent disc — In the presence of chlorine, addition of orthotolidine reagent produces a yellow color. The deepness of the yellow color is proportional to the concentration of the chlorine. The concentration is read directly from the accompanying color disc in terms of milligrams per liter. The orthotolidine method has been discontinued because of the toxic nature of orthotolidine.

Comparator with DPD (N,N-diethyl-p-phenylene-diamine) reagent — Chlorine reacts with DPD by producing a red color. The deepness of the red color corresponds to the concentration of chlorine present expressed in milligrams per liter. For the correct operation of comparator KITs, follow the manufacturer's instructions.

22.1.4 Chlorine Determination by Iodometric Titration Method

Method No. 330.3 (EPA-600/4-79-020).
The minimum detectable concentration is 0.04 mg/L chlorine (Cl_2) if 0.01 N $Na_2S_2O_3$ and a 1000 mL sample is used.

Principle of the method — Correct operation and performance of the KIT measurement has to be checked periodically by a more accurate laboratory method. The iodometric method is a simple oxidation–reduction titrimetric method (see Section 11.2).

Chlorine is a strong oxidant agent the liberates free iodine from potassium iodide (KI) at pH 8 or less. The liberated iodine is titrated with a standard reducing agent, such as sodium thiosulfate ($Na_2S_2O_3$), using starch as an indicator.

$$Cl_2 + KI \rightarrow 2\ KCl + I_2$$

Apparatus

- 2 L Erlenmeyer flask or 2 L beaker for titration
- Microburette
- Stirrer and magnetic stirring bar
- Pipets

Reagents

- Acetic acid, concentrated (glacial).
- Potassium iodide (KI) crystal.
- Sodium thiosulfate, $Na_2S_2O_3$, 0.1 N: 25 g $Na_2S_2O_3.5H_2O$ dissolved and diluted to 1000 mL with DI water. Add a few milliliters of chloroform to minimize bacterial decomposition. After at least 2 weeks storage, standardize the solution with iodate or dichromate method as outlined in Section 6.3.
- Sodium thiosulfate, $Na_2SO_2O_3$, 0.025 N or 0.01 N: dilute 250 mL (for 0.025 N) or 100 mL (for 0.01 N) 0.1 N $Na_2S_2O_3$ to 1000 mL with DI water.
- Potassium biiodate, $KH(IO_3)_2$, 0.025 N or 0.01 N: dilute 250 mL (for 0.025 N) or 100 mL (for 0.01 N) of 0.1 N potassium biiodate (preparation is given in Section 6.3) to 1000 mL with DI water.
- Potassium dichromate, $K_2Cr_2O_7$, 0.025 N or 0.01 N: dilute 250 mL (for 0.025 N) or 100 mL (for 0.01 N) of 0.1 N potassium dichromate (preparation is given in Section 6.3) to 1000 mL with DI water.
- Starch indicator: add 5 g starch in a little water. Pour the suspension into 1 L boiling water. Cool and let to settle overnight. Use the supernate. Preserve with a few drops of toluene.

Standardization of sodium thiosulfate titrant

Standardization with 0.025 N or 0.01 N potassium biiodate, $KH(IO_3)_2$

1. Measure 80 mL DI water into Erlenmeyer flask.
2. Add 1 mL concentrated H_2SO_4 with constant stirring.
3. Add 10 mL 0.025 N (or 0.01 N) potassium biiodate solution.
4. Add 1 g potassium iodide, KI.
5. Titrate with 0.025 N (or 0.01 N) $Na_2S_2O_3$ until the liberated yellow color of iodine is almost discharged.
6. Add 1 mL starch indicator. Color becomes blue.
7. Titrate with 0.025 N (or 0.01 N) $Na_2S_2O_3$ until blue color disappears.

Standardization with 0.025 N or 0.01 N potassium dichromate, $K_2Cr_2O_7$

1. Add to 80 mL DI water 1 mL concentrated H_2SO_4 with constant stirring.
2. Add 10 mL 0.025 N (or 0.01 N) $K_2Cr_2O_3$.
3. Let the mixture stand for 6 minutes in the dark.
4. Titrate with 0.025 N (or 0.01 N) $Na_2S_2O_3$ until the liberated iodine color is almost discharged.
5. Add 1 mL starch indicator; solution becomes blue.
6. Titrate with 0.025 N (or 0.01 N) $Na_2S_2O_3$ until the blue color disappears.

Calculate the normality

$$N = (0.025 \times 10)/mL\ Na_2S_2O_3 \tag{22.1}$$

1 mL of 0.025 N $Na_2S_2O_3$ = 886.3 mg Cl as Cl_2.

$$N = (0.01 \times 10)/\text{mL Na}_2\text{S}_2\text{O}_3 \qquad\qquad (22.2)$$

1 mL of 0.01 N Na$_2$S$_2$O$_3$ = 354.5 mg Cl as Cl$_2$

Titration procedure

1. Measure 5 mL acetic acid glacial into a 2 L Erlenmeyer flask or beaker.
2. Add 1 g potassium iodide (KI) crystal.
3. Add 500 or 1000 mL sample (500 mL with a chlorine range of 1 to 10 ppm and 1000 mL with a chlorine range of 0 to 1 ppm).
4. Place magnetic stirrer.
5. Titrate with 0.01 or 0.025 N sodium-thiosulfate in the presence of starch indicator with the color change from blue to colorless.
6. Use the same volume of DI water as a blank.

Calculation

$$\text{mg/L Cl as Cl}_2 = [(A - B) \times N \times 35.43 \times 1000]/\text{mL sample} \qquad (22.3)$$

where A = thiosulfate used for titration of the sample (mL), B = thiosulfate used for titration of the blank (mL), N = normality of Na$_2$S$_2$O$_3$, and 35.43 = equivalent weight of chlorine.

Quality control — For reference of QC items, calculations, and the accepted criteria, see Section 3.2.7, Section 3.2.9, and Section 16.1.

Method blank — Method blank is performed once for each sample batch.

Duplicate sample — Duplicate samples are analyzed once in every sample batch and once in every 10 samples.

22.2 CHLORIDE

22.2.1 General Discussion

Chloride, in the form of the Cl$^-$ ion, is one of the major inorganic anions in water and wastewater, originated from natural minerals, seawater intrusion, and industrial pollution. In potable water, the salty taste produced by chloride concentrations is variable and depends on the chemical composition of water. Some waters containing 250 mg/L may have a detectable salty taste if the chloride is in the form of sodium chloride (NaCl), but chloride concentration of 1,000 mg/L may not cause a typical salty taste to waters if the salts are calcium and magnesium chlorides (CaCl$_2$ and MgCl$_2$). The recommended chloride content in drinking water is 250 mg/L. The primary drinking water standard does not include chloride in the regulated health-related parameters. Plants are more sensitive to high chloride, and high chloride may harm metallic pipes.

22.2.2 Sample Collection and Handling

Collect representative samples in clean glass or plastic bottles. No special preservation is necessary if the sample is stored. Maximum holding time is 28 days.

22.2.3 Analysis by Mercuric Nitrate Titrimetric Method

Method No. 325.3 (EPA 600/4-79-020). Method No. 9252 (EPA SW-846).

Principle of the method — An acidified sample is titrated with mercuric nitrate, $Hg(NO_3)_2$, in the presence of mixed diphenylcarbazone–bromphenol blue indicator. In the pH range 2.3 to 2.8, diphenylcarbazone indicates the titration end point by formation of a blue-violet complex with the excess mercuric ion.

Interference — Bromide and iodide are titrated with $Hg(NO_3)_2$ in the same manner as chloride. Chromate, ferric, and sulfite ions interfere when present in excess of 10 mg/L. End point detection is difficult for samples that are colored or turbid.

Apparatus

- Erlenmeyer flasks, 250 mL or beakers
- Buret, 25 and 50 mL
- Graduated cylinders, 100, 50 mL
- Volumetric flasks, 1000 mL
- Pipets
- Magnetic stirrer and magnetic stirring bars

Reagents

- Sodium chloride, NaCl, standard 0.0141 *N*: dissolve 0.824 g NaCl (previously dried at 140°C and cooled in desiccator) and dilute to 1000 mL with DI water.

$$1 \text{ mL} = 0.5 \text{ mg Cl}^-$$

- Sodium chloride, NaCl, standard, 0.141 *N*: dissolve 8.240 g NaCl (previously dried at 140°C and cooled in desiccator) and dilute to 1000 mL with DI water.

$$1 \text{ mL} = 5.0 \text{ mg Cl}$$

- Mercuric nitrate, $Hg(NO_3)_2$, titrant 0.0141 *N*: to 2.3 g Hg $(NO_3)_2$ or 2.5 g Hg $(NO_3)_2 \cdot H_2O$ dissolved in 100 mL distilled water, add 0.25 mL concentrated HNO_3 and dilute to 1 L with distilled water. *Store in amber bottle!* 1 mL is equivalent to 0.5 mg Cl⁻.
- Mercuric nitrate, $Hg(NO_3)_2$ titrant, 0.141 *N*: dissolve 25 g $Hg(NO_3)_2 \cdot H_2O$ in DI water, add 5 mL concentrated HNO_3, and dilute to 1000 mL. *Store in amber bottle!*
- Mixed chloride indicator: dissolve 0.5 g crystalline diphenylcarbazone and 0.05 g bromophenol blue powder in 75 mL of 95% ethyl or isopropyl alcohol and dilute to 100 mL with the same alcohol. *Store in amber bottle! Indicator is good for 6 months!*
- 0.1 *N* HNO_3 solution: 6.4 mL concentrated HNO_3 diluted to 1000 mL with DI water.
- 0.1 *N* NaOH solution: 4.0 g NaOH dissolved and diluted to 1000 mL with DI water.
- Reference or QC check standard stock solution, 11,000 mg/L: dissolve 23.1335 g anhydrous KCl and dilute to 1000 mL with DI water. *Solution may be used for at least 6 months!*

$$Cl^- = 11,000 \text{ mg/L}$$

- Reference or QC check standard solution, 110 mg/L: dilute 10 mL of stock reference standard to 1000 mL with DI water. *Solution may be used for at least 6 months!*

$$Cl^- = 110 \text{ mg/L}$$

- Spike stock solution: same as reference stock standard (11,000 mg/L Cl⁻).

Standardization of 0.0141 N Hg(NO₃)₂ with 0.0141 N NaCl standard

1. Measure 10 mL 0.041 N NaCl standard.
2. Add a few drops of mixed chloride indicator. Color is blue-violet.
3. Add dropwise 0.1 N HNO₃ until the blue-violet just turns to yellow.
4. Titrate with 0.0141 N Hg(NO₃)₂ until a blue-violet color persists through the solution.
5. Calculate the exact normality:

$$N = (0.0141 \times 10)/\text{mL Hg(NO}_3)_2 \text{ used}$$

Standardization of 0.141 N Hg(NO₃)₂ with 0.141 N NaCl standard

1. Measure 10 mL 0.141 N NaCl solution into an Erlenmeyer flask.
2. Add five drops of mixed chloride indicator. Color turns to blue-violet.
3. Add dropwise 0.1 N HNO₃ until color changes to yellow.
4. Titrate with 0.141 N Hg(NO₃)₂ solution until a blue-violet color persists throughout the solution.
5. Calculate exact normality by using the following formula:

$$N = (0.141 \times 10)/\text{mL Hg(NO}_3)_2 \text{ used} \qquad \text{(22.4)}$$

Titration procedure

1. Measure 100 mL sample (or suitable aliquot according to the chloride concentration).
2. Add 5 drops of mixed indicator and mix well.
3. If blue-violet color appears, add 0.1 N HNO₃ until the color changes to yellow.
4. If a yellow or orange color forms immediately on addition of mixed indicator, add 0.1 N NaOH solution dropwise until the color changes to blue-violet; then add 0.1 N HNO₃ solution dropwise until the color changes to yellow.
5. Titrate with 0.0141 N Hg(NO₃)₂ until a blue-violet color persists throughout the solution. If the consumed amounts of Hg(NO₃)₂ solution is over 30 mL, use a smaller sample size and retitrate the sample. In the event of a sample containing higher than 2000 mg/L chloride, titrate another aliquot using the 0.141 N Hg(NO₃)₂ titrant or a smaller sample size.

Calculation

$$\text{mg/L Cl}^- = [(A - B) \times N \times 35.450 \times 1000]/\text{mL sample} \qquad \text{(22.5)}$$

where A = titrant for sample (mL), B = titrant for blank (mL), and 35.450 = equivalent weight of Cl.

Quality control — For reference of QC items, calculations, and the accepted criteria, see Section 3.2.7, Section 3.2.9, and Section 6.1.

Method blank — Method blank is to be performed once for each sample batch or once in every 20 samples in a batch, whichever is most frequent.

Reference standard or QC check standard — Analyze immediately after the method blank, once every 10 samples, and once at the end of the sample batch.

Matrix spike/matrix spike duplicate (MS/MSD) — MS/MSDs are performed once for each sample batch or once every 20 samples, whichever is most frequent.

22.3 TOTAL CYANIDE

22.3.1 General Discussion

Cyanide refers to all of the cyanide groups in cyanide compounds that can be determined as the cyanide ion, CN^-. The cyanide ion is a conjugate base of a weak acid, hydrogen cyanide, which is an extremely poisonous gas with an almond odor. Cyanide salts are equally lethal. The cyanide ion is one of the most rapidly working poisons. Lethal doses taken orally act in minutes. Cyanide poisons by asphyxiation, as does carbon monoxide, but the mechanism is different. Instead of preventing the cells from getting oxygen, cyanide interferes with oxidative enzymes, such as cytochrome oxidase, vital to every cell in the use of oxygen. Oxidases are enzymes containing a metal, usually iron or copper. Fe^{3+} ions are essential in the operation of the enzyme system. Cyanide binds tightly to the enzyme cytochrome C and forms stable cyanide complexes with the Fe^{3+} ion and deactivates the enzyme system.

Simple cyanide compounds are inorganic cyanide salts, such as potassium and sodium cyanide, KCN and NaCN, or other metal cyanides that are highly water soluble.

Organically bound cyanides, such as acrylonitrile (cyanoethene) are important industrial chemicals. Cyanides are used in plastics, electroplating, and metallurgy as well as in synthetic fibers and chemicals. In industrial finished wastewaters, cyanides are oxidized with chlorine to cyanates in alkaline media. The generally accepted industrial waste treatment of cyanide compounds is an alkaline titration.

22.3.2 Sample Collection and Handling

Samples may be collected in glass or plastic containers, a minimum of 1 L of capacity. If residual chlorine is present, add 0.6 g ascorbic acid, $C_6H_8O_6$, to remove chlorine. Oxidizing agents, such as chlorine, decompose most cyanides. If sulfide is present, samples must be pretreated in the field or must be taken to the laboratory unpreserved at 4°C for analysis within 24 hours. Sulfide must be checked in the field by lead acetate paper. If sulfide is present, indicated by the developing black color of the lead acetate paper, add cadmium nitrate, $Cd(NO_3)_2$, until yellow precipitation of cadmium sulfide, CdS, appears. The sample should then be filtered and preserved with 2 mL of 10 N sodium hydroxide, NaOH, to pH >12. Transport and store samples at 4°C. Holding time for preserved samples is 14 days.

22.3.3 Determination of Cyanide by Colorimetric Method After Distillation

Method Nos. 9010, 9010 A (EPA SW-846) and Method No. 335.1 (EPA 600/4-79-020).

The colorimetric method is used for concentrations below 1 mg/L of cyanide and has a practical quantitation limit of 0.02 mg/L.

Method description — The cyanide as hydrocyanic acid (HCN) is released from cyanide complexes by means of a reflux-distillation and is absorbed in a scrubber containing sodium hydroxide solution. The cyanide ion in the absorbing solution is then determined colorimetrically. The method is used for total and/or amenable CN in water wastes and leachates. In the colorimetric measurement, the cyanide is converted to cyanogen chloride, (CNCl), by reaction of the cyanide with chloramine-T at a pH <8.0. *CAUTION:* CNCl is a toxic gas; avoid inhalation! CNCl forms a red-blue dye on addition of a pyridine-barbituric acid reagent. The absorbance is read at 578 nm wavelength. The applicable range of the method is 0.2 to 6.0 µg CN.

Interference — All known interferences are eliminated or reduced to a minimum by distillation.

Apparatus — Reflux distillation apparatus (Figure 22.1) includes:

- Boiling flask, 1 L, with inlet tube and provision for a water-cooled condenser.
- Gas absorber, with gas dispersion tube equipped with medium porosity fritted outlet.

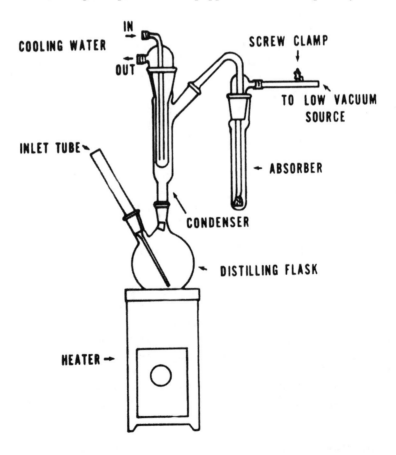

Figure 22.1
Cyanide distillation apparatus.

- Heating element, adjustable.
- Ground glass ST joints, TFE sleeved or with an appropriate lubricant for the boiling flask and condenser.
- Spectrophotometer suitable for measurement at 578 μm.
- Standard laboratory glassware.

Reagents

50% NaOH solution — Can be purchased or prepared in the laboratory as follows: 500 g NaOH add to a 2 L container containing about 700 mL of DI water. Place the flask on a magnetic stirrer and add a magnetic stirring bar. Prepare under laboratory hood! The solution must be stirred while the pellets are dissolving to help disperse the heat generated. When the solid is completely dissolved and the solution is cool, carefully dilute the solution to 1 L with DI water. Mix well. Store in a polyethylene bottle.
    CAUTION: Dissolution of NaOH in water generates extreme heat!

0.4 N NaOH solution — 10 g NaOH dissolved, cooled, and diluted to 1 L.

Magnesium chloride reagent — 510 g $MgCl_2 6H_2O$ dissolved and diluted to 1 L.

Sulfuric acid, H_2SO_4 1+1 — Remember the general laboratory safety rule when diluting sulfuric acid: *Always add the acid to the water!* The acid shall be added in small amounts over a period of time by constant stirring.
    CAUTION: Solution becomes very hot!

Stock CN solution, 1000 mg/L — Dissolve 2.0 g KOH and 2.51 g KCN and dilute to 1 L. Standardize against 0.0192 N $AgNO_3$ solution weekly, because the solution gradually loses strength.
    CAUTION: KCN is highly toxic! Avoid contact or inhalation!

$$1 \text{ mL} = 1 \text{ mg CN}$$

Check exact concentration of the stock CN solution:

- Measure 25 mL CN stock solution into Erlenmeyer flask.
- Add 0.5 mL *p*-dimethylaminobenzal rhodanine indicator.
- Titrate with 0.0192 N $AgNO_3$ until color changes from canary yellow to salmon blue.
- Calculate the concentration of the CN stock solution if

$$1 \text{ mL } AgNO_3 = 1 \text{ mg CN}$$

$$\text{conc. of CN solution, mg/L} = (\text{mL } AgNO_3 \times 1000)/\text{mL CN standard}$$

Intermediate standard CN solution, 10 mg/L — 10 mL stock diluted to 1 L with NaOH solution (2 g/L)

$$1 \text{ mL} = 10 \text{ μg CN } (0.010 \text{ mg})$$

CAUTION: Toxic, avoid ingestion!

Working standard CN solution, 1 mg/L — 10 mL intermediate standard solution diluted to 100 mL with NaOH solution (2 g/L). *Prepare fresh daily!*
    CAUTION: Toxic, avoid ingestion!

$$1 \text{ mL} = 1 \text{ μg CN } (0.001 \text{ mg})$$

AgNO₃, 0.0192 N — Dissolve 3.27 g AgNO₃ and dilute to 1 L.

P-dimethylaminobenzal-rhodanine indicator — Dissolve 20 mg indicator in 100 mL acetone.

Pyridine-barbituric acid reagent — Place 15 g barbituric acid in a 250 mL volumetric flask and add just enough water to wash side of flask and wet barbituric acid. Add 75 mL pyridine and mix. **NOTE:** Pyridine is to be used under a hood! Place the flask on a magnetic stirrer and add a magnetic stirring bar. Stir (under the hood) until the barbituric acid is dissolved. Add 15 mL concentrated HCl and mix. Cool to room temperature and dilute to the mark with water. Mix. Store in a polyethylene bottle at 4°C. *The reagent is stable for 1 month! Discard if precipitate develops.*

Phosphate buffer, sodium dihydrogen phosphate, NaH₂PO₄, 1 M — Dissolve 138 g sodium dihydrogen phosphate dihydrate, $NaH_2PO_4 \cdot H_2O$, and fill up to 1 L with DI water. Refrigerate!

Sodium hydroxide, NaOH, solution 0.05 N — Dissolve 2 g NaOH and dilute to 1 L with DI water.

Chloramine-T solution — Dissolve 1 g chloramine-T and dilute to 100 mL with DI water. Refrigerate until ready to use. Prepare fresh daily!

Reference (independent) standard, 0.01 mg/L — Prepare stock, intermediate, and working standard CN solutions as described previously, but use potassium cyanide (KCN) from different source than that used for the calibration standards. The reference standard is diluted from the working standard (1 mL = 1.0 μg CN) by adding 5 mL working standard solution (5.0 μg CN) to 500 mL DI water. *Prepare working standard and reference (independent) standard solutions fresh daily!*

Sample pretreatment by distillation

1. Add the sample, blank, or standard to the 1 L boiling flask using the following procedures separated by matrix:
   - Aqueous samples: Shake the sample and add 500 mL (diluted if necessary) to the boiling flask.
   - Soil samples: Weigh 15 g sample (use top-loader balance), transfer into the boiling flask and add 500 mL DI water.
   - Waste (nonoily) samples: Weigh 1.0 g sample (use top-loader balance) and transfer into the boiling flask and add 500 mL DI water.
   - Waste (oily) sample: Weigh 1 to 10 g sample (use top-loader balance) and transfer into the boiling flask, and add 500 mL DI water. Add 10 mL of 50% NaOH. Without connecting the condenser, boil the contents of the flask for a minimum of 1 hour. This process should remove all oily components that will interfere with color development.
2. Add 50 mL 1 N NaOH to the gas scrubber.
3. Connect the train (boiling flask air inlet, flask, condenser, gas washer, suction flask trap, and aspirator).
4. Start a new stream of air entering the boiling flask by adjusting the vacuum source. Adjust the vacuum so that approximately two bubbles of air per second enter the boiling flask through the air inlet tube. Maintain air flow throughout the reaction.

5. Slowly add 50 mL 18 N (1+1) sulfuric acid (H_2SO_4) through the air inlet tube. Rinse the tube with distilled water and allow the air flow to mix with the flask contents for 3 minutes.

6. Pour 20 mL of magnesium chloride solution into the air inlet and wash down with distilled water. A precipitate that may form redissolves on heating.

7. Heat the solution to boiling and reflux for at least 1 hour.

8. Turn off heat and continue the airflow for at least 15 minutes.

9. After cooling the boiling flask, disconnect absorber and close off the vacuum source.

10. Drain the solution from the absorber into a 250 mL volumetric flask. Rinse connecting tube between condenser and gas washer with DI water. Add this rinsate also to the 250 mL volumetric flask and dilute to the mark with distilled water. Mix well.

Preparation of calibration curve — Prepare calibration standards from the working standard solution containing 1 µg CN in 1 mL. The calibration standards are not distilled.

1. Blank: 20 mL 0.05 N NaOH in 50 mL volumetric flask.
2. 0.2 µg CN/20 mL: 0.2 mL standard CN + 19.8 mL 0.05 N NaOH in 50 mL volumetric flask.
3. 0.5 µg CN/20 mL: 0.5 mL standard CN + 19.5 mL 0.05 N NaOH in 50 mL volumetric flask.
4. 1.0 µg CN/20 mL: 1.0 mL standard CN + 19.0 mL 0.05 N NaOH, in 50 mL volumetric flask.
5. 2.0 µg CN/20 mL: 2.0 mL standard CN + 18.0 mL 0.05 N NaOH, in 50 mL volumetric flask.
6. 4.0 µg CN/20 mL: 4.0 mL standard CN + 16.0 mL 0.05 N NaOH, in 50 mL volumetric flask.
7. 5.0 µg CN/20 mL: 5.0 mL standard CN + 15.0 mL 0.05 N NaOH, in 50 mL volumetric flask.
8. 6.0 µg CN/20 mL: 6.0 mL standard CN + 14.0 mL 0.05 N NaOH, in 50 mL volumetric flask.
9. Continuing calibration standard (CCS): 20 mL in 50 mL volumetric flask.
10. Calibration verification standard (CVS): 20 mL in 50 mL volumetric flask.

The continuing calibration standard (CCS) is a mid-level calibration standard. The calibration verification standard (CVS) is the reference (independent) standard. The purpose of this standard to verify the curve without any factors from the distillation being considered. *An initial calibration curve must be performed at least once every 3 months!*

Color development

1. To each standard add 4 mL phosphate buffer and mix thoroughly.
2. Add 2 mL chloramine-T and swirl to mix.
3. Immediately add 5 mL pyridine–barbituric acid reagent and swirl gently to mix.
4. Dilute to the 50 mL mark with DI water and mix well by inversion.
5. Transfer solution into a 1 cm cuvette. Wipe the cuvette clean with a kim-wipe.
6. Measure absorbance at 578 nm wavelength after 8 minutes but within 15 minutes from the time of adding the pyridine–barbituric acid reagent. Standardize time for each reading.
7. Use the blank to zero the instrument
8. Plot absorbance of standards against CN concentration and check the correctness of the calibration as discussed in Section 3.2.5, Section 12.3.1, and Section 13.3.2.

Measurement of samples

1. Pour the contents of the scrubber into a 250 mL volumetric flask. Rinse the scrubber with DI water and combine the rinsings in the flask. Bring to volume with DI water.

2. Pipet 20 mL or aliquot of the absorbing solution (distillate) diluted to 20 mL into a 50 mL volumetric flask.

3. Develop the color and measure the absorbance by following steps 1 to 7 in the "Color development" of the calibration standards above.

4. The results are read directly from the calibration curve by using the linear regression calculation as discussed in Section 12.3.1 and Section 13.3.2. Enter the absorbance and read the result in total micrograms.

5. If the absorbance of the sample is greater than that of the largest calibration standard, a smaller aliquot of the scrubber solution must be used.

Calculation

$$mg/L\ CN = (A \times B)/(C \times D) \tag{22.6}$$

where A = micrograms of CN read from calibration curve (50 mL final volume), B = total volume of absorbing solution from the distillation (mL), C = total volume of original sample used for distillation (mL), and D = volume of absorbing solution used in colorimetric test (mL).

Quality control — For reference of QC items, calculations, and the accepted criteria, please see Section 3.2.7, Section 3.2.9, and Section 16.1.

Calibration blank or reagent blank — A reagent blank is analyzed as part of the calibration curve. A reagent blank is 250 mL DI water which is not distilled. A reagent blank must be prepared for each sample set, since this is the blank against which the spectrophotometer is calibrated.

Method blank or preparation blank — A method blank is distilled and analyzed once with each batch or once with every 20 samples, whichever is more frequent. The purpose of the method blank is to demonstrate that there are no positive interferences from the glassware, distillation process, or analytical process. A method blank consists of 500 mL DI water carried through the entire distillation and analytical process.

Continuing calibration standard (CCS) — Analyzed immediately after the calibration to check calibration, and analyzed once every 10 samples.

Calibration verification standard (CVS) or reference (independent) standard (undistilled) — A low-level and a high-level undistilled reference standard is analyzed with every sample batch. The purpose of this standard is to verify the curve without any factors from the distillation being considered. The *low-level standard* is prepared by adding 5 mL of reference (independent) working standard (1 mL = 1 µg CN) to 500 mL DI water, yielding to 5 µg. For the *high level standard,* add 50 mL of reference (independent) working standard (1 mL = 1 µg) to 500 mL DI water, yielding to 50 µg. The recoveries of these standards shall be within the limits of 90 to 110%.

Laboratory control sample (LCS) or reference (independent) standard (distilled) — A high- and a low-level reference standard is distilled and analyzed with every sample batch and every 20 samples, whichever is more frequent. These standards are used to verify the calibration curve and the accuracy of the analyst. The preparation and the concentration of these standards are the same as for the undistilled reference (independent) standards. The accepted limits for the percentage of recovery is determined by individual laboratories.

Matrix spike/matrix spike duplicates (MS/MSDs) — MS/MSDs are performed once for every sample batch or every 20 samples, whichever is more frequent. MS/MSDs are used to demonstrate that the analytes of interest can be recovered and the analysis can be performed with acceptable precision and accuracy. To prepare an MS/MSD, one of the samples from the sample batch is prepared in triplicate, with the second and third aliquots being the MS and the MSD. The MS/MSD consists of 500 mL of sample with 25 mL of reference (independent) working standard, equivalent to 25 µg CN. The MS/MSD is then distilled and analyzed as a routine sample. The percentage of recovery and the RPD limits are different for each laboratory.

22.4 CYANIDES AMENABLE TO CHLORINATION AFTER DISTILLATION

22.4.1 General Discussion

After part of the sample is chlorinated to decompose the cyanides, both the chlorinated and the untreated sample are distilled and analyzed for cyanide content. The difference between the cyanide concentrations in the two samples is expressed as the cyanide amenable to chlorination.

22.4.2 Analysis

Apparatus — See total cyanide determination, Section 22.3.3.

Reagents — See total cyanide determination, Section 22.3.3. In addition, use calcium hypochlorite solution: dissolve 5 g calcium hypochlorite, $Ca(OCl)_2$, in 100 mL DI water. *Store in an amber-colored glass bottle in the dark. Prepare monthly!*

Procedure

1. Divide sample into two equal parts.
2. To one part add dropwise calcium hypochlorite solution while agitating and maintaining pH between 11 and 12 by adding NaOH solution. Test for chlorine by a strip of KI-starch paper. A distinct blue color indicates sufficient chlorine.
3. Add 0.5 g crystalline sodium thiosulfate, $Na_2S_2O_3 \cdot 5H_2O$, to reduce residual chlorine. Check again with KI-starch paper the presence of residual chlorine. There should be no color change. Add 0.1 g thiosulfate in excess.
4. Distill both chlorinated and untreated sample portion and analyze as described under total cyanide determination, Section 22.3.3.

Calculation

$$\text{mg/L CN amenable to chlorination} = A - B \qquad (22.7)$$

where A = CN in the unchlorinated sample portion (mg/L) and B = CN in the chlorinated sample portion (mg/L).

Chapter 23

NITROGEN COMPOUNDS AND ANALYSIS OF AMMONIA NITROGEN

23.1 NITROGEN COMPOUNDS AND NITROGEN CYCLE

Nitrogen compounds are members of the inorganic *nutrient* group (nitrogen- and phosphorus-containing compounds), and the wide variety of forms of nitrogen in environmental samples have great interest. Major pollution sources of nutrients are surface and subsurface agricultural and urban drainage, animal waste runoff, as well as domestic and industrial waste effluents.

Nitrogen gas makes up 78% (v/v) of dry air. Nitrogen forms part of many organic molecules, usually amino acids (the building block of proteins) and the genetic materials RNA (ribonucleic acid) and DNA (deoxyribonucleic acid). However, plants and animals cannot use nitrogen in the form of atmospheric nitrogen (N_2). To be usable, it must first be converted to ammonia (NH_3) and nitrate (NO_3). The conversion of atmospheric nitrogen to usable forms of nitrogen is called *nitrogen fixation*, which is associated with bacterial activity. One nitrogen-fixing bacteria, called *Rhizobium*, invades the roots of leguminous species (peas, beans, alfalfa, clover, and others). The roots form tiny nodules where the nitrogen fixation takes place. Bacteria species, as *Azotobacter*, are ready to fix atmospheric nitrogen directly in the soil. The fixed nitrogen is taken by plants and then synthesized to amino acids, proteins, DNA, and RNA. Animals, in turn, receive the needed nitrogen by eating plants and other animals.

Nitrogen-rich wastes from plants and animals will be decomposed by certain types of bacteria and converted to ammonia (NH_3). This process is called *ammonification*.

Ammonia oxidizes to nitrite (NO_2) by oxidizing bacteria species *Nitrosomonas*, and the nitrite oxidizes to nitrate (NO_3) by other oxidizing bacteria called *Nitrobacter*. The process is called *nitrification*. Nitrates decompose to nitrites, convert into a gas, nitrous oxide (N_2O), by bacterial activity (*Pseudomonas* and others), and are released into the atmosphere. The process called *denitrification*.

In various forms, nitrogen travels from air to plant to animal and than back to soil and atmosphere in a never-ending cycle, called the *nitrogen cycle* (Figure 23.1).

259

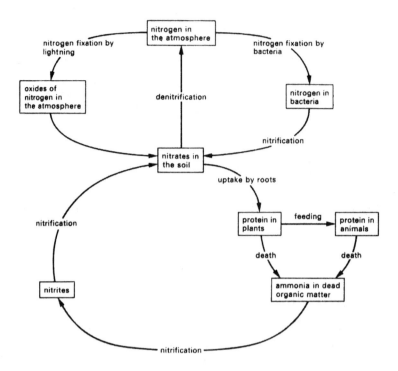

Figure 23.1
The nitrogen cycle.

23.2 AMMONIA-NITROGEN (NH$_3$-N)

23.2.1 General Discussion

Ammonia is mostly produced by decomposition of organic nitrogen-containing compounds and by the hydrolysis of urea (NH$_2$-CO-NH$_2$). During chlorination, ammonia combines with chlorine to form chloramine, as combined residual chlorine.

Ammonia is naturally present in surface and ground waters. Ammonia concentration is usually low in groundwater, because it adheres to soil particles and does not leach easily.

23.2.2 Sample Collection and Storage

Collect samples in plastic or glass container. Add sodium thiosulfate, Na$_2$S$_2$O$_3$, to destroy residual chlorine immediately after collection to prevent its reaction with ammonia.

Most reliable results are obtained on fresh samples; if prompt analysis is impossible, preserve samples with 0.8 mL concentrated H$_2$SO$_4$ to pH <2 and store at 4°C. Holding time is 28 days. If acid preservation is used, neutralize samples with NaOH immediately before making the determination.

23.2.3 Ammonia-Nitrogen Determination by Nesslerization

Method No. 4500-NH$_3$C (*Standard Methods*, 17th ed., 1989) and Method No. 350.2 (EPA-600/4-79-020).

Principle of the method — Nesslerization can be used directly, or after distillation of the sample. Nessler reagent gives a yellow to brown color with ammonia, depending on the concentration. The Nessler method is sensitive to 20 µg/L NH_3-N under optimum conditions and may be used for up to 5 mg/L NH_3-N. The yellow color characteristic to low ammonia nitrogen concentration (0.4 to 5 mg/L) can be measured with acceptable sensitivity within the wavelength region of 400 to 425 nm when a 1 cm light path is available. A light path of 5 cm extends measurements into the nitrogen concentration range of 5 to 60 µg/L.

Nessler reagent may respond in optimum conditions to 1 µg NH_3-N per 50 mL. This represents 20 µg/L NH_3-N. However, the reproducibility below 100 µg/L may be erratic.

Interferences — Interferences are eliminated by preliminary distillation. Addition of EDTA reagent or Rochelle salt solution inhibits precipitation of residual calcium and magnesium ions in the presence of the alkaline Nessler reagent.

Apparatus

- Distillation apparatus
- pH meter
- Spectrophotometer, for use at 400 to 425 nm and providing a light path of 1 cm or longer
- Standard laboratory glassware

Reagents

Sodium tetraborate solution, 0.025 M — Dissolve 9.5 g sodium tetraborate decahydrate ($Na_2B_4O_7 \cdot 10H_2O$) in DI water, dilute to 1 L, and mix.

Borate buffer solution — Add 88 mL 0.1 N NaOH solution to 500 mL approximately 0.025 M sodium tetraborate ($Na_2B_4O_7$) solution, dilute to 1 L, and mix.

Sodium hydroxide solution, 1 N — Dissolve 40 g NaOH in DI water under laboratory hood with constant stirring. After solution has cooled to room temperature, fill up to 1 L with DI water and mix. ***CAUTION:*** Reaction to NaOH with water releases heat!

Sulfuric acid solution, 1 N — Add 28 mL concentrated H_2SO_4 to about 500 mL DI water. After solution has cooled to room temperature, dilute to 1 L.
   REMEMBER: Always add acid to water, never add water to acid! Mixing sulfuric acid with water generates heat!

Sulfuric acid, 0.04 N — Dilute 1 mL concentrated H_2SO_4 to 1 L with DI water.

Absorbent solution (plain boric acid) — Dissolve 20 g boric acid (H_3BO_3) in DI water, dilute to 1 L, and mix.

Rochelle salt solution — Dissolve 50 g potassium sodium tartrate tetrahydrate ($KNaC_4H_4O_6 \cdot 4H_2O$) in 100 mL DI water. Remove ammonia by boiling off 30 mL of solution. After cooling, dilute back to 100 mL.

EDTA reagent — Dissolve 50 g disodium ethylenediaminotetraacetate dihydrate (EDTA) in 60 mL DI water containing 10 g NaOH. If necessary, apply gentle heat to complete dissolution. Cool to room temperature and dilute to 100 mL.

Nessler reagent —

1. Dissolve 100 g mercuric iodide (HgI_2) and 70 g potassium iodide (KI) in a small quantity of water.
2. Dissolve 160 g NaOH in 500 mL water with stirring. After dissolution, cool to room temperature.
3. Add slowly with stirring solution 1 to solution 2 and dilute to 1 L.

Store in rubber-stoppered borosilicate glass bottle and protect from sunlight. The reagent is stable up to 1 year under normal laboratory conditions.
  CAUTION: Toxic! Take care to avoid ingestion!

Stock ammonia solution, 1000 mg/L NH_3-N —

1. Dry ammonium chloride (NH_4Cl) in oven at 100°C.
2. From this dried anhydrous ammonium chloride (NH_4Cl), measure 3.819 g, dissolve, and dilute to 1 L with analyte-free water. Transfer the solution in an amber bottle and preserve it with several drops of chloroform. *Store in a refrigerator!*

$$1 \text{ mL} = 1.00 \text{ mg } NH_3\text{-N} = 1.22 \text{ mg } NH_3$$

Standard ammonia solution, 10 mg/L NH_3-N — Dilute 10 mL stock ammonia solution to 1 L with analyte-free water. Preserve the solution with several drops of chloroform. *Store in an amber bottle in refrigerator.*

$$1 \text{ mL} = 10 \text{ μg } NH_3\text{-N} = 12.2 \text{ μg } NH_3$$

Reference (independent) stock, 1000 mg/L NH_3-N — Preparation is the same as for ammonia stock solution, but the NH_4Cl should be originated from another source. Preserve with several drops of chloroform. *Store in amber bottle in refrigerator.*

Reference (independent) intermediate standard, 10 mg/L NH_3-N — Dilute 10 mL of reference (independent) stock solution to 1 L with analyte-free water. Preserve with several drops of chloroform. *Store in an amber bottle in refrigerator!*

Reference (independent) standard, 0.5 mg/L NH_3-N — Pipet 5 mL of reference (independent) intermediate standard into a 100 mL volumetric flask and dilute to 100 mL with analyte-free water. *Prepare fresh daily.*

Preliminary distillation — The sample is buffered at pH 9.5 with a borate buffer and distilled into boric acid solution if the nesslerization method is used, and into 0.04 N H_2SO_4 solution if the ion selective method is used to determine ammonia in the distillate.

Preparation of equipment

1. Add 500 mL ammonia-free water and 20 mL borate buffer to a distillation flask.
2. Adjust pH to 9.5 with 6 N NaOH solution and add a few glass beads or boiling chips.

3. Steam out the distillation apparatus until distillate shows no traces of ammonia (check the color by adding Nessler reagent).

Distillation

1. Measure 500 mL sample or a portion diluted to 500 mL if the ammonia nitrogen concentration is possibly high, or measure 1000 mL sample if the ammonia nitrogen concentration is less than 0.1 mg/L.
2. Measure the pH of the sample and if necessary neutralize to approximately pH 7 with acid or base, using a pH meter.
3. Add 25 mL borate buffer solution and adjust pH to 9.5 with $6\,N$ NaOH using a pH meter.
4. Disconnect steaming-out flask and immediately connect the flask with sample to the distillation apparatus.
5. Measure 50 mL boric acid solution (if nesslerization method follows distillation) or 50 mL $0.04\,N$ H_2SO_4 solution (when ion-selective electrode method used) into a 500 mL Erlenmeyer flask. This is the receiving flask for the distillate.
6. Distill at a rate of 6 to 10 mL/min with a tip of the delivery tube submerged to at least 2 cm below the surface of the receiving solution. Collect at least 200 mL distillate.
7. Lower collected distillate free of contact with delivery tube and continue distillation for 1 or 2 minutes to clean condenser and delivery tube.
8. Dilute the distillate to 500 mL with ammonia-free water.

Preparation of calibration curve

1. Prepare calibration curve in 50 mL final volume with a range of 0.1 to 1.0 mg/L or 10 to 50 µg/50 mL NH_3-N.
   - 50 mL DI water   Blank
   - 1 mL standard ammonia solution diluted to 50 mL 0.2 mg/L
   - 2 mL standard ammonia solution diluted to 50 mL 0.4 mg/L
   - 3 mL standard ammonia solution diluted to 50 mL 0.6 mg/L
   - 4 mL standard ammonia solution diluted to 50 mL 0.8 mg/L
   - 5 mL standard ammonia solution diluted to 50 mL 1.0 mg/L
2. Transfer blank and standards, following with 50 mL continuing calibration standard (CCS) (middle-range standard) and 50 mL calibration verification standard (CVS) (QC-check standard) into 250 mL Erlenmeyer flasks marked accordingly.
3. Add to each flask 2 drops of Rochelle salt reagent. Mix well.
4. Add 1 mL Nessler reagent and mix well.
5. Let reaction proceed for 10 minutes after the addition of Nessler reagent.
6. Read the absorbance at 425 nm using reagent blank for zeroing the instrument.
7. Prepare calibration curve, and check its acceptance by calculating the correlation coefficient and the percentage recovery of the CCS and CVS. For reference, please check Section 3.2.5 and Section 12.3.

Sample measurement

1. In the case of the acceptance of the calibration, analyze samples by measuring 50 mL sample or smaller sample size diluted to 50 mL final volume.
2. Use the same procedure for color development as prescribed above for calibration standards, steps 2 to 6, with the following exception: if using undistilled samples, add 1 mL

Nessler reagent, and for distilled samples, add 2 mL Nessler reagent for color development.

3. Calculate the concentration of NH_3-N and report in mg/L. The ammonia concentration measured is calculated by linear regression calculation (Section 12.2.3).

Calculation

Direct nesslerization, without distillation

$$\text{mg/L } NH_3\text{-N} = (\mu g \ NH_3\text{-N measured} \times 1000)/\text{sample mL} \qquad (23.1)$$

Nesslerization after distillation

$$\text{mg/L } NH_3\text{-N} = [A/(\text{mL sample})] \times [B/C] \qquad (23.2)$$

where A = NH_3-N measured (μg), B = total volume distillate collected (mL), including absorbent acid solution, and C = volume of distillate taken for nesslerization. If the result has to be reported as ammonia as NH_3, convert ammonia nitrogen as NH_3-N by using the factor of 1.22. For example, if ammonia nitrogen NH_3-N is 0.6 ppm, the corresponding ammonia, NH_3, is 0.73 ppm.

Quality control — For reference of QC items, calculations, and the accepted criteria, see Section 3.2.7, Section 3.2.9, and Section 16.1.

Initial calibration check — The continuing calibration standard (CCS) (middle-range calibration standard) is analyzed immediately after the calibration curve to check calibration. The recovery must be within ±10%.

Continuing calibration check — Continuing calibration standard (CCS) is analyzed once every 10 samples. The purpose of the CCV is to verify that the calibration curve is constant through the run.

Reference (independent) standard — This is analyzed once every 10 samples. Since the reference standard is prepared from a source different from the CCS and the calibration, the reference check is used to verify that the CCS and calibration standards are actually at the concentrations claimed by the analyst.

Matrix spike/matrix spike duplicate (MS/MSD) — Performed once for every sample batch or once every 20 samples. For MS/MSDs, a sample is analyzed in triplicate with 50 μL of the 1000 mg/L stock ammonia solution added to two of the aliquots in 100 mL volumetric flasks. Enough sample is to be added so that the total volume is 100 mL. Spike concentration is 0.5 mg/L NH_3-N. The MS/MSDs are treated as samples and taken through the entire analytical process.

23.2.4 Ammonia Determination by Ion-Selective Electrode

Method No. 350.3 (EPA-600/4-79-020).

Principle of the method — Ammonia is determined potentiometrically using an ion-selective ammonia electrode and a pH meter having an expanded MV scale or a specific ion meter. The ammonia electrode uses a hydrophobic gas-permeable membrane to separate the sample solution from an ammonium chloride internal solution. Ammonia in the sample diffuses through the membrane and changes the internal solution pH that is sensed by a pH electrode. The constant level of chloride in the internal solution is sensed by a chloride-selective ion electrode which acts as a reference electrode. Ion-selective electrodes are discussed in Section 13.2.3.

This method is applicable to measurement of ammonia concentration of potable and surface waters, and domestic and industrial wastes. Color and turbidity have no effect on the measurement. Standards and samples should be the same temperature. The measurement is affected by high concentration of dissolved ions. Therefore, when analyzing high salt content samples, standards must be made up with synthetic ocean water (SOW).

Interferences — Amines give positive interference. Mercury and silver interfere by complexing with ammonia.

Apparatus

- pH meter with expanded MV scale or specific ion meter
- Ammonia-selective electrode, ORION 95-10
- Magnetic stirrer (thermally insulated) and Teflon coated
- Stirring bars

Reagents

Sodium hydroxide, NaOH 10 N — Add 400 g NaOH to 800 mL DI water and a stirring bar. Place the bottle on a magnetic stirrer and with constant stirring dissolve NaOH. After solution cools, dilute to 1 L. Store in polyethylene bottle. *CAUTION:* This quantity of NaOH reaction with water releases extreme heat!

Ammonia stock solution — Same as in Section 23.2.3.

Ammonia standard solution — Same as in Section 23.2.3.

Synthetic ocean water (SOW)

Sodium chloride, NaCL	24.50 g/L
Magnesium chloride, $MgCL_2$	5.20 g/L
Sodium sulfate, Na_2SO_4	4.00 g/L
Calcium chloride, $CaCL_2$	1.16 g/L
Potassium chloride, KCl	0.70 g/L
Sodium bicarbonate, $NaHCO_3$	0.20 g/L
Potassium bromide, KBr	0.10 g/L
Boric acid, H_3BO_3	0.03 g/L
Strontium chloride, $SrCL_2$	0.03 g/L
Sodium fluoride, NaF	0.003 g/L

Reference (independent) stock solution, 1000 mg/L NH_3-N — Same as in Section 23.2.3.

Reference (independent) intermediate standard, 10 mg/L NH₃-N — Same as in Section 23.2.3.

Reference (independent) standard, 0.5 mg/L NH₃-N — Same as in Section 23.2.3.

Measurement by ion-analyzer — A variety of instruments are available for the test. Read carefully and follow the instructions of the manufacturer before starting analysis. The following procedure is suitable for the *ORION 901 ion analyzer.*

Instrument setup — Ammonia electrode plugs into Sense "A" connection, reference plugs into "Ref A." (There is no separate reference electrode: the ammonia electrode is a combination electrode, the reference electrode is built in.) Place the electrode into 1000 mg/L NH₃-N stock solution for at least 30 minutes or until the reading stabilizes.

Slope determination

1. Put instrument into MV mode.
2. Measure 100 mL distilled water into a 150 mL beaker, and put a stirring bar in it.
3. Immerse electrode (previously rinse with distilled water).
4. Mix solution with a magnetic stirrer. Do not stir so rapidly that air bubbles are sucked into the solution, because they will become trapped on the electrode membrane. Maintain the same stirring rate and a temperature of about 25°C throughout calibration and testing.
5. Add 2 mL of 10 *N* NaOH, to raise pH to above 11.
6. Pipet 1 mL of 1000 mg/L NH₃-N standard into the beaker by volumetric pipet to maintain accurate measurement! Stir and wait until MV reading is stable. Record MV reading.
7. Pipet 10 mL of 1000 mg/L NH₃-N standard into the same solution with the same volumetric accuracy as in step 6. Stir and wait until MV reading is stabilized. Record mV reading.
8. The difference between the two MV readings is the value of the slope. Acceptance slope should be between –54 to –60.
9. Adjust the slope switch to the determined slope value. Set the sign to minus (–)!

Standardization

1. Prepare standards, 100 mL, that bracket expected sample concentration.
   - 1.0 mg/L NH₃-N: pipet 10 mL standard ammonia solution (10 mg/L) into 100 mL volumetric flask and dilute to the mark with DI water.
   - 10 mg/L NH₃-N: pipet 0.1 mL standard ammonia solution (1000 mg/L) into 100 mL volumetric flask and dilute to the mark or use directly the 10 mg/L standard ammonia solution.
   - 0.1 mg/L NH₃-N: pipet 1.0 mL standard ammonia solution (10 mg/L) into 100 mL volumetric flask and dilute to 100 mL with DI water.
   - 0.5 mg/L NH₃-N: pipet 5.0 mL standard ammonia solution (10 mg/L) into 100 mL volumetric flask and dilute to the mark.
   - The 0.1 and 0.5 mg/L standards are used for checking the accuracy of low-concentration measurements.
2. Set instrument into **CONC** mode.
3. Pour the first standard, 1.0 mg/L, into a clean beaker, add the stirring bar, immerse electrode, and place on stirrer.
4. Add 2 mL of 10 *N* NaOH. Do not add NaOH until the electrode has been immersed, and measure, because ammonia may be lost!

5. Adjust concentration on the panel to the desired concentration (1.0) and press **CLEAR/READY MV** and wait for a stable reading.

6. Press **SET CONC,** reading should be 1.0.

7. Set the second standard 10 mg/L by pouring into the beaker, add stirring bar, immerse electrode.

8. Add 2 mL 10 N NaOH, stir, and read. If the reading is not exactly 10.0, adjust the slope switches until the reading of 10 appears. Record the changed slope reading.

9. Measure 100 mL of reference (independent) standard into a beaker, immerse electrode, stir.

10. Add 2 mL 10 N NaOH and read. If your value is not giving an accurate reading, or far from the true value, stop, find the problem, and correct it.

Sample measurement

1. If the reading of the reference standard value is acceptable, follow the procedure by measuring samples the same way as the standards were measured.

2. If any sample shows a low reading, check the performance of the instrument in this range, by measuring 0.1 and 0.5 mg/L standards. If needed, standardize the instrument again by using 0.1 to 1.0 mg/L range.

3. If the sample reading is higher than the calibrated range, dilute the sample and calculate the final value by multiplying the reading with the dilution factor.

Measurement by pH meter — An initial calibration curve is prepared and analyzed daily. Prepare calibration standards by dilution from ammonia standard solution in 100 mL final volume. The concentration range of the standards is variable, depending on the analyzed samples.

Preparation of calibration curve

1. Prepare standards in concentrations of 0.1, 0.2, 0.4, 0.6, 0.8, 1.0, 1.2, 1.6, and 2.0 mg/L, by adding 1, 2, 4, 6, 8, 10, 12, 15, and 20 mL ammonia standard solution (1 mL = 10 µg NH_3-N) into 100 mL volumetric flask and diluting to the volume with analyte-free water. A blank is 100 mL analyte-free water.

2. Place 100 mL standard solution in a 150 mL beaker. Immerse electrode in standard of lowest concentration and mix with a magnetic stirrer. Maintain a same stirring rate and a temperature of about 25°C throughout calibration and testing.

3. Add 2 mL 10 N NaOH solution and keep electrode in the solution until a stable millivolt (mV) reading is obtained. Record millivolt reading. *Do not add NaOH solution before immersing electrode, because ammonia may be lost from a basic solution!*

4. Repeat procedure with the remaining standards, proceeding from the lowest to the highest concentration.

5. Using semilogarithmic graph paper, plot ammonia concentration in NH_3-N mg/L on the log axis vs. millivolts on the linear axis starting with the lowest concentration at the bottom of the scale.

6. Using a calculator with logarithmic regression capabilities, enter the millivolts as the independent (x) variable and the log or natural log (ln) of the concentration as the dependent (y) variable. Perform the linear regression and obtain a value for the regression factor (correlation coefficient). This value must be >0.9950. See Section 12.3.2.

Slope check — Immediately after the calibration curve is generated, a slope check must be performed at two different points on the curve to verify the electrode sensitivity. The ideal difference in millivolts for a tenfold difference in concentration is −59. The slope is checked between 0.1 and 1.0 mg/L and again 0.2 and 2.00 mg/L. Calculate the slope according to the formula

$$S = mV_{0.1} - mV_{1.0} \text{ and } S = mV_{0.2} - mV_{2.0}$$

The difference in millivolt readings must be between 50 and 60.

Measurement of samples — Measure samples in the same way as standards (steps 2 and 3).

Calculation — By using ion analyzer instrument, the reading directly gives milligram per liter values. For diluted samples, use dilution factor for correct result. By using pH meters, calibration curves should be prepared, and the logarithmic regression calculation should be used for checking the curve and calculating the results (see Section 12.3.2).

Quality control — For reference of QC items, calculations, and the accepted criteria see Section 3.2.7, Section 3.2.9., and Section 16.1.

Method blank — A method blank is analyzed once with every sample batch or once every 20 samples, whichever is more frequent. If distilled samples are being analyzed, the blank must also be distilled. Undistilled samples must be accompanied by an undistilled bank.

Matrix spike/matrix spike duplicates (MS/MSD) — A MS/MSD is performed once for each sample batch or once every 20 samples, whichever is more frequent. To prepare an MS/MSD one of the samples in the batch is prepared in triplicate, with the second and third aliquots being the MS and the MSD. The MS/MSD consists of 0.1 mL (100 μl) of the 1000 mg/L stock ammonia solution pipetted into a 100 mL sample with a micropipet. The added spike value is 1.0 mg/L NH_3-N. The MS/MSDs are analyzed as routine samples.

Reference (independent) standard — Preparation is listed in Section 22.2.3. It is analyzed once every sample batch and once in every 10 samples.

23.2.5 Ammonia Determination by Titrimetric Method

For high ammonia concentrations, a distillation and titration technique is preferred. The titrimetric method is used only on samples that have been carried through preliminary distillation (Section 22.2.3).

Apparatus

- Distillation apparatus (Section 22.2.3)
- Standard titrimetric glassware

Reagents

- Sulfuric acid solution, 0.02 *N*: See Section 20.1.3. 1 mL = 280 μg N.
- Mixed indicator: Dissolve 0.2000 g methyl red indicator in 100 mL 95% ethyl or isopropyl alcohol. Dissolve 0.1000 g methylene blue in 50 mL ethyl or isopropyl alcohol. Combine the two solutions. *Prepare fresh monthly!*

- Indicating boric acid solution: Dissolve 20 g boric acid (H_3BO_3) in ammonia-free DI water, add 10 mL mixed indicator solution, and dilute to 1 L. *Prepare fresh monthly!*

Procedure

- Distillation: Proceed as described in Section 22.2.3, using indicating boric acid solution (instead of the plain boric acid) as absorbent for the distillate.
- Titration: Titrate ammonia in the distillate with standard 0.02 N H_2SO_4 titrant until indicator turns to pale lavender.

Calculation

$$NH_3\text{-}N \text{ mg/L} = [(A - B) \times N \times 14 \times 1000]/\text{sample size (mL)} \qquad (23.3)$$

If the sulfuric acid normality is exactly 0.02, the following formula can be used as an alternative.

$$NH_3\text{-}N \text{ mg/L} = [(A - B) \times 280]/\text{sample size (mL)} \qquad (23.4)$$

where A = volume (mL) of titrant used for sample titration, B = volume (mL) of titrant used for blank titration, N = exact normality of the sulfuric acid, 14 = equivalent weight for nitrogen, and 280 = $0.02 \times 14 \times 1000$.

23.2.6 Automated Phenate Method

Method No. 350.1 (EPA-600/4-79-020).

Principle of the method — The method is used to determine ammonia nitrogen in potable, surface, and saline waters. This method is also applicable to domestic and industrial wastewaters over a range of 0.02 to 2.0 mg/L when photometric measurement is made at 630 to 660 nm in a 15 mm or 50 mm tubular flow cell. Determine higher concentrations by diluting the sample. Ammonia in the sample is reacted with alkaline phenol and hypochlorite to form indophenol blue. The blue color formed is intensified with sodium nitroferricyanide and read at 640 nm. The intensity of the color is directly proportional to the concentration of ammonia in the sample.

Interferences — Precipitation of calcium and magnesium ions are eliminated by the addition of a combined potassium sodium tartrate/sodium citrate complexing reagent. Variation of acidity or alkalinity among samples will affect color intensity. This is eliminated by preservation with 5 mL sulfuric acid per 1 L sample. Turbid samples are filtered prior to determination to remove interferences. Samples with background absorbance at the analytical wavelength may also be interferred with.

Apparatus, reagents, and analysis — Use an automated continuous flow analytical instrument, such as Technicon Autoanalyzer II, Alpkem RFA system, or equivalent. Prepare reagents, set up manifold, and complete system along with the detailed analysis depending on instrumentation. Consult manufacturer's manual.

23.2.7 Ammonia Determination in Soil, Sludge, or Sediment Samples

Corps of Engineers Method 3-154.

Principle of the method — It is a preparation method for ammonia applicable to soils and sediments. The practical quantitation limit (PQL) is 15 ppm (mg/kg) for a 1.00 g sample. The PQL should be calculated each time the sample set is run.

The wet sediment or soil is made into a slurry, acidified, and added to clean distillation apparatus where the ammonia is distilled from the sample and trapped in a boric acid solution. The distillate is then analyzed using one of the listed EPA methods.

Interferences — This method removes the exchangeable ammonia, making it similar to the cation exchange capacity procedure. This procedure is more thorough in its recovery of ammonia, due to the presence of the phosphate buffer. This buffer neutralizes the sulfuric acid used to preserve the slurry. Digestion of the sample in the presence of acid may convert organic nitrogen to ammonia, resulting in positive interference. Other interferences are unknown at this time.

Sample preparation

1. Weigh 1.0 (±0.5) g of wet sediment or soil sample into a 200 mL Erlenmeyer flask.

2. Add 50 mL DI water and 3 to 4 drops of concentrated sulfuric acid to preserve the sample. Preservation is good for 24 hours.

3. After sample preservation is completed, assemble the distillation apparatus.

4. To the side-arm flask, add 500 mL reagent water, 10 mL phosphate buffer, and several Hengar granules. Insert the glass stopper into the opening in the top of the flask. Turn on the water for the condensers, and turn on the heating mantles to their highest settings. Heat the contents of the flask to boiling, and continue the distillation until the volume of the distillate is in excess of 100 mL.

5. Turn off the heating mantle, and let the apparatus cool.

6. Carefully remove the side-arm flask from the distillation apparatus and dispose of its contents.

7. Transfer the slurry from the Erlenmeyer flask into the 1000 mL side-arm flask. 450 mL DI water should be used in this step so that the resultant volume in the side-arm flask is 500 mL. Add several Hengar granules to the flask.

8. Turn the heating mantles on to their highest setting. Heat to boiling and allow to boil for 2 to 3 minutes; this will remove some interferants such sulfide and volatile organics. During this 2 to 3 minute period, place a 500 mL Erlenmeyer flask containing 50 mL boric acid trapping solution (or indicating boric acid solution if the titrimetric method is used to determine ammonia in the distillate) under the condenser. Ensure that the bottom edge of the Tygon tubing extension is below the surface of the boric acid solution; this is essential for recovery of ammonia from the sample.

9. After the 2 to 3 minute initial boiling period has elapsed, add 15 mL of phosphate buffer solution using a Mohr pipet. Continue the distillation until the volume of the distillate, including the boric acid solution, is in excess of 350 mL but less than 500 mL.

10. When the distillation is completed, remove the glass stopper from the top of the side-arm flask, turn off the heating mantles, and turn off the water for the condensers. Remove the Erlenmeyer collection flask, taking care not to spill any distillate.

11. Transfer the distillate to a 500 mL volumetric flask and dilute to volume with analyte-free water.

12. Determine the ammonia content of the distillate according to one of the methods described in Sections 23.2.3, 23.2.4, 23.2.5, or 23.2.6.

Calculation — Use the following formula to convert the concentration to milligrams per kilogram:

$$\text{Conc}_{\text{solids}} = (\text{Conc}_{\text{mg/L}} \times \text{FV})/\text{weight of sample (g)} \tag{23.5}$$

where $\text{Conc}_{\text{solids}}$ = concentration of NH_3-N in the sample, mg/kg (wet base), $\text{Conc}_{\text{mg/L}}$ = concentration determined in milligrams per liter, and FV = final volume after distillation (mL).

Use the following formula to convert the milligrams per kilogram wet base concentration to the milligrams per kilogram dry base:

$$\text{Conc}_{\text{solids}} = \text{conc}_{\text{wet base}}/\text{decimal fraction of dry weight \%} \tag{23.6}$$

where $\text{Conc}_{\text{solids}}$ = concentration of NH_3-N in the sample, mg/kg (dry base), $\text{Conc}_{\text{wet base}}$ = concentration of NH_3-N in the sample, mg/kg (wet base), and dry weight % = 100% − moisture %.

Calculations for solid matrixes are presented in Section 15.2.6.

Chapter 24

Analysis of Nitrite, Nitrate, and Total Kjeldahl Nitrogen

24.1 NITRITE–NITROGEN, NO$_2$-N

24.1.1 General Discussion

Nitrite is the intermediate oxidation state of nitrogen, both in the oxidation of ammonia to nitrate and in the reduction of nitrate. Such oxidation and reduction occur in wastewater treatment plants, water distribution systems, and natural waters. Nitrite can enter water supply systems through its use as a corrosion inhibitor in industrial process waters.

A small amount of sodium nitrite (NaNO$_2$) is added to meats such as cold cuts and frankfurters, because it is a good preservative. When the nitrite ion reaches the stomach, the high concentration of hydrochloric acid there converts the nitrite (NO$_2^-$) to nitrous acid (HNO$_2$), which can react with secondary amines in the digestive tract to produce N-nitrosamine. N-Nitrosamines are carcinogens. The problem is not easily solved and the question still exists: can we find another preservation technique for meat products, or accept the risk of using this chemical?

24.1.2 Sample Collection and Handling

Collect samples in plastic or glass bottles, preserve samples by cooling at 4°C. Make the determination promptly on fresh samples to prevent bacterial conversion of NO$_2^-$ to NO$_3^-$ or NH$_3$. Holding time for samples is 28 days. When sample is preserved with acid, NO$_3^-$ and NO$_2^-$ cannot be determined as individual species.

24.1.3 Determination of NO$_2$-N by Colorimetric Method

Method No. 354.1 (EPA-600/4-79-020).

Principle of the method — Nitrite (NO$_2^-$) is determined through formation of a reddish-purple azo dye produced at pH 2.0 to 2.5 by coupling sulfanilamide with N-(1-naphthyl)-ethylenediamine dihydrochloride (NED dihydrochloride). The applicable range for spectrophotometric measurements is 10 to 1000 μg/L NO$_2$-N (0.01 to 1.0 mg/L). The color system obeys Beer's Law up to 180 μg/L N with a 1 cm lightpath at 543 nm. Higher concentrations can be determined by diluting samples. Never use acid-preserved samples to analyze for NO_2-N.

Interferences — Sb^{3+}, Au^{3+}, Bi^{3+}, Fe^{3+}, Pb^{2+}, Hg^{2+}, Ag^{1+} and metavanadate (VO_3^{2-}) ions should be absent, because they interfere by precipitation under test conditions. Cu^{2+} ions may cause low results by catalyzing decomposition of the diazonium salt. Colored ions should be absent because they alter the developing color. Suspended solids should be removed prior to testing by filtrating the sample through 0.45 μm pore-size membrane filter.

Apparatus — Spectrophotometer, for use with 543 nm wavelength, providing a light path of 1 cm or longer.

Reagents

Color reagent — To 800 mL distilled water add 100 mL 85% phosphoric acid and 10 g sulfanilamine. After sulfanilamine dissolves completely, add 1 g NED dihydrochloride and mix to dissolve, then dilute to 1 L with DI water. Solution is stable for about a month when stored in a dark bottle in a refrigerator.

Stock NO_2-N solution, 100 mg/L — Use 0.1493 g dried sodium nitrite (dried in desiccator for 24 hours) dissolved and diluted to 1 L. Preserve with 1 mL chloroform. Store at 4°C. This solution is stable for 3 months.

$$1 \text{ mL} = 0.10 \text{ mg } NO_2\text{-N (1 mL} = 100 \text{ μg)}$$

Standard NO_2-N solution, 1 mg/L — Use 10 mL stock solution diluted to 1000 mL. Prepare daily!

$$1 \text{ mL} = 0.001 \text{ mg } NO_2\text{-N (1 mL} = 1 \text{ μg)}$$

Reference (independent) stock solution, 1000 mg/L — In a 1 L volumetric flask, dissolve 6.072 g of dry potassium nitrite, KNO_2 (dried in a desiccator for 24 hours), in 800 mL DI water. Dilute to volume with DI water and mix well. Preserve the solution with 2 mL/L chloroform. This standard is stable for 3 months. Store at 4°C.

Reference (independent) intermediate standard, 10 mg/L — Pipet 10 mL of reference (independent) stock solution into 1 L volumetric flask, fill up to volume with DI water, and mix. The solution is stable for 1 month. Store at 4°C.

Reference (independent) standard, 0.05 mg/L — Pipet 2.5 mL reference (independent) intermediate standard solution into 500 mL volumetric flask. Fill up to volume and mix well. Prepare fresh daily.

Procedure

1. Calibration standards, final volume 50 mL (0.01 to 0.2 mg/L).

Standard NO_2-N (mL)	Dilute to (mL)	NO_2-N (mg/L)
0	50	0 (blank)
0.5	50	0.01
1.0	50	0.02
2.0	50	0.04
4.0	50	0.08
5.0	50	0.10
10.0	50	0.20

2. Add 2 mL color reagent to each standard. Mix well.

3. Allow color to develop at least 15 minutes. (pH should be between 1.5 to 2.0).

4. Read the absorbance at 540 nm against reagent blank.

5. Prepare calibration curve and check its acceptance by calculating the correlation coefficient.

6. Check the calibration curve by a continuing calibration standard (CCS), followed by the analysis of a calibration verification standard (CVS) (reference or independent standard). See Sections 3.2.5 and 12.3.2.

7. In the case of accepted calibration, analyze the samples using 50 mL or aliquots diluted to 50 mL, as described for the standards, step 2 to step 4.

Calculation — Compute sample concentration directly from the curve by using the linear regression calculation (See Section 12.3.2).

Quality control — For reference of QC items, calculations, and acceptance criteria, see Sections 3.2.7, 3.2.9, and 16.1.

Initial calibration verification — The continuing calibration standard (CCS) is analyzed immediately after the calibration curve is established.

Continuing calibration verification — A continuing calibration standard (CCS) is analyzed once every 10 samples. The purpose is to verify that the calibration curve is constant throughout the run.

Reference (independent) standard or calibration verification standard (CVS) — Analyzed after the calibration curve has been established to check calibration, and after each 10 samples.

Duplicates — Duplicates are useful to establish the preciseness of the analysis (Section 3.2.5).

24.2 NITRATE–NITROGEN, NO_3-N

24.2.1 General Discussion

In water supplies, nitrate-nitrogen owes its origin to several possible sources, including the atmosphere, legume plants, plant debris, animal excrement, and sewage, as well as nitrogenous fertilizers and some industrial wastes. Since the atmosphere consists of about 78% by volume of nitrogen, some of that present in water originates from this source, but most is generated by the decay of organic matter, and from industrial and agricultural chemicals. Bacteria-decomposed organic matter, such as sewage and excrement, and these complex proteins change to ammonia (NH_3), then nitrite (NO_2), and finally nitrate (NO_3). Since the various forms are highly soluble in water, they are easily leached downward from the soil by infiltrating water and may quickly reach the water table. Much of that present in the soil, however, is used by plants, since nitrate is the major nutrient for vegetation and is essential to all forms of life. In addition to decaying organic matter, fertilizers are a major source of nitrate in water supplies. The concentration of nitrate in both surface and groundwater can increase to alarming amounts.

Nitrate–nitrogen (NO_3-N) concentration greater than 10 mg/L has been known to cause infant methemoglobinemia, a disease characterized by cyanosis, a bluish coloration of the skin. Nitrate inactivates hemoglobin into methemoglobin and in this form it becomes unfit to transport oxygen. The sickness is especially prevalent in infants in the first 3 months. This disease may occur when a nursing child consumes either formula prepared with nitrate-containing water or milk directly from the mother that contains a large amount of nitrate.

Some evidence indicates that high concentrations of nitrate in drinking water for livestock has resulted in abnormally high mortality rates in baby pigs and calves and abortion in brood animals.

Shallow groundwater supplies are susceptible to pollution by nitrates. The concentration of nitrate in groundwater can fluctuate greatly from one season to the next. During the wet season, particularly while plants are still dormant, infiltrating rain or snowmelt may leach large amounts of nitrate to the water table. Most of it is either used by plants during the growing season or remains in the ground because of a soil-moisture deficiency, but after crops are harvested in the fall, nitrate may again leach down to pollute shallow aquifers.

Nitrate pollution of some surface and groundwater has become the major problem in some agricultural areas. Although fertilizers have been implicated in such pollution, there is evidence that feedlots are the major source of nitrate pollution. The growth of livestock populations and the concentration of livestock in feedlots have aggravated the problem. Such concentrations of cattle, coupled with the fact that a steer produces approximately 18 times as much waste material as a human, have resulted in high levels of water pollution in rural areas with small human populations.

Nitrate in farm wells is a common and especially damaging manifestation of nitrogen pollution from feedlots, because of the susceptibility of ruminant animals to nitrate poisoning.

24.2.2 Sample Collection and Handling

Start determination promptly after sample collection. If storage is necessary, store for up to 24 hours at 4°C. For longer storage, preserve with 2 mL concentrated H_2SO_4 per liter and store at 4°C. When sample is preserved with acid, NO_3^- and NO_2^- cannot be determined as individual species.

24.2.3 Nitrate Analysis by Ion-Selective Electrode Method

Method No. 4500-NO_3-D (*Standard Methods*, 17th ed., 1989).

Principle of the method — The NO_3^- ion electrode is a selective sensor that develops a potential across a thin, porous inert membrane that holds in place a water-immiscible liquid ion exchanger. The electrode responds to NO_3^- ion activity (see Section 12.3).

Apparatus

- pH meter or ion-selective meter.
- Double-junction reference electrode, ORION 90-02 or equivalent. NO_3^- ion electrode, ORI-ON 93-07, or CORNING 476134, or equivalent. Carefully follow manufacturer's instructions regarding care and storage.
- Magnetic stirrer.
- TFE-coated stirring bars.

Reagents

Stock NO_3-N solution, 1000 mg/L — Dry potassium nitrate (KNO_3) at 105°C for 24 hours. Dissolve 7.218 g in distilled water and dilute to 1 L. Preserve with 2 mL chloroform per liter. The preserved solution is stable for at least 6 months. Store in refrigerator.

$$1 \text{ mL} = 1 \text{ mg} (1000 \text{ µg}) NO_3\text{-N}$$

Standard NO_3-N solution, 10 mg/L — Measure 10 mL stock NO_3-N solution and dilute to 1 L. Preserve with 2 mL of chloroform per 1 L. This solution is stable for 3 months. Store in refrigerator.

$$1 \text{ mL} = 0.01 \text{ mg} (10 \text{ µg}) NO_3\text{-N}$$

Ionic strength adjustor (ISA) — Dissolve 26.4 g ammonium sulfate, $(NH_4)_2SO_4$, and dilute to 100 mL with DI water.

Reference electrode filling solution — 2 mL ISA diluted to 100 mL.

Reference (independent) stock solution, 100 mg/L NO_3-N — In a 1 L volume flask, dissolve 0.6068 g anhydrous sodium nitrate, $NaNO_3$, dilute to volume with DI water, and mix. Preserve the solution with 2 mL/L chloroform. Store in refrigerator. The solution is stable for 3 months.

Reference (independent) standard, 0.5 mg/L — Pipet 5 mL of reference (independent) stock solution into 1 L volumetric flask, dilute to volume, and mix. Prepare daily.

Procedure

Prepare reference electrode — Fill with the reference electrode filling solution.

Check nitrate electrode's sensing tip — Check for cleanness, and change the sensing tip if necessary.

Instrument setup — The following instructions are suitable for ORION 901 Ion Analyzer: Connect nitrate electrode into "Sense A" and reference electrode into "Ref A." Shorting strap should be across "Sense B" and "Ref B."

Slope determination

1. Instrument in MV mode.

2. Measure 100 mL distilled water into 150 mL beaker and add a stirring bar.

3. Add 2 mL of ISA solution and stir with a magnetic stirrer. The stirring rate and temperature (25°C) should be constant during standardization and measurement.

4. Place electrodes into the solution.

5. Pipet 1 mL of 1000 ppm standard, stir.

6. Read MV and record.

7. Pipet 10 mL of 1000 ppm standard to the same sample, while stirring.

8. Read MV and record.

9. The difference between the two readings is the value of the slope. Correct operation gives –54 ± 1 MV of slope.

10. Adjust the slope switch to the slope value with minus (–) sign.

Standardization

1. Prepare 100 mL of two standards with concentration range expected of the samples. The two standards differ by a factor of 10. For example 1.0 ppm and 10.0 ppm. 1 mg/L = 10 mL standard solution (10 ppm) diluted to 100 mL. 10 mg/L = 1 mL stock solution (1000 ppm) diluted to 100 mL.

2. Turn instrument to **CONC** mode.

3. Measure 100 mL 1 mg/L standard, add stirring bar, place on stirrer.

4. Add 2 mL ISA and immerse electrodes.

5. Flip the concentration switch to the accepted concentration, 1.0 mg/L.

6. Press **CLEAR/READ MV.**

7. Press **SET CONC** button. Reading should give the correct concentration.

8. Pour the 100 mL of second standard (10 mg/L) into a beaker, add the stirring bar.

9. Add 2 mL ISA and measure. If the value is not exactly the desired value, adjust by changing the slope until the reading is exactly the prepared value of the standard. Record the new slope.

10. If the changed slope value is acceptable, measure the reference or independent standard. The percentage of recovery of its true value should be within the laboratory established acceptance limit.

Measurement of samples

1. Obtain 100 mL sample.

2. Add 2 mL ISA, stir, and read directly in the concentration mode.

3. If the reading is higher than the calibration range, sample should be diluted, and the reading multiplied by the dilution factor.

Quality control — For reference of QC items, calculations, and accepted criteria see Section 3.2.7, Section 3.2.9, and Section 16.1.

Method blank — A method blank is analyzed once every sample batch or once every 20 samples, whichever is more frequent.

Reference (independent) standard — Analyzed immediately after instrument calibration and once every 10 samples. The recovery of its value must be within the in-house generated limits.

Duplicate — Analyzed once in every sample batch. The calculated precision value must be within the in-house established limits.

24.2.4 Nitrite + Nitrate–Nitrogen Determination by Cadmium Reduction Method

Principle of the method — A filtered sample is passed through a column containing granulated copper-cadmium. This process reduces nitrate (NO_3^-) to nitrite (NO_2^-), which is determined by diazotizing with sulfanilamide and coupling with N-(1-naphthyl)ethylenediamine dihydrochloride to form a highly colored azo dye which is measured spectrophotometrically. Knowing the NO_3-N + NO_2-N mg/L value and the separately analyzed NO_2-N concentration, the NO_3-N is then calculated by difference.

Interferences

- Turbidity can interfere with color development. Samples containing turbidity should be filtered through 0.45 μm filter prior to analysis.
- Iron, copper, and other metals with concentrations of several milligrams per liter can cause low results. EDTA eliminates this problem by complexing metal ions.
- Residual chlorine will interfere by oxidizing the cadmium column, reducing its efficiency. Remove residual chlorine by adding a dechlorinating agent, for example sodium thiosulfate ($Na_2S_2O_3$).
- Samples containing oil and grease will coat the cadmium reduction column and interfere with its functioning. The oil and grease can be removed by extraction with an organic solvent such as hexane prior to analysis.
- Suspended materials must be filtered prior to analysis, since they will clog the cadmium reduction column.

Apparatus

- Reduction column: purchase or construct from a 100 mL pipet by removing the top portion. See Figure 24.1.
- Spectrophotometer for use at 540 nm, providing a light path of 1 cm or longer.

Reagents

- Cadmium granules, 40 to 50 mesh size.
- Copper–cadmium granules: 25 g cadmium (Cd) granules (new or used) are cleaned with 6 N HCl and rinsed with distilled water. The color of the Cd should be silver. Swirl Cd with 100 mL of 2% copper sulfate ($CuSO_4$) solution for 5 minutes, or until blue color partially fades. Decant and repeat with fresh $CuSO_4$ until a brown colloidal precipitate forms. Gently flush with DI water to remove all precipitated copper. The color of the Cd should be black.
- Ammonium chloride–EDTA solution (stock): Dissolve 13 g ammonium chloride (NH_4Cl) and 1.7 g EDTA in 900 mL DI water. Adjust pH to 8.5 with concentrated ammonium hydroxide (NH_4OH) and dilute to 1 L.
- Ammonium chloride–EDTA, diluted solution: Dilute 300 mL stock NH_4Cl–EDTA to 500 mL with DI water.

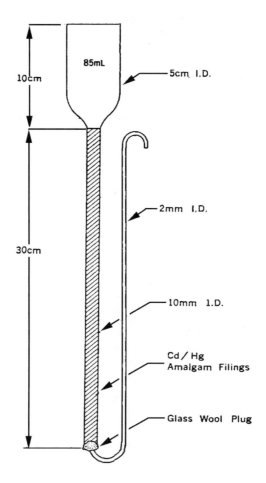

Figure 24.1
Reduction column.

- Color reagent: See NO_2-N test in Section 24.1.3.
- Stock NO_3-N solution: Prepare as directed in Section 24.2.3.
- Standard NO_3-N solution: See Section 24.2.3.
- Stock NO_2-N solution: See Section 24.1.3.
- Standard NO_2-N solution: See Section 24.1.3.
- Hydrochloric acid, HCl 6 N: 500 mL concentrated HCl and 500 mL DI water (1+1 dilution).
- Copper sulfate ($CuSO_4$) solution, 2%: Dissolve 20 g $CuSO_4$ or 31.32 g $CuSO_4 \cdot 5H_2O$ and dilute to 1 L.
- Reference (independent) standard: See Section 24.2.3.
- Cadmium reduction column check standard, 0.5 mg/L NO_2-N: In a 100 mL volumetric flask add 500 μL (0.5 mL) of the nitrite stock solution (100 mg/L), dilute to the volume, and mix.

Procedure

Preparation of reduction column

1. Insert a glass wool plug into bottom of the reduction column and fill with water.

2. Add sufficient Cu-Cd granules to produce a column 18.5 cm long. Maintain water level above Cu-Cd granules to prevent entrapment of air.

3. Wash column with 200 mL diluted NH_4Cl–EDTA solution.

4. Activate column by passing through it, at 7 to 10 mL/min, at least 100 mL of a solution composed of 25 mL of 1 mg/L NO_3-N standard and 75 mL of NH_4Cl–EDTA solution.

Preparation of calibration standard — Using the standard NO_3-N (10 mg/L) standard solution, prepare standards in the range of 0.05 to 1.0 mg/L NO_3-N as follows:

Standard NO_3-N (mL)		NO_3-N (mg/L)
0	100 mL DI water	0
0.5	diluted to 100 mL	0.05
1.0	diluted to 100 mL	0.10
2.0	diluted to 100 mL	0.20
5.0	diluted to 100 mL	0.50
10.0	diluted to 100 mL	1.00

Reduction of standards

1. To 25 mL blank and standards add 75 mL NH_4Cl–EDTA solution and mix.

2. Pour mixed standard into column and collect at a rate of 7 to 10 mL/min. Discard first 25 mL and collect the rest in reaction flask. *There is no need to wash column between standards!*

Color development and measurement of standards

1. As soon as possible, but no more than 15 minutes after reduction, measure 50 mL reduced standard and add 2 mL color reagent and mix.

2. Between 10 minutes and 2 hours afterward, measure absorbance at 543 nm wavelength against reagent blank.

3. Compare at least one NO_2-N standard to a reduced NO_3-N standard at the same concentration to verify reduction column efficiency. Reactivate Cu-Cd granules as described earlier, when the efficiency of reduction falls below about 75%.

4. When all the standards are treated similarly, establish the calibration curve, calculate its correlation coefficient, and verify it by a continuing calibration standard (CCS) and a calibration verification standard (CVS). The curve is valid until it fails acceptance criteria. See Section 3.2.5 and Section 12.3.

Treatment of samples

1. Turbid samples should be filtered through 0.45 nm membrane filter prior to analysis.

2. pH of the samples should be between 7 and 9; adjust if necessary by using pH meter. This pH range ensures a pH of 8.5 after adding NH_4Cl–EDTA solution.

3. Sample reduction is the same as for standards by using 25 mL sample or a portion diluted to 25 mL.

4. Color development and measure is the same as described for standards.

5. Calculate the concentration of the sample by using the linear regression calculation concerning the curve. If any dilution was necessary prior to analysis, multiply the value with the dilution factor for the final result. Report as the sum of NO_3-N and NO_2-N, unless the concentration of NO_2-N is separately determined and subtracted.

Quality control — For reference of QC items, calculations, and the accepted criteria, see Section 3.2.7, Section 3.2.9, and Section 16.1.

Method blank — A method blank is to be performed at least once every sample batch or once every 20 samples, whichever is more frequent.

Continuing calibration standard (CCS) — This is analyzed immediately after the calibration curve to check calibration and once every 10 samples.

Calibration verification standard (CVS) or reference (independent) standard — This is analyzed once every 10 samples. It verifies that the calibration standards are actually at the concentrations claimed by the vendor/or the analyst. The reference standard must fall within the in-house limits.

Cadmium reduction column check standard — A mid-level nitrite, NO_2-N, standard is analyzed once immediately preceding the analysis of samples and once at the end of the run. This standard verifies the reduction capacity of the column.

Matrix spike/matrix spike duplicate (MS/MSD) — This is performed once for every sample batch or once every 20 samples. MS/MSDs are used to demonstrate that the analytes of interest can be recovered and analysis of the analyte can be performed with acceptable accuracy and precision. For MS/MSDs, a sample is analyzed in triplicate with 50 μL (0.05 mL) of the nitrate stock solution (1000 mg/L) added to one of the two aliquots in 100 mL volumetric flasks. The concentration of the spike is 0.5 mg/L.

24.2.5 Automated Cadmium Reduction Method

Method No. 353.2 (EPA-600/4-79-020).

Follow general procedure described by the manufacturer of the continuous flow analytical instrument, such as AutoAnalyzer II, Technicon Instrument Corp., or Alpkem RFA, Alpkem Corp., or equivalent. Quality control requirement is the same as mentioned in Section 24.4.2.

24.2.6 Determination of Nitrate–Nitrogen in Soils and Sediments

Corps of Engineering Method No. 3-183 is applicable to soils and sediments. The method is a separation technique based on the high-solubility nitrates and nitrites in water, but should be considered operationally defined. The practical quantitation limit is 10 mg/kg for a 1.00 g sample size. The PQL should be calculated each time the sample set is run.

Principle of the method — The wet sediment or soil is made into a slurry, acidified, heated for 15 minutes, and centrifuged to separate the liquid and solid. The liquid phase is then analyzed using the EPA cadmium reduction method. This method is only valid for the combined nitrate plus nitrite number. Interferences commonly found in sediments and soils cause the nitrite to convert to nitrate.

Interferences — High concentration of water soluble forms of sulfate and phosphate will be extracted with the nitrate and cause a low result in the colorimetric analysis.

Procedure

1. Weigh 1.00 (±0.5) g of wet sediment into a 200 mL Erlenmeyer flask.
2. Add 50 mL of DI water and 3 to 4 drops of concentrated sulfuric acid to preserve the sample. Preservation is good for 24 hours.
3. Add 50 mL to the acidified slurry and boil the sample for 15 minutes. The procedure is operationally defined, so the heating time and temperature must be the same for all samples and standards.
4. Transfer the sample to centrifuge tubes and centrifuge the slurry at 2000 rpm for 5 to 10 minutes. Decant the liquid into a 200 mL volumetric flask.
5. Add 50 mL of DI water to the solid in the centrifuge tube. Mix thoroughly. Centrifuge again for the 5 to 10 minutes at 2000 rpm.
6. Decant the wash into the 200 mL volumetric flask.
7. Repeat the washing procedure a second time and add the wash to the 200 mL volumetric flask. Dilute the sample to volume if necessary with DI water.
8. Filter the sample through a 0.45 μm membrane filter.
9. Analyze the sample according to the EPA Method No. 353.1 or 353.2.

24.3 TOTAL KJELDAHL NITROGEN (TKN) AND ORGANIC NITROGEN

24.3.1 General Discussion

TKN is referred to as organically bound nitrogen, but analytically ammonia and organic nitrogen can be determined together and called total Kjeldahl nitrogen (TKN).

Organic nitrogen includes such natural materials as proteins and peptides, nucleic acids and urea, and numerous synthetic materials. Typical organic nitrogen concentrations vary from a few hundreds of ppb in some lakes to more than 20 ppm in raw sewage. Organic nitrogen value is calculated as (TKN) – (NH_3-N).

24.3.2 Sample Collection and Handling

The most reliable results are obtained from fresh samples. If any immediate analysis is not possible, preserve samples by acidifying to pH <2 with H_2SO_4 and storing at 4°C. Samples may be collected in plastic or glass bottles.

24.3.3 Analysis of Total Kjeldahl Nitrogen (TKN)

Method No. 351.3 (EPA-600/4-79-020).

Principle of the method

Total Kjeldahl nitrogen (TKN) is the sum of free ammonia and organic nitrogen compounds. Organic nitrogen is the difference obtained by subtracting NH_3-N from TKN. The sample is heated in the presence of concentrated H_2SO_4, K_2SO_4, and $HgSO_4$, and evaporated until SO_3 fumes are obtained, and the solution becomes colorless or pale yellow. The residue is cooled, diluted, and made alkaline. The ammonia is distilled and determined by nesslerization, titration, potentiometric, or automated RFA (rapid flow analyzer) method.

Apparatus

Digestion apparatus

- Traditional Kjeldahl digestion system with 800 mL Kjeldahl flasks
- Fume removal system
- Instead of the traditional digestion system, a block digestor (Technicon, Labconco, Lachate or equivalent) may be used with digestion tubes and cold fingers.
- Non-selenized Hengar granules

Distillation apparatus — Distillation apparatus used with traditional Kjeldahl. With the block digester, use rapid still distillation unit.

Apparatus used for ammonia determination — See Chapter 23.

Reagents — All reagents used for the determination of ammonia nitrogen, in Sections 23.2.3, 23.2.4, 23.2.5, and 23.2.6. In addition the reagents listed below are used. All chemicals should be ultra-pure and free from nitrogen!

Mercuric sulfate solution — Dissolve 8 g red mercuric oxide (HgO) in 50 mL 1:4 H_2SO_4 (400 mL water + 100 mL H_2SO_4). Dilute to 100 mL with distilled water.

Digestion reagent — Dissolve 134 g potassium sulfate (K_2SO_4) in 600 mL water and 200 mL concentrated H_2SO_4. Add with stirring 25 mL mercuric sulfate solution and dilute this combined solution to 1 L with DI water. Keep at a temperature close to 20°C to prevent crystallization.

Sodium hydroxide, NaOH 10 N — Place a 2 L beaker or Erlenmeyer flask on a magnetic stirrer under laboratory hood. Measure 400 g NaOH and 600 mL DI water into the container with a TFE-coated stirring bar. Stir constantly while dissolving.
  CAUTION: Reaction of NaOH with water generates extreme heat! After solution cools to room temperature, dilute to 1 L with DI water. Mix well and store in plastic container.

Boric acid, 2% solution — Dissolve 20 g boric acid, HBO_3, in water and dilute to 1 L.

Sodium hydroxide–sodium thiosulfate reagent — Dissolve 500 g NaOH and 25 g $Na_2S_2O_3$ · $5H_2O$ in DI water and dilute to 1 L.

Glutamic acid stock standard, 200 mg/L as N — In a 500 mL volumetric flask, dissolve 1.05 g oven-dried glutamic acid in 200 mL DI water and fill up to 500 mL with DI water. Preserve with several drops of chloroform ($CHCl_3$).

Glutamic acid working standard, 1 mg/L — In the digestion tube, add 0.1 mL (100 µL) of glutamic acid stock standard solution and dilute to 20 mL with DI water and with 0.5% sulfuric acid for the automated analysis.

Ammonium chloride spike stock solution, 200 mg/L — In a 500 mL volumetric flask, dissolve 0.3819 g oven-dried NH_4Cl in 200 mL DI water. Dilute to volume. Preserve with a few drops of chloroform ($CHCl_3$).

Digestion and distillation by Kjeldahl apparatus

1. Selection of sample volume and sample preparation for water and for sludge and sediment samples.

   a. Place a volume of well-mixed sample into a 800 mL Kjeldahl flask. Sample size selection is based on the possible concentration of organic nitrogen (25 mL to 300 mL. If smaller sample size is used, dilute sample to 300 mL with DI water. Neutralize to pH 7.00 and dechlorinate with a dechlorinating agent, such as sodium thiosulfate.

   b. For sludge and sediment samples, weigh wet sample in a weighing boat and transfer content into the Kjeldahl flask. Determination of TKN in dried sludge and sediment samples is not accurate, because drying results in loss of ammonium salts. Determine moisture and dry weight of the sample in a separate sample portion.

2. Digestion: Add 50 mL digestion solution and a few glass beads or boiling chips. After mixing, heat under a hood or with suitable ejection equipment to remove acid fumes. Boil briskly until the volume is reduced to about 25 to 50 mL. Then continue to digest for an additional 30 minutes. As digestion continues, colored or turbid samples will turn clear or straw-colored. After digestion, let solution cool and dilute back to 300 mL with DI water and mix. Add few drops of phenolphthalein indicator. Tilt flask and carefully add 50 mL sodium hydroxide- thiosulfate reagent. Add more NaOH if red phenolphthalein color fails to appear at this stage. Do not mix until connected to distillation apparatus. Connect flask to steamed-out distillation apparatus and shake flask well to insure complete mixing. A black precipitation of HgS will form and the pH should exceed 11.0.

3. Distillation: Distill and collect 200 mL distillate below surface of 50 mL of 2% boric acid solution. Extend the tip of the condenser well below the level of boric acid solution and do not let the temperature in condenser rise above 29°C. Lower the collected distillate free of contact with the delivery tube and continue distillation 1 to 2 minutes.

4. Ammonia determination: Determine NH_3-N in the distillate by titrimetric, colorimetric or potentiometric method.

Digestion with block digester and distillation with rapid still

1. Digestion: Measure 20 mL sample, or an aliquot diluted to 20 mL and place in a digestion tube. Add 5 mL digestion solution and mix. Place tubes in digestion that has been preheated to 200°C. Set low temperature at 200°C for 1 hour. Set high temperature at 380°C. After the temperature reaches 380°C, the time should be set for 30 minutes. Longer time and higher temperature may result in complete loss of the acid. Cool and add 25 mL DI water, and mix.

2. Distillation: Distill on rapid still into 50 mL 2% boric acid solution. Collect about 125 mL distillate. For correct distillation with the rapid still, follow manufacturer's instruction.

3. Determination of ammonia: Determine NH_3-N in the distillate by titrimetric, colorimetric, or potentiometric method.

Calculation

1. After nesslerization (23.2.3) finish:

$$\text{TKN mg/L} = [(A \times 1000)/B] \times [C/D] \qquad (24.1)$$

where A = NH_3-N from curve (mg), B = total distillate collected including the H_3BO_3, (mL), C = distillate taken for nesslerization (mL), and D = original sample taken (mL).

2. After titrimetric (23.2.5) finish, add 3 drops of mixed alkalinity indicator to the distillate and titrate the ammonia with 0.02 N H_2SO_4. See Alkalinity, page XXX, Chapter 20.

$$TKN, mg/L = [(A - B) \times N \times F \times 1000]/S \qquad (24.2)$$

where A = H_2SO_4 used for sample titration (mL), B = H_2SO_4 used for blank titration (mL), N = normality of acid, F = milliequivalent weight of N (14 mg), and S = sample digested (mL).

If the H_2SO_4 is exactly 0.02 N:

$$TKN\ mg/L = [(A - B) \times 280]/S \qquad (24.3)$$

where A = H_2SO_4 used for sample titration (mL), B = H_2SO_4 used for blank titration (mL), $0.02 \times 14 \times 1000 = 280$, and S = sample digested (mL).

3. Potentiometric finish (Section 23.2.4).

Quality control — For reference of QC items, calculations, and acceptance criteria, see Section 3.2.7, Section 3.2.9, and Section 16.1.

Digested blank or method blank — This is to be performed in duplicate at least once every digestion batch. This blank is to consist of 0.5% H_2SO_4, which is the matrix of the preserved sample.

QC check standard or reference standard (glutamic acid standard) — This standard is digested in duplicate with the sample batch, and analyzed every 10 samples.

Matrix spike/matrix spike duplicate (MS/MSD) — MS/MSDs are performed once for every sample batch or once every 20 samples, whichever is more frequent. For MS/MSDs, a sample is analyzed in triplicate with 100 μL (0.1 mL) of the 100 mg/L ammonia nitrogen standard added to 20 mL sample aliquots if the block digestor method used, and 500 μL (0.5 mL) of 100 mg/L ammonia nitrogen standard added to 100 mL sample aliquots if the Kjeldahl flask digestion method is applied. This process results in a spike of 0.5 mg/L N. The MS/MSDs are treated as samples taken through the entire digestion and analytical process.

Chapter 25

DETERMINATION OF PHOSPHORUS

25.1 PHOSPHATE-PHOSPHORUS AS PO$_4$-P

25.1.1 General Discussion

Phosphorus is present in nature only in its compound forms. They occur in the animal, plant, and mineral kingdoms, almost always as one or another variation of the phosphates, which contain the PO$_4^{3-}$ ion. Phosphates are an important part of DNA and RNA, and phosphorus is found in cell membranes as phospholipid. The most important diphosphates and triphosphates in the body are called adenosine diphosphate (ADP) and adenosine triphosphate (ATP). They store and transfer energy. The body stores energy by forming P-O-P bonds. When it needs to use the energy, it hydrolyzes, so the energy is given off.

Tripolyphosphate ion as sodium salt (Na$_5$P$_3$O$_{10}$) is the chief phosphate in laundry detergents. The trouble with phosphates in laundry detergents is that they support the life of algae, so when phosphate-containing wastewater gets into rivers, lakes, and bays, they become unfit for aquatic life (except algae) or for human recreation. Therefore, the sale of phosphate-containing laundry products has been banned in many areas.

Phosphates belong to the group called *nutrients*. Nutrients flow in cycles, as carbon cycle (photosynthesis and respiration), nitrogen cycle (see Chapter 23), and phosphorus cycle. Phosphates are slowly dissolved (leached) from rocks by rain and carried to waterways. Dissolved phosphates are incorporated by plants and passed to animals by the food web. Phosphorus reenters the environment directly by animal excretum and by detritus decay. Each year, large quantities of phosphates are washed into the oceans, where much of it settles to the bottom and is incorporated into the marine sediments. Sediments may release some of the phosphate needed by aquatic organisms, and the rest may become buried.

Phosphate is the major component of fertilizers. By applying excess fertilizer, farmers may alter the phosphate cycle.

In natural waters and wastewaters, phosphorus occurs as phosphates. These are classified as ortho-, condensed, organically bound, and total phosphates. *Orthophosphates* are inorganic, water-soluble phosphates applied to agricultural or residential cultivated land as fertilizer, and carried into surface waters with storm runoff. *Condensed phosphates* are used extensively in the treatment of boiler waters and added to some water supplies during treatment to prevent scale formation and inhibit corrosion. *Organic phosphates* are formed primarily by biological processes. They are contributed to sewage by body wastes and food residues and can also be formed from orthophosphates in biological treatment processes or by receiving water biota. *Total phosphates* include all of the above-mentioned forms of phosphorus and are reported as total phosphorus.

The reporting form of phosphorus may be as phosphate (PO_4) or as phosphate-phosphorus (PO_4-P). The conversion factor is 3.08. For example, if total phosphate as PO_4 is 4.2 mg/L, the corresponding report of total phosphate-phosphorus as PO_4-P will be 1.36 mg/L.

25.1.2 Sampling and Storage

Collect samples in plastic or glass bottles, preserve with H_2SO_4 to pH <2, and store at 4°C until analysis. Holding time for hydrolyzable and total phosphates is 28 days. If the request is dissolved orthophosphates, sample filter through 0.45 μm membrane filter on site and cool at 4°C without acid preservation. Holding time is only 48 hours. In the case of total dissolved phosphate analysis, filter sample on site through 0.45 μm membrane filter, acidify with H_2SO_4 to pH <2 and cool to 4°C until analysis. Holding time is 48 hours.

25.2 DETERMINATION OF PHOSPHATE, PO_4-P

The method applies to drinking, surface, and saline waters, domestic and industrial wastes. The method is specific for the orthophosphate ion. Preliminary acid hydrolysis at boiling temperature converts condensed phosphates to orthophosphates. The preliminary digestion process converts the condensed and polyphosphates and the organic phosphates to orthophosphate for quantitation.

It is imperative that all glassware be meticulously cleaned with 50% hydrochloric acid and rinsed with DI water prior to usage! This includes all glassware used in the generation of reagents and standards. Whenever possible, standards and samples should be prepared in dedicated phosphate glassware.

25.2.1 Acid Hydrolysis for Condensed Phosphates

Acid hydrolysis converts condensed and polyphosphates to orthophosphates, ready for determination by the colorimetric method.

Apparatus — Autoclave or pressure cooker capable of operating at 110°C, 5 psi pressure or 116°C, 10 psi pressure, or 121°C, 15 psi pressure (psi = pounds per square inches).

Reagents

- Phenolphthalein indicator: Dissolve 2.5 g phenolphthalein in 250 mL 95% ethanol or isopropyl alcohol and add 250 mL DI water. Mix well.
- Strong acid solution: Slowly add 300 mL concentrated H_2SO_4 to about 600 mL DI water with stirring. When cool, add 4 mL concentrated HNO_3 and dilute to 1 L with DI water.
- Sodium hydroxide, NaOH 6 *N*: Add 240 g NaOH in a beaker with 600 mL DI water. Add a TFE stirring bar. Prepare under laboratory hood on a magnetic stirrer with constant stirring. *CAUTION:* Reaction generates extreme heat! After dissolution, let to cool at room temperature, and fill up to 1 L with DI water.

Procedure

1. To 100 mL sample or a portion diluted to 100 mL, add 0.05 mL (about one drop) phenol-phthalein indicator solution. If a red color develops, add strong acid solution dropwise to just discharge the color. Then add 1 mL more.
2. Boil gently for at least 90 minutes, adding distilled water to keep the volume between 25 and 50 mL. Alternatively, heat for 30 minutes in an autoclave or pressure cooker at 10 to 15 psi.
3. Cool, neutralize to a faint pink color with NaOH solution, and restore to the original 100 mL volume with DI water.

25.2.2 Persulfate Digestion Method for Total Phosphates

The digestion process converts the condensed or polyphosphates and the organic phosphates to orthophosphate for quantitation. The digestion process must oxidize the organic matter effectively to release total phosphorus as orthophosphate. The digestion method used is the ammonium persulfate method. Once the sample is digested, it is analyzed by the ascorbic acid colorimetric method. The range of the sample is 0.01 to 1.00 mg/L phosphorus.

Apparatus

- Hot plate: A 30–50 cm heating surface is adequate.
- Autoclave: See under acid hydrolysis, Section 25.2.1.
- Glass scoop: To hold required amount of persulfate crystals.

Reagents

- Phenolphthalein indicator: See Section 25.2.1, Acid Hydrolysis.
- Sulfuric acid solution: Carefully add 300 mL concentrated H_2SO_4 solution to 600 mL water. When solution is cool, fill up to 1 L with DI water.
- Ammonium persulfate $((NH_4)_2S_2O_8)$ or potassium persulfate $(K_2S_2O_8)$ crystals.
- Sodium hydroxide, NaOH 1 N: Dissolve 40 g NaOH in 600 mL DI water with constant stirring. After solution is cool, dilute to 1 L with DI water.

Procedure

Digestion using hot plate

1. Measure 50 mL or a suitable portion of thoroughly mixed sample into dedicated phosphate Erlenmeyer flask.
2. Add 0.05 mL (1 drop) of phenolphthalein indicator solution. If a red color develops, add H_2SO_4 solution dropwise to just discharge the color, then add 1 mL more.
3. Add 0.4 g ammonium persulfate $(NH_4)_2S_2O_8$ or 0.5 g potassium persulfate, $K_2S_2O_8$.
4. Boil gently on a preheated hot plate for 30 to 40 minutes or until a final volume of 10 mL reached.
5. Cool and dilute to 30 mL with DI water, add 0.05 mL (1 drop) of phenolphthalein indicator solution, and neutralize to a faint pink color with NaOH.

6. Make up to 100 mL with DI water. In some samples a precipitate may form at this stage, but do not filter. The precipitate is possible calcium phosphate precipitate, and will dissolve in acidic conditions during the following analytical process.

Digestion using autoclave

1. Heat for 30 minutes in autoclave at 10 to 15 psi pressure.
2. Measure 100 mL well mixed sample into dedicated phosphate Erlenmeyer flask. Cover the mouth of the flasks with aluminum foil.
3. Follow steps 2 and 3 of the "Digestion using hot plate" section above.
4. Heat for 30 minutes in autoclave at 10 to 15 psi pressure.
5. When the pressure gauge on the front reads zero, carefully open the door to the autoclave and allow the samples to cool.
6. Add 0.05 mL (1 drop) phenolphthalein indicator solution, and neutralize to a faint pink color with NaOH.

Digestion of soil samples

1. Weigh 5 g air-dried soil sample, transfer into HCl washed Erlenmeyer flask, and add 50 mL DI water.
2. Shake for 30 minutes.
3. Filter through Whatman No. 42 filter paper, wash the soil with additional aliquots of DI water, collect the filtrates, and dilute it to a final volume of 100 mL. Cover the mouth of the flask with aluminum foil and follow steps 3 to 6 of the digestion technique described above in "Digestion using autoclave."
4. Determine phophate phosphorus in the filtrate (Section 25.2.3 or Section 25.2.4 below).
5. Calculate the total phosphorus content of the soil, as shown in Section 15.2.6.

25.2.3 Determination of Phosphorus by Ascorbic Acid Method

Method No. 365.2 (EPA-600/4-79-020).

Principle of the method — Ammonium molybdate and antimony potassium tartrate react in an acid medium with dilute solution of phosphorus, to form an antimony-phospho-molybdate complex. This complex is reduced to a blue-colored complex by ascorbic acid. The blue color is proportional to the phosphorus concentration. The range of the sample is 0.01 to 1.00 mg/L phosphorus.

Interferences — Arsenates react with the molybdate reagent to produce a blue color similar to that formed with phosphate. A concentration of arsenic of 0.1 mg/L interferes. Iron will use some of the reducing agent, giving a negative interference. Turbidity can interfere with color development. Samples containing turbidity should be filtered through 0.45 μm filters only after the digestion procedure. Sample color that absorbs in the photometric range will also interfere.

Concentration ranges

- 0.30 to 2.0 mg/L P range with 0.5 cm light path

- 0.15 to 1.30 mg/L P range with 1.0 cm light path
- 0.01 to 0.25 mg/L P range with 5.0 cm light path

Apparatus

- Spectrophotometer for use 880 nm
- Acid-washed glassware
- General laboratory glassware

Reagents

Sulfuric acid, H_2SO_4 5 N — Add about 300 mL DI water to a volumetric flask and add 70 mL concentrated H_2SO_4. Mix well. When the temperature of the mixture is at room temperature, dilute to 500 mL with DI water and mix.

Potassium antimonyl tartrate solution — Dissolve 1.3715 g potassium antimonyl tartrate, $K(SbO)C_4H_4O_6 \cdot 1/2 H_2O$, and dilute to 500 mL. Store in a glass-stoppered bottle. The solution is good for 6 months.

Ammonium molybdate solution — Dissolve 20 g ammonium molybdate, $(NH_4)_6Mo_7O_{24} \cdot 4H_2O$, in DI water and dilute to 500 mL. Store in a glass-stoppered bottle.

Ascorbic acid, 0.01 M — Dissolve 1.76 g ascorbic acid in 100 mL water. The solution is stable for 1 week at 4°C.

Combined reagent — Mix the above reagents in the following proportions for 100 mL of combined reagent. Let all reagents reach room temperature before they are mixed and mix in the order given! Mix after addition of each reagent!

- 50 mL 5 *N* H_2SO_4.
- 5 mL potassium antimonyl tartrate
- 15 mL ammonium molybdate
- 30 mL ascorbic acid

The reagent is stable for 4 hours!

Stock phosphate solution, 50 mg/L phosphorus — Dissolve 0.2195 g anhydrous potassium dihydro-phosphate (KH_2PO_4), dried at 105°C, and dilute to 1 L. Preserve with 2 drops of chloroform. The solution is good for 3 months. Store in refrigerator.

$$1 \text{ mL} = 50 \text{ µg PO}_4\text{-P}$$

Standard phosphate solution, 2.5 mg/L P — Dilute 50 mL stock phosphate solution to 1000 mL with DI water. The solution is good for 1 week. Store in refrigerator.

$$1 \text{ mL} = 2.5 \text{ µg PO}_4\text{-P}$$

Reference (independent) stock solution, 25 mg/L — Obtain potassium dihydro-phosphate (KH_2PO_4) from other source as used for standard preparation. Dissolve 0.1098 g anhydrous (previously dried at 105°C) potassium dihydrophosphate KH_2PO_4 in about 500 mL DI water contained in a 1 L volumetric flask. Dilute to volume with DI water and mix well. Store solution in a refrigerator.

Reference (independent) standard, 0.5 mg/L — Pipet 2 mL reference (independent) stock solution into a 100 mL volumetric flask and dilute to the volume with DI water. Prepare fresh daily.

Procedure

Preparation of the calibration curve

1. Measure from the standard phosphate solution (2.5 mg/L) the following volumes and dilute to 50 mL with DI water:

    0 mL standard diluted to 50 mL = Blank

    2 mL standard diluted to 50 mL = 0.10 mg/L

    4 mL standard diluted to 50 mL = 0.20 mg/L

    8 mL standard diluted to 50 mL = 0.40 mg/L

    10 mL standard diluted to 50 mL = 0.50 mg/L

    15 mL standard diluted to 50 mL = 0.75 mg/L

    20 mL standard diluted to 50 mL = 1.00 mg/L

2. Add to the blank and to each standard 8 mL combined reagent and mix thoroughly.
3. After at least 10 minutes but no more than 30 minutes, measure absorbance at 880 nm wavelength against reagent blank.
4. Prepare calibration curve by plotting absorbance vs. phosphate concentration. Calculate the correlation coefficient to verify proper calibration. Its value must be >0.9950.
5. Check this initial calibration with continuing calibration standard (CCS), which is a middle-range standard, followed by calibration verification standard (CVS), which is a middle-range standard, but prepared from another source than the calibration standards (reference/independent standard).
6. Once the calibration curve has been established, its criteria should be verified by testing at least one standard and one CVS standard with each set of samples. See Section 3.2.5 and Section 12.3.

Measurement of samples

1. Measure 50 mL sample or smaller sample size diluted to 50 mL.
2. Add to 50 mL sample one drop of phenolphthalein indicator. If red color develops, add 5 N H_2SO_4 until the color is just discharged.
3. Add 8 mL combined reagent and mix well.
4. After at least 10 minutes, but no more than 30 minutes, measure absorbance at 880 nm against reagent blank. Because the color at first develops progressively and later fades, maintain equal timing conditions for samples and standards.
5. For highly colored or turbid waters, prepare a blank by adding all reagents except ascorbic acid and antimonyl potassium tartrate. Subtract blank absorbance from the absorbance of the samples.

Calculation — Obtain phosphate concentration from the calibration curve by using linear regression calculation. See Section 12.3.2.

25.2.4 Automated Ascorbic Acid Reduction Method

Method No. 365.4 (EPA-600/4-79-020).

Once the sample is digested it may be analyzed on a continuous flow analyzer, such as ALPKEM RFA, Technicon, or equivalent. Follow the general procedure described by the manufacturer.

Quality control — For reference of QC items, calculations, and acceptance criteria, see Section 3.2.7, Section 3.2.9, and Section 16.1.

Continuing calibration standard (CCS) — This is the initial calibration verification with a middle range of working calibration standard. It is analyzed immediately after the calibration curve to check calibration. It is also analyzed after every 10 samples to verify that the calibration curve is constant throughout the run. The recovery of the CCS must be within ±10%.

Calibration verification standard or reference (independent) standard — Since the reference standard is prepared from a source different from calibration standards, it is used to verify that these standards are at the concentration claimed by the analyst. Analyze once every 10 samples. The reference standard analytical value must be within the in-house established acceptance limits.

Method (digestion) blank — Performed at least once with every digestion batch. The total phosphorus content of the blank must not exceed the practical quantitation limit (PQL) of 0.01 mg/L.

Calibration blank — Used to prepare together with the calibration curve and once in sample batch to zero the instrument. It is not digested!

Matrix spike/matrix spike duplicate (MS/MSD) — Performed once for every sample batch or once every 20 samples, whichever is most frequent. For MS/MSDs, the sample is analyzed in triplicate with 1 mL of the 25 mg/L independent standard added to two of the aliquots (50 mL) in flasks. The added spike value is 0.5 mg/L phosphorus.

Chapter 26

DETERMINATION OF SULFIDE

26.1 ANALYSIS OF HYDROGEN SULFIDE

26.1.1 General Discussion

Sulfide is often present in groundwater, especially in hot springs. Groundwater that contains hydrogen sulfide (H_2S) gas is easily recognized by its rotten-egg odor. As little as 0.05 mg/L of H_2S in cold water is noticeable and the odor from 1 mg/L is definitely offensive. Its common presence in wastewaters comes partly from the decomposition of organic matter, sometimes from industrial wastes, but mostly from bacterial reduction of sulfate. Hydrogen sulfide escaping into the air from sulfide-containing wastewater causes odor nuisance. H_2S is very toxic and has claimed the lives of numerous people working in sewers. It reacts with metals to form insoluble sulfides, and makes corrosivity a problem for all metallic pipes. It has caused serious corrosion on concrete sewers by biologically oxidizing to H_2SO_4 on the pipe wall.

From the analytical standpoint, three categories of sulfide are distinguished in water and wastewaters. *Total sulfide* includes dissolved H_2S and HS^-, as well as acid soluble metallic sulfides present in suspended matter. *Dissolved sulfide* is that remaining after suspended solids have been removed by flocculation and settling. *Un-ionized hydrogen sulfide* may be calculated from the concentration of dissolved sulfide, the sample pH, and the practical ionization constant of H_2S.

26.1.2 Sample Collection and Handling

Samples must be collected with the minimum of aeration. Water samples must be preserved with zinc acetate and strong base (NaOH) to bring the pH above 12. Use 2 mL of 2 N zinc acetate solution and 1 mL 6 N NaOH per 1 L sample. Add preservatives into the sampling bottle before filling it with sample. The sample bottle must be filled completely, stoppered with no air bubbles, and mixed by rotating. The sulfide will precipitate to the bottom as zinc sulfide (ZnS).

26.1.3 Sulfide Determination by Iodometric Method

Method No. 376.1 (EPA-600/4-79-020).

Principle of the method — This method is applicable to all waste materials, or effluents, with the condition that the waste does not form an explosive mixture when mixed with acid. Sulfide is separated from the sample by reaction of zinc acetate. The remaining zinc sulfide is then oxidized to sulfur with excess iodine, and the unreacted iodine is back-titrated with sodium thiosulfate (back-titration is discussed in Section 9.3). The resultant number is for sulfide (S^{2-}) and is used to calculate un-ionized hydrogen sulfide (H_2S). The last section of the method calculates the un-ionized hydrogen sulfide.

Interferences — This method separates out potential interferences, reducing substances that react with iodine, such as sulfite, thiosulfate, and some organic compounds that decompose in acid and may form sulfur dioxide. Care should be taken to minimize contact with the air to avoid loss of volatile sulfide and to avoid reactions with oxygen that convert the sulfide to sulfur compounds not detected by this method.

Apparatus

- Glass fiber filter paper, Type A/E, Gelman p/n 61631
- General laboratory glassware

Reagents

Carbonate stabilized water — Adjust the pH of DI water to between 7.0 and 9.0 with sodium carbonate, Na_2CO_3.

Hydrochloric acid, HCl 6 N — To 500 mL water add 500 mL concentrated HCl solution. Mix well.

Iodine solution, 0.1 N — For preparation and standardization, see Section 6.3.4.

Iodine solution, 0.025 N — Dilute 250 mL of iodine solution to 1 L with DI water.

Sodium thiosulfate, $Na_2S_2O_3$ 0.1 N — For preparation and standardization, see Section 6.3.5.

Sodium thiosulfate, $Na_2S_2O_3$ 0.025 N — Dilute 250 mL of $Na_2S_2O_3$ 0.1 N solution to 1 L with DI water. Standardize this solution daily in accordance with the direction given for the standardization of 0.1 N sodium thiosulfate, using 0.025 N $K_2Cr_2O_7$ or iodate. Standardization is detailed in Section 22.1.4.

Starch indicator solution — To 5 g starch add a little cold water and mix it to a paste. Add this mixture to 1 L of boiling DI water, stir, and let settle overnight. Use the clear supernatant. Preserve by adding either 1.3 g salicylic acid, 4 g $ZnCl_2$, or a combination of sodium propionate and 2 g sodium azide to 1 L starch solution.

Stock sulfide solution — Into a 500 mL volumetric flask, dissolve 2.01 g of sodium sulfide nonahydrate, $Na_2S \cdot 9H_2O$ (stored in a sealed can with desiccant) in 400 mL with pH-adjusted DI water (see above). Dilute to volume with DI water and mix. This solution contains 570 mg/L H_2S and should be made fresh daily, and must be standardized prior to use. Standardization should be performed in triplicate. Directions for standardization follow.

1. Pipet 10 mL 0.025 N iodine solution into a 250 Erlenmeyer flask and acidify with 2 mL 6 N HCl.

2. Pipet 3 mL of sulfide stock solution, just prepared, into the flask and mix.

3. Dilute the mixture to 50 mL with DI water and mix again.

4. Titrate the pale yellow color using 0.025 N sodium thiosulfate as a titrant.

5. Add three drops of starch indicator and titrate until blue color disappears. Approximately 6.0 mL of the titrant will be required.

6. Calculation:

$$\text{mg/L } S^{2-} = [(v_1 - v_2) \times N \times 16 \times 1000]/\text{mL sulfide standard}$$

where v_1 = volume of iodine solution added, v_2 = volume of sodium thiosulfate titrant used, N = normality of sodium thiosulfate, 16 = equivalent weight of sulfide, and 1000 = conversion factor (mL/L).

Quality control check standard (QCCS) 34.2 mg/L S^{2-} — Add 15 mL of the 570 mg/L sulfide stock solution (see above) to 250 mL of DI water in a volumetric flask. True value is 34.2 mg/L. Prepare daily!

Reference check stock solution, 285 mg/L S^{2-} — Dissolve 1.01 g $Na_2S \cdot 9H_2O$ (stored in sealed can with desiccant) in 450 mL DI water in a 500 mL volumetric flask. Dilute to the volume with pH-adjusted DI water (see above) and mix. This solution contains 285 mg/L hydrogen sulfide. Standardize 5.0 mL aliquot of the standard using the same standardization procedures as for the stock sulfide solution. Approximately 6.5 mL of the 0.025 N sodium thiosulfate titrant should be required.

Reference check standard solution — Dilute 1.0 mL of the reference check stock solution to a final volume of 25 mL. The value is calculated according to the concentration of the standardized stock solution. Prepare daily!

Sodium hydroxide, NaOH 6 N — Dissolve 24 g NaOH pellets in approximately 70 mL of DI water with constant stirring. Cool and dilute to 100 mL with DI water.

   CAUTION: The reaction of NaOH with water is exothermic; solution becomes extremely hot!

Zinc acetate solution — Dissolve 220 g zinc acetate dihydrate, $Zn(C_2H_3O_2)_2 \cdot 2H_2O$, in 870 mL DI water; this makes 1 L solution.

Procedure

1. This step is necessary only if the sample was not preserved in the field!
   - To a 1000 mL sample, add sufficient 6 N NaOH to raise the pH above 9.0. For ideal water this will be approximately 1 mL. Mix thoroughly without aeration.
   - Add 1.5 mL of zinc acetate solution and mix thoroughly without aeration, and let settle.
   - Add an additional 0.5 mL of zinc acetate solution and mix thoroughly. If additional precipitation is not observed, then proceed to the next step. If this additional precipitate is observed, then continue adding 0.5 mL aliquots until all the sulfide has been precipitated. A small excess of zinc acetate is to be expected.

2. Filter the sample through a glass fiber filter and discard the filtrate.

3. Return the filter to the original sample container. Resuspend the precipitate in 500 mL of carbonate stabilized water and proceed with the titration.

4. Measure from a buret standardized 0.025 N iodine solution into the sample container in 5 mL increments until there is an excess of the amount needed to oxidize the sulfide (indicated by the amber color of the iodine remaining in the flask). The amount of iodine used to begin with is determined by the analyst.

- If titrating a standard or other known amount, use 15 mL of iodine.
- If the sample emits the odor of H_2S, probably start with 20 mL of iodine.
- If the sample is unknown and odor-free, the analyst should start with 5 mL of iodine. It must be added in 5 mL aliquots until the amber color persists.

5. Record the amount of iodine used.
6. Add 2 mL 6 N HCl.
7. Titrate the solution in the flask with standard 0.025 N sodium thiosulfate solution until the amber color fades to yellow.
8. Add enough starch indicator for the solution to turn dark blue and titrate until the blue disappears.
9. Record the volume of titrant used.

Calculation

$$mg/L\ S^{2-} = [(A \times B) - (C \times D)] \times 16,000/mL\ sample$$

where A = iodine solution (mL), B = normality of iodine, C = $Na_2S_2O_3$ (mL), D = normality of $Na_2S_2O_3$, 16 = Eqw of S^{2-}, 16,000 = 16 × 1000, and 1000 = mL/L.

Quality control

Method blank — A method blank is analyzed once with every sample batch or once every 20 samples, whichever is most frequent. It is 200 mL DI water.

Matrix spike/matrix spike duplicate (MS/MSD) — **NOTE:** This method consumes the entire liter of sample. If spikes are required, three 1 L bottles must be provided! The sample is analyzed in triplicate with 30 mL of the sulfide stock standard (see above) added to two of the sample aliquots. MS/MSDs are treated as samples and taken through all steps of the analytical process! (Concentration of the spike depends on the real concentration of the stock solution. If the value is 285 mg/L, the added spike is 0.285 × 30 = 8.55 mg/L.)

Quality control check standard (QCCS) — The QCCS is to be processed through the entire analytical procedure with each sample batch. It is titrated at the beginning of the batch and after every 10 samples.

Reference check standard (RCS) — The RCS is to be titrated only. The described pretreatment prior to titration (see above "Procedure" steps 1 to 3) is not needed. Prepared by diluting 1.0 mL of the reference check stock solution to a final volume of 25 mL. Titrate immediately after preparation. The entire 25 mL aliquot is titrated. (11.4 mg/L).

26.1.4 Calculation of Un-Ionized Hydrogen Sulfide

H_2S and HS^-, which together constitute dissolved sulfide, are in equilibrium with hydrogen ions:

$$H_2S \Leftrightarrow H^+ + HS^-$$

The ionization constant of H_2S is used to calculate the distribution of dissolved sulfide between the two forms. The practical constant written in logarithmic form, as pK', is used. The constant varies with temperature and ionic strength of the solution. The ionic strength effect can be estimated most easily from the conductivity. Table 26.1 gives the pK' values for various temperatures and conductivities. The temperature effect is practically linear from 15°C to 35°C.

1. From sample pH and appropriate pK' value, calculate pH – pK'.
2. From Figure 26.1, read proportion of dissolved sulfide present as H_2S (left-side scale on figure).
3. Let this proportion equal J.
4. Calculate un-ionized H_2S expressed as S^{2-} as follows:

$$\text{Un-ionized } H_2S \text{ as } S^{2-} \text{ mg/L} = J \times \text{dissolved sulfide}$$

26.2 DETERMINATION OF TOTAL SULFIDE IN LIQUID AND SOLID WASTES

26.2.1 General Discussion

This method is applicable to all aqueous and solid waste materials, or effluents, with the condition that the waste does not form an explosive mixture when mixed with acid.

Total sulfide is usually defined as the acid-soluble fraction of a waste. It is this parameter that is addressed by this method. For wastes where only metal sulfides are suspected, the acid insoluble fraction should be performed. That methodology is not addressed here.

The method is applicable to determining sulfide levels between 2.0 mg/kg and 400 mg/kg or 1.0 mg/L and 2000 mg/L.

TABLE 26.1

Values for pK', Logarithm of Practical
Ionization Constant for Hydrogen Sulfide

Conductivity at 25°C (μmhos/cm)	pK' at Given Temperature		
	20°C	25°C	30°C
0	—	7.03	—
100	7.08	7.01	6.94
200	7.07	7.00	6.93
400	7.06	6.99	6.92
700	7.05	6.98	6.91
1200	7.04	6.97	6.90
2000	7.03	6.96	6.89
3000	7.02	6.95	6.88
4000	7.01	6.94	6.87
5200	7.00	6.93	6.86
7200	6.99	6.92	6.85
10,000	6.98	6.91	6.84
14,000	6.97	6.90	6.83
22,000	6.96	6.89	6.82
50,000	6.95	6.88	6.81

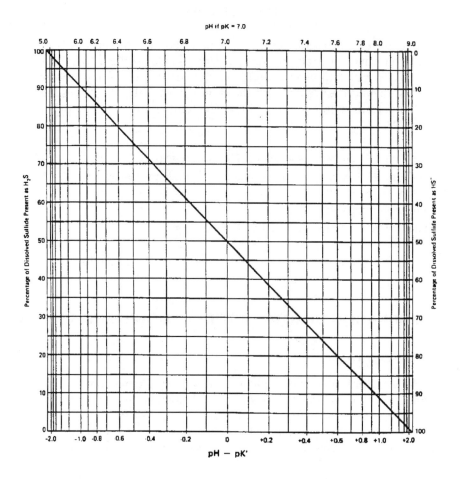

Figure 26.1
Proportions of H₂S and HS⁻ in dissolved sulfide.

26.2.2 Sample Collection and Handling

Solid samples are not preserved but are stored at 4°C. Liquid samples are collected and preserved as described in Section 26.1. Samples must be analyzed within 7 days of sampling.

26.2.3 Distillation Followed by Iodometric Titration

Method No. 9030A (EPA SW-846).

Principle of the method — Sulfide is separated from the sample by the addition of H_2SO_4. The sample is heated to 70°C and the resulting hydrogen sulfide is distilled under acidic conditions and carried by a stream of nitrogen into zinc acetate gas scrubbing bottles, where it is precipitated as zinc sulfide. This sulfide is oxidized to sulfur with a known excess amount of iodine. Then the excess iodine is determined by titration with a standard solution of sodium thiosulfate until the blue iodine-starch complex disappears.

Interferences — Reduced sulfur compounds, such as sulfite and hydrosulfite, decompose in acid and may form sulfur dioxide. This may react with the iodine solution and positively interfere. Formaldehyde is used in the gas scrubbing bottles to remove this interference. Care should also be taken to minimize contact with the air to avoid the loss of volatile sulfide and to avoid reactions with oxygen that convert the sulfide to sulfur compounds not detected by this method.

Apparatus and materials — Same as listed in Section 26.1.3 and the following.

- Round-bottom flask: Three neck with 24/40 ground glass joints, 500 mL
- Gas scrubber with fritted inlet, 125 mL
- Addition funnel with 24/40 ground glass joints, 250 mL
- Teflon sleeves, 24/40
- Magnetic stirrer/hot plate and oval-shaped magnet
- Gas manifold: $1/8$ in. copper tubing with micro-metering needle valves to control N_2 flow rate
- Flexible tubing (tygon or equivalent), $1/8$ in., to connect the manifold to the gas inlet on the round-bottom flask
- Flexible tubing (tygon or equivalent), $1/8$ in., to connect the scrubber bottle to the flask outlet
- UHP-grade N_2 with a two-stage regulator attached to the manifold
- Top-loading balance capable of weighing to the nearest 0.01 g

Reagents — Same as in Section 26.1.3 and zinc acetate solution, about 0.5 *M:* Dissolve 110 g zinc acetate dihydrate in 200 mL DI water in a 1 L volumetric flask. Add 1 mL concentrated HCl. Fill up to 1 L with DI water. Also, Nitrogen gas, UHP (ultra-high purity).

Distillation procedure

1. Determine the approximate amount of H_2SO_4 required to adjust a measured amount of the sample to a pH <1. Place a known amount of sample into a disposable beaker. Stir the mixture and record the measured pH. Slowly add sulfuric acid until the pH is <1. *This procedure should be carried out in a fume hood!* Discard this preliminary sample!

2. Into each gas scrubbing bottle, pipette 10 mL of 0.5 *M* zinc acetate, 5 mL of 37% formaldehyde, 100 mL DI water, and 7 mL 1 *N* NaOH.

3. Turn on the nitrogen at the source, regulate the pressure to approximately 7.5 psig.

4. Proceed from steps 5 to 13 below for each unit individually. Ensure that there is excess gas flow for each unit before proceeding to the next unit. After flow for each unit has been established, adjust the flow to 25 mL/min using a bubble flow meter at the outlet of the gas scrubber bottles.

5. If the matrix of the sample is soil, place on a top-loading balance and accurately weigh a 15.0 g (±0.5 g) portion into the gas evolution flask (round-bottom reaction flask).

6. Add the volume of H_2SO_4 calculated (step 1) plus an additional 50 mL into the dropping funnel. The stopcock must be closed.

7. Place an oval-shaped, Teflon-covered stirring bar into the flask. Assemble the three-neck 50 mL flask, gas inlet tube, and exhaust tube.

8. Connect the gas evolution flask, dropping funnel, and gas scrubbing bottles. Secure all fittings joints with plastic joint clamps and monitor for leaks.

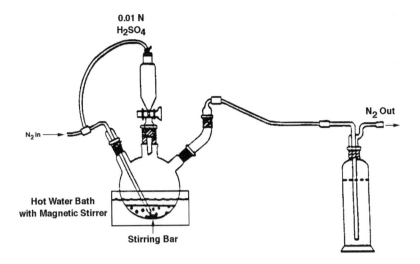

Figure 26.2
Gas evolution apparatus to determine reactive sulfide released from wastes.

9. If the sample matrix is aqueous, place 200 mL of the sample, measured in a graduated cylinder, into the gas evolution flask through the side arm where the gas inlet attaches.

10. If the sample matrix is solid, add 200 mL DI water to the weighed sample already in the flask.

11. Start with the unit closest to the nitrogen cylinder. Establish that there is flow in each unit by observing bubbles in the gas scrubber bottles. Adjust the flow to 25 mL/min for each unit. Turn on the magnetic stirrer and adjust it to a level sufficient to keep solid samples from settling.

12. Purge system for 15 minutes with nitrogen to remove all oxygen from the system.

13. Heat sample by bringing the temperature to 70°C.

14. After 70°C has been attained, open the dropping funnel to a position that will allow a flow of sulfuric acid of approximately 5 mL/min. Allow the sulfuric acid to drain into the sample flask until the acid level in the dropping flask reaches the index mark on the funnel. At that point, close the funnel to stop the flow. Do not let the funnel empty completely.

15. Purge, stir, and maintain a temperature of 70°C for a total of 90 minutes from start to finish. Monitor the flow to ensure that it remains at 25 mL/min. This is necessary because high samples will form zinc sulfide, which will precipitate and may clog the frit, requiring the needle valve to be opened to maintain flow in the unit.

16. After 90 minutes have elapsed, disconnect the gas scrubbing bottles from the exhaust tube for each unit. Then, shut off nitrogen supply, magnetic stirrers, and hot plates. Use the power strip switch to shut off the heaters so that they will not have to be readjusted each time.

Titration

1. Pipet standardized 0.025 N iodine solution into a 500 mL Erlenmeyer flask until there is an excess of the amount needed to oxidize the sulfide. Add 2 mL of 6 N HCl and bring the volume to 100 mL with DI water. The amount of added iodine should be estimated according to the character of the sample.

   • For standard or other known amount, use 15 mL of iodine.

   • If the sample emits the odor of hydrogen sulfide, start with 20 mL of iodine.

- If the sample is unknown and odor-free, start with 5 mL of iodine at the beginning.
- Record the added milliliters of iodine solution.

2. Using a long-necked funnel with a rubber hose attached that extends below the surface of the iodide solution, pour the gas scrubbing bottle solutions into the flask. If at any time in transferring the scrubbing solutions or rinsing the bottles the amber color of the iodine solution fades or disappears, more 0.025 N iodine solution must be added until the amber color is stable. The total amount of added 0.025 N iodine solution must be recorded.

3. If there appears to be any zinc sulfide remaining, prepare a rinse solution of a known amount of standard iodine, 1 mL of 6 N HCl and DI water to rinse the remaining zinc sulfide from the gas scrubbing bottles into the Erlenmeyer flask.

4. Titrate the solution in the flask with standard 0.025 N sodium thiosulfate solution until the amber color fades to yellow. Add enough starch indicator for the solution to turn dark blue and titrate until the blue disappears. Record the volume of titrant used.

Calculation

$$\text{mg/L sulfide} = [[(A \times B) - (C \times D)] \times 16{,}000]/V] \times [250/\text{wt}_{sample} \text{ or } v_{sample}]$$

where A = iodine added (mL), B = normality of iodine, C = sodium thiosulfate titrant (mL), D = normality of the sodium thiosulfate titrant, 16,000 = 16 × 1000 (16 is the equivalent weight of sulfide and 1000 is the factor, mL/L), V = amount of sample aliquot titrated (mL), 250 = amount of the scrubbing solution (mL), wt_{sample} = weight of the sample, as g (solid matrix), and v_{sample} = volume of the sample, as mL (liquid matrix).

Calculation for standards is the same as in Section 26.1.3.

Quality control — Same as in Section 26.1.3.

Chapter 27

DETERMINATION OF SULFITE AND SULFATE

27.1 ANALYSIS OF SULFITE, SO_3^{2-}

27.1.1 General Discussion

Sulfite (SO_3^{2-}) ions may occur in boilers and boiler feedwaters treated with sulfite for dissolved oxygen control. Excess sulfite in boiler waters is deleterious, because it lowers pH and promotes corrosion. Sulfite in natural waters or wastewaters is a result of industrial pollution, and in a treatment plant, dechlorination with sulfur dioxide. Control of sulfite ions in wastewater treatment and discharge may be environmentally important, principally because of its toxicity to fish and other aquatic life and its rapid oxygen demand.

27.1.2 Sample Collection and Handling

Collect fresh samples, taking care to minimize contact with air. No preservation is necessary; analyze samples immediately after collection.

27.1.3 Iodometric Titration Method

Method No. 377.1 (EPA-600/4-79-020). Method detection limit: 2 mg/L.

Principle of the method — An acidified sample is titrated with potassium iodine–iodate titrant. Free iodine, liberated by the iodine–iodate reagent, reacts with SO_3^{2-}. The titration, the end point is signalled by the blue color resulting from the first excess of iodine reacting with a starch indicator.

The method is applicable to drinking waters, sewage, and industrial wastes.

Interferences — The temperature of the sample must be below 50°C. Care must be taken to allow as little contact with air as possible. For example, do not filter the sample. Keep the buret tip below the surface of the sample.

Other oxidizable substances, such as organic compounds, ferrous iron, and sulfide, are positive interferences. Sulfide may be removed by adding 0.5 g of zinc acetate and analyzing the supernatant of the settled sample.

Nitrite gives a negative interference by oxidizing sulfite when the sample is acidified; this is corrected by either using a proprietary indicator which eliminates nitrite or by adding sulfamic acid. Copper and possibly other heavy metals catalyze the oxidation of sulfite; EDTA is used to complex metals.

A blank must be run to correct for interferences present in the reagents.

Apparatus — General laboratory glassware is needed, including a volumetric flask, 1000 and 250 mL, volumetric pipettes, a buret, and a Erlenmeyer flask.

Reagents

- Sulfuric acid, H_2SO_4 1+1.
- Potassium iodide–iodate (KI-KIO$_3$) titrant, 0.0125 N: Dissolve 0.4458 g anhydrous potassium iodate, KIO$_3$, primary standard-grade (dried at 120°C for 4 hours and stored in desiccator), 4.25 g potassium iodide, KI, and 0.310 g sodium bicarbonate, NaHCO$_3$, in DI water and dilute to 1 L. 1 mL of this titrant is equivalent to 500 μg SO_3^{2-}. 1 mL = 500 μg or 0.5 mg SO_3^{2-}.
- Sulfamic acid (NHSO$_3$H) crystals.
- Starch indicator: Make a paste from 5 g starch in a little cold water. Add the suspension to the 1 L of boiling water. Stir and let settle overnight. Use the clear supernatant. Preserve by adding either 1.3 g salicylic acid, 4 g $ZnCl_2$, or a combination of 4 g sodium propionate and 2 g sodium azide to 1 L starch solution.
- EDTA reagent: Dissolve 2.5 g disodium EDTA in 100 mL DI water.
- Stock reference (independent) standard, 200 mg/L: Weigh 3.150 g ±0.005 g anhydrous sodium sulfite, Na$_2$SO$_3$, into a 1 L volumetric flask with approximately 800 mL DI water. Mix to dissolve and bring to volume with DI water.
- Working reference (independent) standard, 16 mg/L: In a 250 mL volumetric flask filled with approximately 200 mL of DI water, add 2.0 mL of stock independent standard with volumetric pipet. Swirl to mix and bring to volume with DI water. Prepare daily.

Procedure

1. Place 1 mL 1+1 H_2SO_4 in a titration vessel.
2. Add 0.1 g sulfamic acid crystals.
3. Add 50 to 100 mL sample.
4. Add 1 mL starch (approximately five drops).
5. Titrate with potassium iodide–iodate titrant until a faint permanent blue color develops. View the color change against a white background. Record the volume of the titrant.

Calculation

$$\text{mg/L } SO_3^{2-} = [(A - B) \times N \times 40 \times 1000]/\text{mL sample} \qquad (27.1)$$

where A = titrant for sample (mL), B = titrant for blank (mL), N = normality of KI-KIO$_3$, and 40 = equivalent weight of SO_3^{2-} (80/2).

Quality Control

Method (reagent) blank — A method blank is analyzed once with every sample batch or once every 20 samples, whichever is more frequent. In this titrimetric procedure, the method blank measures the amount of titrant used to reach the end point in the absence of a sample. This value is subtracted from the amount of titrant used to reach the end point of a sample.

Duplicate samples — One duplicate sample must be analyzed once with every sample batch or once every 20 samples, whichever is more frequent. Duplicate samples are used to calculate precision.

Reference (independent) standard — A reference standard is titrated immediately after the blank, once every 10 samples, and once at the end of the sample batch. It is a sample taken from the working reference standard solution, with the value of 16 mg/L.

27.2 DETERMINATION OF SULFATE, SO_4^{2-}

27.2.1 General Discussion

Many sulfate compounds are soluble in water. Most of them originate from the oxidation of sulfate ores, gypsum ($CaSO_4 \cdot 2H_2O$) and anhydride ($CaSO_4$), oxidation of pyrite (fool's gold), which is an iron sulfide (FeS_2), and the existence of industrial wastes.

Sulfate content of a groundwater usually does not exceed 100 mg/L. Higher sulfate content is expected in groundwater close to deposits of sedimentary rocks. Groundwater may contain other minerals such as magnesium sulfate heptahydrate, $MgSO_4 \cdot 7H_2O$, (Epsom salt) and sodium sulfate decahydrate, $Na_2SO_4 \cdot 10H_2O$ (Glauber's salt). These salts in sufficient quantities are responsible for a bitter taste in water, which is detectable over 500 mg/L sulfate content. For people not accustomed to drinking high-sulfate water (over 600 mg/L), these salts may act as a laxative. It is mainly the reason that the maximum contaminant level (MCL) for sulfate is 250 mg/L. Since sulfate compounds are quite mobile in aquatic systems, there are few circumstances where concentrations will naturally decrease. One exception, however, is sulfate reduction by bacteria. Sulfate-reducing bacteria derive energy from the oxidation of organic compounds, obtaining oxygen from sulfate ions. The bacterial reduction of sulfates produces H_2S gas as a by-product. If sufficient iron is present under moderate reducing conditions, iron sulfides may also be precipitated.

Much of the H_2S gas escapes directly to the atmosphere if reduction is by bacteria in the soil.

Sulfate is one of the major dissolved constituents in rain. The amount is related to natural and human pollution of the atmosphere. This soluble substance may be derived from dust particles, oxidation of sulfur dioxide (SO_2), and hydrogen sulfide (H_2S) gases released from the decomposition of organic material, volcanic discharges, forest fires, and bacterial decay. Anthropogenic (human pollution) sources of the sulfur oxides are of major concern, because they are concentrated in urban and industrial regions, causing local levels to be quite high. About 70% of anthropogenic SO_2 comes from electric power plants, most of which burn coal. In the atmosphere, SO_2 reacts with oxygen and water to produce sulfuric acid (H_2SO_4) as a toxic pollutant with far-reaching effects. Acid rain and snow are formed when two pollutant gases, sulfur and nitrogen oxides, combine with water. Sulfur oxides form sulfuric acid, nitrogen oxides form nitric acid. Both are powerful acids. They may accumulate in clouds and fall in rain and snow. The process is called *wet deposition*.

Sulfur and nitrogen oxide gases also form sulfate and nitrate particulates. These may settle out from the atmosphere onto surfaces, as plants and other solid surfaces. These particles are called *dry deposits*, and they combine with water to form acids.

27.2.2 Sample Collection and Handling

In the presence of organic matter, certain bacteria may reduce SO_4^{2-} to S^{2-}. To avoid this, store samples at 4°C for a maximum of 28 days.

25.2.3 Sulfate Determination with Turbidimetric Method

Method No. 375.4 (EPA-600/4-79-020). Minimum detectable concentration is 1 mg/L SO_4^{2-}.

This method is applicable to groundwater, drinking and surface waters, domestic and industrial wastes, sediments, and soils.

Principle of the method — Sulfate ion is converted to a barium sulfate suspension under controlled conditions. The resulting turbidity is determined by a spectrophotometer (or nephelometer) and compared to a curve prepared from standard sulfate solution. Solid samples are extracted into water, filtered, and then analyzed.

Interferences — Color and turbidity due to the sample matrix can cause positive interferences which must be accounted for by blank subtraction when necessary. Silica in concentration over 500 mg/L will interfere.

Apparatus

- Magnetic stirrer: Use magnets of identical shape and size. The exact speed is not critical, but keep it constant for each run of samples and standards and adjust it to prevent splashing.
- Spectrophotometer for use at 420 nm.
- Stopwatch.
- Measuring spoon, capacity 0.2 to 0.3 mL.
- General laboratory glassware.

Reagents

- Conditioning reagent: Slowly add 30 mL of concentrated HCl to 300 mL DI water, 100 mL 95% ethanol or isopropyl alcohol (IPA), and 75 g NaCl in a container. Add 50 mL glycerol and mix. For complete dissolution, use of a magnetic stirrer is helpful.
- Barium chloride, $BaCl_2$, crystals, 20 to 30 mesh.
- Standard sulfate (SO_4^{2-}) Solution, 100 mg/L: Dissolve 0.1479 g anhydrous sodium sulfate (Na_2SO_4) in DI water and dilute to 1 L.

$$1 \text{ mL} = 0.1 \text{ mg } SO_4^{2-}$$

- Stock reference standard, 100 mg/L: Dissolve 0.1479 g of anhydrous sodium sulfate (Na_2SO_4) (from other source than the calibration standard) into 800 mL DI water in a 1 L volumetric flask. Swirl and dilute to 1 L with DI water.
- Working reference standard, 20 mg/L: In a 100 mL volumetric flask, add about 50 mL DI water and transfer by volumetric pipet 20 mL of stock reference standard. Dilute to the volume and mix well.

Procedure

Extraction of solid samples

1. Into a clean 250 mL plastic container with a threaded top, weigh 10.0 (±0.5 g) of sample on a top-loading balance. Add 100 mL DI water and close the cover tightly.
2. Repeat step 1 for all samples and place containers on the automatic shaker for 30 minutes.
3. After shaking, filter out the solids. Rinse the filter cakes with three small aliquots of water. Discard the filters.
4. Adjust the final volumes of the filtrates to 100 mL in volumetric flasks.

Preparation of calibration standards — Prepare calibration standards by diluting standard sulfate solution (100 mg/L). All standards are prepared in 100 mL volumetric flasks. Measure standards and dilute to 100 mL with DI water to obtain 5, 10, 20, 30, and 40 mg/L sulfate standards. Use 100 mL DI water as a blank. Above 40 mg/L, accuracy decreases and $BaCl_2$ suspensions lose stability. Prepare the calibration standards as follows:

Volume of standard	Concentration SO_4^{2-}
0 mL diluted to 100 mL	Blank
5 mL diluted to 100 mL	5 mg/L
10 mL diluted to 100 mL	10 mg/L
20 mL diluted to 100 mL	20 mg/L
30 mL diluted to 100 mL	30 mg/L
40 mL diluted to 100 mL	40 mg/L

Formation and measurement of barium sulfate turbidity

1. Transfer blank and standards into 250 mL Erlenmeyer flasks.
2. With a volumetric pipet, add exactly 5 mL conditioning reagent.
3. Mix in a stirring apparatus.
4. While the solution is being stirred, add a measuring spoonful of $BaCl_2$ crystals and begin timing immediately.
5. Stir exactly 1 minute at constant speed.
6. Read the absorbance at 420 nm at 30 second intervals for 4 minutes.
7. Record the maximum reading obtained in the 4-minute period.
8. Prepare calibration curve by plotting sulfate standard concentrations against absorbance readings.
9. Using a calculator with linear regression capabilities, calculate the correlation coefficient (R value). The correlation coefficient must be greater than 0.995. If the R value is less than 0.995, the calibration standards must be remade and a new curve analyzed.

Analysis of samples

1. Place 100 mL sample (or soil extract), or a suitable portion diluted to 100 mL, into a 250 mL Erlenmeyer flask.
2. Follow steps 2 to 7 in the previous section, "Formation and measurement of barium sulfate turbidity."

Correction of sample color and turbidity — Run a sample blank using the procedure without addition of $BaCl_2$. Subtract reading from sample reading. Use this corrected reading to calculate the concentration of the sample.

Calculation — Calculate sample result from the calibration curve by using linear regression calculation. In the case of any dilution, correct the result by the dilution factor. For liquid samples, result is expressed as sulfate as SO_4^{2-} mg/L. For solid samples, result expressed as sulfate as SO_4^{2-} mg/kg on dry bases.
   For calculation of the result for solid samples, see Section 15.2.6.

$$mg/kg \text{ on wet base} = (mg/L \times \text{final volume})/g \text{ sample}$$

$$mg/kg \text{ on dry base} = mg/kg \text{ on wet base}/f$$

where f = decimal fraction of dry weight.

Quality control — For reference of QC items, calculations, and acceptance criteria, see Section 3.2.7, Section 3.2.9, and Section 16.2.

Method blank — A method blank is to be performed at least once every sample batch or every 20 samples, whichever is more frequent.

Continuing calibration standard (CCS) — This is a 30 mg/L calibration standard, analyzed immediately after the calibration curve to check calibration, and also analyzed once every 10 samples to verify that the calibration curve is constant throughout the run. Recovery of this standard should be within ±10% of the value.

Reference (independent) standard — This is analyzed once every 10 samples. Since the reference standard is prepared from a source different from the calibration standards and CCS, the reference standard is used to verify that calibration standards are actually at the concentrations claimed by the analyst. Acceptable limits are established by individual laboratories.

Matrix spike/matrix spike duplicates (MS/MSD) — MS/MSDs are performed once for every sample batch or once every 20 samples, whichever is more frequent. A sample is analyzed in triplicate with 10 mL of stock sulfate standard. Spike value is 10 mg/L sulfate. Acceptance limit for percentage of recovery is established by individual laboratories.

27.2.4 Automated Methylthymol Blue (MTB) Method

This automated method is applicable to drinking and surface waters, and domestic and industrial wastes. Samples can be analyzed in the range of 5.0 mg/L to 50 mg/L. This range can be expected by dilution.

Principle of the method — The sample is first passed through a sodium form of cation-exchange resin to remove multivalent metal ions, which interfere with the analysis. The sample containing sulfate is then reacted with an alcohol solution of barium chloride and methylthymol blue (MTB) at a pH of 2.5 to 3.0, so that excess barium reacts with MTB. The uncomplexed MTB color is gray; if it is all chelated with barium, the color is blue. Initially, the barium and MTB are equimolar and equivalent to 300 mg/L of sulfate; thus the amount of uncomplexed MTB is equal to the sulfate present.

Interferences — The ion exchange column eliminates interferences from multivalent cations. Samples with pH below 2 must be neutralized, because high acid concentrations elute cations from the ion exchange resin. Turbid samples must be filtered or centrifuged.

Apparatus, materials, reagents, and procedure — According to the available automated continuous flow analyzer, such as Technicon II, Alpkem RFA, etc. Consult manufacturer's manual.

Quality control — Same as for Section 27.2.3.

Chapter 28

CATION EXCHANGE CAPACITY

28.1 GENERAL DISCUSSION

Cation exchange capacity (CEC) is a reversible chemical reaction. The cations held on the surface of the soil minerals and within the crystal framework of some mineral species and those which are a part of certain organic compounds can be reversibly replaced by those of salt solutions and acids. Cation exchange capacity is expressed in milliequivalents (meq) per 100 g of soil. It is defined as the sum of the exchangeable cations of a soil. The advantages of the ammonium acetate CEC method described belfow are that it is easy to use, it is highly buffered so the soil remains neutral throughout the leaching process, and the ammonium ion is easy to determine analytically.

28.1.1 Sample Collection and Storage

Samples must be collected in soil jars and kept at 4°C until the aliquots are removed for air drying.

28.2 CATION EXCHANGE CAPACITY BY THE AMMONIUM ACETATE METHOD

MSA 57-2.1.2.

Principle of the method — An air-dried, sieved soil sample is soaked overnight with ammonium acetate. It is then filtered and leached three times with ammonium acetate, three times with 1 N ammonium chloride, and once with 0.25 N ammonium chloride, and finally washed with isopropyl alcohol. The filter cake is then leached with 225 mL of acidified NaCl. The ammonium content is determined by one of the method listed in Sections 23.2.3, 23.2.4, 23.2.5, and 23.2.6. The practical quantitation limit (PQL) is established by the method. The applicable range of this method is 0.2 meq/100 mL based on a 10 g sample.

Interferences — Some exchangeable ions are more easily replaced than others due to the salt solution and the characteristics of the soil. In soils with high organic content or 1:1 type clay material such as kaolin or halloysite, the ammonium acetate method will yield somewhat lower results than the barium acetate method. This is due to the less complete replacement of absorbed hydrogen and aluminum ions by ammonium acetate.

Some common materials in soils such as $CaCO_3$ and $CaSO_4$ are soluble in some of the extractants. The results obtained by calcareous soils will be too low, because the dissolved calcium present in the ammonium acetate solution prevents complete saturation of exchange positions with ammonium.

In soils containing vermiculite clay, interlayer cations such as Ca^{2+}, Mg^{2+}, Na^+, and H^+ can be replaced with ammonium, but the ammonium so fixed is not replaceable by the method.

Apparatus and materials

- Desiccator with color-coded desiccant
- Top-loading balance
- Buchner funnels
- pH paper, full range
- Glassware:
  - Volumetric flask, 2 L and 1 L
  - Filter flask, 1 L
  - Volumetric pipets, 50 mL, 25 mL, 10 mL, 5 mL, 3 mL
  - Erlenmeyer flasks, 500 mL
- Parafilm
- Filter paper, Whatman No. 41 or equivalent

Reagents

- Ammonium hydroxide, concentrated.
- Ammonium acetate, 1 N: In a 2 L volumetric flask, dilute 114 mL of glacial acetic acid with DI water to a volume of approximately 1 L. Add 138 mL concentrated ammonium hydroxide. Add DI water to approximately 1950 mL. Check the pH with pH paper, add more ammonium hydroxide until the pH is 7.00, and dilute the solution to 2 L with DI water.
- Isopropyl alcohol, 99%.
- Ammonium chloride, 1 N neutral: In a 1 L volumetric flask, add approximately 500 mL DI water, and dissolve 53.49 g of ammonium chloride in it. Adjust the pH to 7.0 with ammonium hydroxide, and dilute to the volume with DI water and mix.
- Ammonium chloride, 0.25 N neutral: In a 1 L volumetric flask, add approximately 500 mL DI water. Dissolve 13.37 g ammonium chloride. Adjust the pH to 7.00 with ammonium hydroxide. Fill up to the volume with DI water and mix.
- Hydrochloric acid, 1 N: In a 1 L volumetric flask, pipet 83 mL of concentrated HCl, dilute to volume, and mix well.
- Sodium chloride, acidified, 10%: In a 2 L volumetric flask, add approximately 1 L DI water. Add 200 g NaCl and 10 mL 1 N HCl. Mix well and dilute to volume with DI water, mix.

Procedure

1. Dry the soil sample in the desiccator until all the moisture is gone. Grind the sample in a mortar and pestle until all the clumps are broken and the particles are finely divided.

2. On a top-loader balance, weigh 10.00 ± 0.5 g and record the weight and transfer into one 500 mL Erlenmeyer flask.

3. Add 250 mL neutral ammonium acetate, 1 N solution to the flask.

4. Cover the flask with "parafilm" and mix the contents thoroughly.

5. Let the sample sit overnight.

6. Filter the soil using a Buchner funnel and No. 41 filter paper. Do not let the soil dry.

7. Rinse the soil 3 times with 25 mL aliquots of neutral 1 N ammonium chloride.

8. Rinse the soil 3 times with 25 mL aliquots of neutral 0.25 N ammonium chloride.

9. Wash the soil with 150 to 200 mL of 99% isopropyl alcohol.

10. Dispose of the liquids and rinse the vacuum flask with DI water.

11. Wash the ammonium-saturated soil with 10% acidified NaCl solution until 225 mL have passed through the sample. Add the NaCl in small portions, allowing each portion to pass through soil before adding the next portion. Store liquid in wide mouth soil jars in the refrigerator until time for the analysis.

12. Analyze the ammonia with one of the methods described in Sections 23.2.3, 23.2.4, and 23.2.5. The samples must be diluted (about 1:50) before analysis.

Calculation — The result is calculated as milliequivalents per 100 g soil:

$$CEC = Reading \times DF \times 1/14.0067 \times 0.225 \times W_{soil} \qquad (28.1)$$

1. Reading $\times$ dilution factor $= $ mg/L NH_3-N.

2. mg/L convert to meq/L $=$ mg/L $\times$ factor (1/14.067 for N).

3. meq/L convert to meq/225 mL (10 g soil).

4. meq/10 g soil convert to meq/100 g soil.

For example:

1. Reading is 0.5 mg/L N, dilution factor is 50, therefore $0.5 \times 50 = 25$ mg/L NH_3N.

2. Factor to convert mg/L to meq/L for N is $1/14.067 = 0.07108$. 25 mg/L $\times 0.07108 = 1.777$ meq/L N.

3. 1.777 meq/L convert to meq/225 mL. $1.777 \times 0.225 = 0.3998$ meq/225 mL (10 g soil).

4. 0.3998 meq/10 g soil, 3.998 meq/100 g soil.

Quality control

Method blank — The method blank is to be analyzed once with each sample batch. The MB is an aliquot of the reagents used to soak the soil and the leaching solutions used to wash the filter cake.

Duplicate — One duplicate sample must be analyzed once with every sample batch, once every 10 samples, or once a month, whichever is more frequent.

Appendix A

LABORATORY FIRST AID

First aid is an emergency procedure to be followed until the physician arrives: stop bleeding, start breathing, treat for poisoning, treat for shock, and then care for the wound(s) — in that order.

GENERAL RULES

1. The prime rule — a difficult one to obey — is *keep calm.*
2. Keep crowds away.
3. Do nothing else unless you are certain of the proper procedure.
4. The right person to administer first aid is a *trained first aider.*
5. Have someone CALL FOR PROFESSIONAL ASSISTANCE IMMEDIATELY. Emergency phone numbers must be posted near to the phone: the number of the nearest hospital and poison-control center.

IMPORTANT STEPS

1. *Call an ambulance;* state type of accident, its location, type of injuries, and the number of persons injured.
2. *Stop any bleeding,* as directed under "Wounds and Fractures" below.
3. *Restore breathing* by administering artificial respiration.
4. *Treat for poisoning* if necessary.
5. *Treat for physical shock,* as directed under "Shock" below, in all injuries.
6. *Remove the patient from a hazardous environment* involving spillage of chemicals or high concentrations of noxious gases, vapors, or fumes.
7. *Wear proper protective clothing and a respirator* so as not to expose yourself to the environment.
8. *Never give liquids to an unconscious person* or to persons with abdominal or lower chest injuries.
9. *Do not move* a person with possible broken bones or possible head or internal injuries unless fumes or fire necessitate it. Then, improvise a splint to support and prevent aggravating the fracture.

FIRST AID FOR THERMAL BURNS

1. Smother flames by rolling the victim in a coat or blanket. If clothing adheres to the burned skin surface, do not attempt to remove it, but carefully cut away the clothing around the burned area.

2. A slight burn should be immersed in cold water, with ice added, to relieve pain.

3. For first- or second-degree burns with no open wounds, apply cold-water compresses.

4. For open wounds or third-degree burns, apply a dry sterile dressing and transport the patient as soon as possible. Fasten dressings securely but not tightly.

5. Do not use ointments on severe burns. A physician will apply the proper medication after the degree and extent of the burn have been determined.

FIRST AID FOR CHEMICAL BURNS

1. Remove contaminated articles of clothing as well as the source of contamination. If necessary, wear protective clothing and respiratory equipment during this process so as not to contaminate yourself.

2. Flush the contaminated skin area with large quantities of water for at least 15 minutes.

3. Do not use oils, fats, or sodium bicarbonate on the burned area unless specifically advised to do so by a physician.

4. Do not apply salves or ointments, since these may increase skin absorption of the noxious chemical.

5. Soap may be used, especially where phenol (carbolic acid) and its derivatives are in the contaminant.

6. The patient should be taken to a hospital immediately for further treatment.

FIRST AID FOR WOUNDS AND FRACTURES

1. If bleeding is copious, it must be controlled before other aid can be given. Apply a large compress with direct pressure on the wound. If the wound is on an extremity, pressure can be applied to one of the two pressure points shown in Figure A-1.

2. When there are chest and abdominal injuries, cover the wound(s) with a sterile dressing moistened with physiological saline solution. Do not attempt to replace protruding viscera.

3. If the cut is slight and bleeding is not profuse, remove all foreign material (glass, dirt, etc.) projecting from the wound. Removal is best accomplished by careful washing with soap and water.

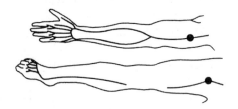

Figure A-1
Pressure points on arm and leg.

4. Apply an antiseptic to all parts of the cut and to approximately $1/2$ in. of skin around the cut.

5. All wounds should be securely, but not tightly, bandaged.

6. In cases of puncture wounds (from broken thermometers, glass tubing, etc.), the patient should be sent to a hospital or physician's office. Danger of foreign material in the wounds and inability to reach the bottom of the wound with antiseptic make this action mandatory.

7. In cases of possible fracture, do not move the patient unless fumes or fire necessitates it. Treat for bleeding and shock and leave splinting to a physician. When necessary to transport the victim to treatment, improvise a splint support to prevent aggravating the fracture in transit.

FIRST AID FOR SHOCK

Shock occurs to some extent in all injuries, varies with the individual, and can cause death. Some easily recognized symptoms are pallor, cold and moist skin with perspiration on the forehead and palms of the hands, nausea, shallow breathing, and trembling.

1. Place the patient in a reclining position, with the head lower than the body. (In cases of severe hemorrhage of the head, fractured skull, or stroke, the head should be elevated.)

2. Control any bleeding.

3. Wrap a cold patient in blankets, and elevate the patient's legs if there are no broken bones.

4. If the patient is overheated, try to reduce body temperature by sponging with cold water. The objective is to attain and maintain normal body temperature.

5. Keep the patient's airway open.

6. If vomiting occurs, turn the head to one side so that the neck is arched.

7. If there is no bleeding, rub the extremities briskly toward the heart to restore circulation, and give stimulants either by mouth or inhalation (inhalation only, if the patient is unconscious).

8. A stimulant cannot be given until bleeding is controlled and should not be given at all in cases of fractured skull, abdominal injuries, or stroke. The preferred liquid stimulant is a formula consisting of 1 teaspoon salt and $1/2$ teaspoon of sodium bicarbonate in a quart of warm water; even plain warm water will do, however. For inhalation, use aromatic spirits of ammonia, dilute ammonia, dilute acetic acid, or amyl nitrite on a cloth.

9. Do not give alcoholic liquids, and never administer any liquids while a person is unconscious.

FIRST AID FOR ELECTRIC SHOCKS

1. Shut off the current and then cautiously remove the wire or other contact while protecting yourself with an insulator. Use heavy rubber or asbestos gloves. Do not use a dry stick or dry towel!

2. Keep the patient warm by covering with blankets.

3. Start artificial respiration immediately! Efforts for revival should be continued for at least 4 hours, or until a physician certifies death, even though there is no sign that the patient is regaining consciousness.

FIRST AID FOR INGESTED POISONS

1. If the patient is conscious, give two to four glasses of water (or milk if water is unavailable) immediately.
2. Call the local poison-control center and then call an ambulance.
3. Induce vomiting *except* when poisoning is due to phenothiazine, strong acids, strong alkalies, cyanide, strychnine, gasoline, kerosene, or other hydrocarbons, or when the patient is already having convulsions. Have the patient place an index finger far back on the tongue and stroke it from side to side (do this for the patient if necessary). Emetics such as ipecac, a teaspoonful of powdered mustard in sufficient warm water to make a paste, clear warm water, or even soap suds can be used. Repeat dosage of the emetic until the vomited fluid is clear. (Ipecac is the best emetic to use.)
4. When poisoning is due to strong acids or strong bases, give the patient a glass of milk.
5. For cyanide, use artificial respiration if the patient is not breathing. Maintain consciousness by allowing inhalation of amyl nitrite for 20 or 30 seconds out of each minute.
6. For strychnine, give 1 oz of powdered charcoal and keep the patient quiet. If the patient is having convulsions, treat for shock while waiting for the ambulance.
7. For methyl alcohol poisoning, give the patient beer to drink.
8. To prevent shock, keep the patient warm.

FIRST AID FOR INHALED POISONS

1. Call an ambulance.
2. Wear protective clothing and respiratory gear. Remove the patient to fresh air immediately. Give oxygen if necessary, using an inhalator to avoid overdosage.
3. Allow the patient to rest. At any sign of cessation of breathing, begin artificial respiration.
4. Treat for shock; keep the patient warm.
5. Note that carbon monoxide poisoning often mimics food poisoning when it is not severe enough to produce unconsciousness.

ARTIFICIAL RESPIRATION

If breathing stops because of electrocution, sedative poisoning, gas poisoning, or suffocation, start artificial respiration immediately. It is important that artificial respiration, when needed, be started quickly. A patient can die within 3 minutes. Irreversible brain damage can also occur. As soon as possible, send for a physician. Mouth-to-mouth respiration is shown in Figure A-2.

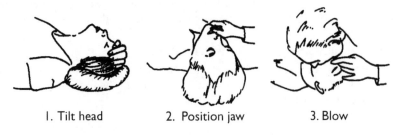

1. Tilt head 2. Position jaw 3. Blow

Figure A-2
Mouth-to-mouth method of artifical respiration.

1. Place the patient face up and loosen tight clothing.

2. Hold the lower jaw up and pinch the patient's nostrils together to close them. Open your mouth wide and place it lightly over the patient's mouth.

3. While breathing into the patient, watch the chest rise to make sure the air passage is clear. Remove your mouth, turn your head to the side, and listen for the return rush of air that indicates air exchange. Repeat the blowing effort. Blow vigorously at the rate of about 12 breaths per minute for an adult. Children require a rate of 20 breaths per minute.

4. Plastic mouth-to-mouth resuscitators are available; these remove the danger of the tongue blocking the air passage. However, don't use one unless you are thoroughly familiar with it.

5. A revived victim should be kept as quiet as possible until regular breathing is restored. Keep the patient from becoming chilled and otherwise treat for shock. Continue artificial respiration until the patient begins to breathe unaided, or until a physician states that death has occurred. Because respiratory and other disturbances may develop as an aftermath, a doctor's care is necessary during the recovery period.

FIRST AID FOR FOREIGN OBJECT IN THE EYE

1. If a piece of glass or any other foreign body flies into an eye, do not attempt to remove it. Cover the eye with a sterile gauze pad and rush the patient to a hospital or doctor's office immediately. Have someone alert the hospital or doctor that you are on the way, so that treatment can be given as soon as possible. It is most important that the patient keep hands away from the eye!

2. If any kind of chemical splashes in the eye, the recommended treatment is a minimum 10-minute washing of the eye with water. An eye-washing fountain is best for this purpose; a stream of water from a slowly running faucet or even water from a washbottle will also work. Cover the eye with a sterile gauze pad and then get the patient to a hospital or doctor's office. It is most important to remove or dilute any chemical in the eye immediately!

Appendix **B**

SALINITY DETERMINATION BY HYDROMETRIC METHOD

Salinity is an important measurement in the analysis of certain industrial wastes and seawater. It is defined as the total solids in water after all carbonates have been converted to oxides, all bromides and iodides have been replaced by chloride, and all organic matter has been oxidized. It is numerically smaller than the filtrable residue and usually is reported as grams per kilogram or parts per thousand (‰).

Salinity can be determined by measuring specific gravity with a hydrometer, correcting for temperature, and converting specific gravity to salinity at 15°C by means of density salinity tables.

The values for converting hydrometer readings at certain temperatures to density at 15°C are given in Table B-1. Determine salinity from Table B-2. Locate corrected density and read salinity as parts per thousand (‰).

TABLE B-1

Values for Converting Hydrometer Readings at Certain Temperatures to Density at 15°C

Observed Reading	Temperature of Water in Jar (°C)												
	−2.0	−1.0	0.0	1.0	2.0	3.0	4.0	5.0	6.0	7.0	8.0	9.0	10.0
0.9960													
0.9970													
0.9980													
0.9990	−1	−2	−3	−4	−5	−5	−6	−6	−6	−6	−6	−5	−5
1.0000	−2	−3	−4	−5	−5	−6	−6	−6	−6	−6	−6	−5	−5
1.0010	−3	−4	−4	−5	−6	−6	−6	−7	−7	−6	−6	−6	−5
1.0020	−3	−4	−5	−6	−6	−7	−7	−7	−7	−7	−6	−6	−5
1.0030	−4	−5	−6	−6	−7	−7	−7	−7	−7	−7	−6	−6	−5
1.0040	−4	−5	−6	−7	−7	−7	−8	−8	−7	−7	−7	−6	−6
1.0050	−5	−6	−6	−7	−8	−8	−8	−8	−8	−7	−7	−6	−6
1.0060	−6	−6	−7	−8	−8	−8	−8	−8	−8	−8	−7	−6	−6
1.0070	−6	−7	−8	−8	−8	−8	−8	−8	−8	−8	−7	−7	−6
1.0080	−7	−8	−8	−9	−9	−9	−9	−9	−8	−8	−7	−7	−6
1.0090	−7	−8	−9	−9	−9	−9	−9	−9	−9	−8	−8	−7	−6
1.0100	−8	−9	−9	−10	−10	−10	−10	−9	−9	−8	−8	−7	−6
1.0110	−9	−9	−10	−10	−10	−10	−10	−10	−9	−9	−8	−7	−6
1.0120	−9	−10	−10	−10	−10	−10	−10	−10	−10	−9	−8	−7	−7
1.0130	−10	−10	−11	−11	−11	−11	−11	−10	−10	−9	−8	−8	−7
1.0140	−10	−11	−11	−11	−11	−11	−11	−11	−10	−10	−9	−8	−7
1.0150	−11	−11	−12	−12	−12	−12	−11	−11	−10	−10	−9	−8	−7
1.0160	−12	−12	−12	−12	−12	−12	−12	−11	−11	−10	−9	−8	−7
1.0170	−12	−12	−12	−13	−13	−12	−12	−12	−11	−10	−9	−8	−7
1.0180	−13	−13	−13	−13	−13	−13	−12	−12	−11	−10	−9	−8	−7
1.0190	−13	−13	−14	−14	−13	−13	−13	−12	−12	−11	−10	−9	−8
1.0200	−14	−14	−14	−14	−14	−13	−13	−12	−12	−11	−10	−9	−8
1.0210	−14	−14	−14	−14	−14	−14	−13	−13	−12	−11	−10	−9	−8
1.0220	−15	−15	−15	−15	−15	−14	−14	−13	−12	−11	−10	−9	−8
1.0230	−15	−15	−15	−15	−15	−15	−14	−13	−12	−12	−10	−9	−8
1.0240	−16	−16	−16	−16	−15	−15	−14	−14	−13	−12	−11	−10	−8
1.0250	−16	−16	−16	−16	−16	−15	−15	−14	−13	−12	−11	−10	−8
1.0260	−17	−17	−17	−16	−16	−16	−15	−14	−13	−12	−11	−10	−8
1.0270	−18	−17	−17	−17	−17	−16	−15	−14	−14	−12	−11	−10	−9
1.0280	−18	−18	−18	−17	−17	−16	−16	−15	−14	−13	−11	−10	−9
1.0290	−19	−18	−18	−18	−17	−17	−16	−15	−14	−13	−12	−10	−9
1.0300	−19	−19	−19	−18	−18	−17	−16	−15	−14	−13	−12	−10	−9
1.0310	−20	−19	−19	−19	−18	−17	−16	−16	−15	−13	−12	−10	−9

Observed Reading	Temperature of Water in Jar (°C)											
	11.0	12.0	13.0	14.0	15.0	16.0	17.0	18.0	18.5	19.0	19.5	20.0
0.9960												
0.9970												
0.9980							3	4	5	6	7	8
0.9990	−4	−3	−2	−1	0	1	3	4	5	6	7	8
1.0000	−4	−3	−2	−1	0	1	3	4	5	6	7	8
1.0010	−4	−3	−2	−1	0	1	3	4	5	6	7	8
1.0020	−4	−3	−2	−1	0	1	3	4	5	6	7	8
1.0030	−4	−3	−2	−1	0	1	3	4	5	6	7	8
1.0040	−5	−4	−3	−1	0	2	3	5	6	6	7	8

TABLE B-1 *(continued)*

Values for Converting Hydrometer Readings at Certain Temperatures to Density at 15°C

Observed Reading	Temperature of Water in Jar (°C)											
	11.0	12.0	13.0	14.0	15.0	16.0	17.0	18.0	18.5	19.0	19.5	20.0
1.0050	−5	−4	−3	−1	0	2	3	5	6	7	8	9
1.0060	−5	−4	−3	−1	0	2	3	5	6	7	8	9
1.0070	−5	−4	−3	−2	0	2	3	5	6	7	8	9
1.0080	−5	−4	−3	−2	0	2	3	5	6	7	8	9
1.0090	−5	−4	−3	−2	0	2	3	5	6	7	8	9
1.0100	−5	−4	−3	−2	0	2	3	5	6	7	8	9
1.0110	−5	−4	−3	−2	0	2	3	5	6	7	8	9
1.0120	−6	−4	−3	−2	0	2	3	5	6	7	8	9
1.0130	−6	−4	−3	−2	0	2	4	5	6	7	8	10
1.0140	−6	−4	−3	−2	0	2	4	5	6	8	9	10
1.0150	−6	−4	−3	−2	0	2	4	5	6	8	9	10
1.0160	−6	−5	−3	−2	0	2	4	6	7	8	9	10
1.0170	−6	−5	−3	−2	0	2	4	6	7	8	9	10
1.0180	−6	−5	−3	−2	0	2	4	6	7	8	9	10
1.0190	−6	−5	−3	−2	0	2	4	6	7	8	9	10
1.0200	−6	−5	−3	−2	0	2	4	6	7	8	9	10
1.0210	−6	−5	−3	−2	0	2	4	6	7	8	9	10
1.0220	−7	−5	−3	−2	0	2	4	6	7	8	9	11
1.0230	−7	−5	−4	−2	0	2	4	6	7	8	9	11
1.0240	−7	−5	−4	−2	0	2	4	6	7	8	10	11
1.0250	−7	−5	−4	−2	0	2	4	6	7	8	10	11
1.0260	−7	−5	−4	−2	0	2	4	6	7	9	10	11
1.0270	−7	−5	−4	−2	0	2	4	6	7	9	10	11
1.0280	−7	−6	−4	−2	0	2	4	6	8	9	10	11
1.0290	−7	−6	−4	−2	0	2	4	6	8	9	10	11
1.0300	−7	−6	−4	−2	0	2	4	6	8	9	10	12
1.0310	−8	−6	−4	−2	0	2	4					

Observed Reading	Temperature of Water in Jar (°C)												
	20.5	21.0	21.5	22.0	22.5	23.0	23.5	24.0	24.5	25.0	25.5	26.0	26.5
0.9960											19	20	21
0.9970			10	11	12	14	15	16	17	18	19	20	22
0.9980	9	10	11	12	13	14	15	16	17	18	19	21	22
0.9990	9	10	11	12	13	14	15	16	17	18	19	21	22
1.0000	9	10	11	12	13	14	15	16	17	19	20	21	22
1.0010	9	10	11	12	13	14	15	17	18	19	20	21	23
1.0020	9	10	11	12	13	14	16	17	18	19	20	22	23
1.0030	9	10	11	12	13	15	16	17	18	19	21	22	23
1.0040	9	10	11	12	14	15	16	17	18	20	21	22	23
1.0050	10	11	12	13	14	15	16	17	19	20	21	22	24
1.0060	10	11	12	13	14	15	16	18	19	20	21	23	24
1.0070	10	11	12	13	14	15	17	18	19	20	21	23	24
1.0080	10	11	12	13	14	16	17	18	19	20	22	23	24
1.0090	10	11	12	13	15	16	17	18	19	21	22	23	25
1.0100	10	11	12	14	15	16	17	18	20	21	22	24	25
1.0110	10	12	13	14	15	16	17	19	20	21	22	24	25
1.0120	10	12	13	14	15	16	18	19	20	21	23	24	25
1.0130	11	12	13	14	15	16	18	19	20	22	23	24	26
1.0140	11	12	13	14	15	17	18	19	20	22	23	24	26

TABLE B-1 *(continued)*

Values for Converting Hydrometer Readings at Certain Temperatures to Density at 15°C

Observed Reading	Temperature of Water in Jar (°C)												
	20.5	21.0	21.5	22.0	22.5	23.0	23.5	24.0	24.5	25.0	25.5	26.0	26.5
1.0150	11	12	13	14	16	17	18	20	21	22	23	25	26
1.0160	11	12	13	14	16	17	18	20	21	22	24	25	26
1.0170	11	12	13	15	16	17	18	20	21	22	24	25	27
1.0180	11	12	14	15	16	17	19	20	21	23	24	25	27
1.0190	11	12	14	15	16	18	19	20	21	23	24	26	27
1.0200	11	13	14	15	16	18	19	20	22	23	24	26	27
1.0210	12	13	14	15	17	18	19	21	22	23	25	26	27
1.0220	12	13	14	15	17	18	19	21	22	23	25	26	28
1.0230	12	13	14	16	17	18	20	21	22	24	25	26	28
1.0240	12	13	14	16	17	18	20	21	22	24	25	27	28
1.0250	12	13	15	16	17	18	20	21	23	24	25	27	28
1.0260	12	13	15	16	17	19	20	22	23	24	26	27	29
1.0270	12	14	15	16	17	19	20	22	23	24	26	27	29
1.0280	12	14	15	16	18	19	20	22	23	25	26	28	29
1.0290	13	14	15	16	18	19	21	22	23				
1.0300	13	14	15	16	18								
1.0310													

Observed Reading	Temperature of Water in Jar (°C)												
	27.0	27.5	28.0	28.5	29.0	29.5	30.0	30.5	31.0	31.5	32.0	32.5	33.0
0.9960	23	24	25	27	28	29	31	32	34	35	37	38	40
0.9970	23	24	26	27	28	30	31	33	34	36	37	39	40
0.9980	23	25	26	27	29	30	31	33	34	36	38	39	41
0.9990	24	25	26	28	29	30	32	33	35	36	38	39	41
1.0000	24	25	26	28	29	31	32	34	35	37	38	40	41
1.0010	24	25	27	28	30	31	32	34	35	37	39	40	42
1.0020	24	26	27	28	30	31	33	34	36	37	39	41	42
1.0030	25	26	27	29	30	32	33	35	36	38	39	41	42
1.0040	25	26	28	29	30	32	33	35	36	38	40	41	43
1.0050	25	26	28	29	31	32	34	35	37	38	40	42	43
1.0060	25	27	28	30	31	32	34	36	37	39	40	42	44
1.0070	26	27	28	30	31	33	34	36	38	39	41	42	44
1.0080	26	27	29	30	32	33	35	36	38	39	41	43	44
1.0090	26	28	29	30	32	33	35	36	38	40	41	43	45
1.0100	26	28	29	31	32	34	35	37	38	40	42	43	45
1.0110	27	28	30	31	32	34	36	37	39	40	42	44	45
1.0120	27	28	30	31	33	34	36	37	39	41	42	44	46
1.0130	27	29	30	32	33	35	36	38	39	41	43	44	46
1.0140	27	29	30	32	33	35	36	38	40	41	43	45	46
1.0150	28	29	31	32	34	35	37	38	40	42	43	45	47
1.0160	28	29	31	32	34	35	37	39	40	42	44	45	47
1.0170	28	30	31	33	34	36	37	39	40	42	44	46	47
1.0180	28	30	31	33	34	36	38	39	41	42	44	46	48
1.0190	29	30	32	33	35	36	38	39	41	43	44	46	48
1.0200	29	30	32	33	35	37	38	40	41	43	45	47	48
1.0210	29	31	32	34	35	37	38	40	42	43	45	47	49
1.0220	29	31	32	34	36	37	39	40	42	44	45	47	49
1.0230	30	31	33	34	36	37	39	41	42	44	46	47	49
1.0240	30	31	33	34	36	37	39	41	42	44	46	48	49

TABLE B-1 *(continued)*

Values for Converting Hydrometer Readings at Certain Temperatures to Density at 15°C

Observed Reading	Temperature of Water in Jar (°C)												
	27.0	27.5	28.0	28.5	29.0	29.5	30.0	30.5	31.0	31.5	32.0	32.5	33.0
1.0250	30	31	33	35	36	38	39	41	43	44	46	48	50
1.0260	30	32	33	35	37	38	40	41	43	45	46	48	50
1.0270	30	32	34	35	37	38	40						
1.0280	31	32											
1.0290													
1.0300													
1.0310													

Note: Add tabular values to the last decimal of observed reading. For example, an observed reading of 1.0000 at 10°C is converted to 1.0000 + (–0.0005) or 0.0005 at 15°C.

TABLE B-2

Corresponding Densitites and Salinities

Density	Salinity	Density	Salinity	Density	Salinity	Density	Salinity
0.9991	0.0	1.0036	5.8	1.0081	11.6	1.0126	17.5
0.9992	0.0	1.0037	5.9	1.0082	11.8	1.0127	17.7
0.9993	0.2	1.0038	6.0	1.0083	11.9	1.0128	17.8
0.9994	0.3	1.0039	6.2	1.0084	12.0	1.0129	17.9
0.9995	0.4	1.0040	6.3	1.0085	12.2	1.0130	18.0
0.9996	0.6	1.0041	6.4	1.0086	12.3	1.0131	18.2
0.9997	0.7	1.0042	6.6	1.0087	12.4	1.0132	18.3
0.9998	0.8	1.0043	6.7	1.0088	12.6	1.0133	18.4
0.9999	0.9	1.0044	6.8	1.0089	12.7	1.0134	18.6
1.0000	1.1	1.0045	6.9	1.0090	12.8	1.0135	18.7
1.0001	1.2	1.0046	7.1	1.0091	12.9	1.0136	18.8
1.0002	1.3	1.0047	7.2	1.0092	13.1	1.0137	19.0
1.0003	1.5	1.0048	7.3	1.0093	13.2	1.0138	19.1
1.0004	1.6	1.0049	7.5	1.0094	13.3	1.0139	19.2
1.0005	1.7	1.0050	7.6	1.0095	13.5	1.0140	19.3
1.0006	1.9	1.0051	7.7	1.0096	13.6	1.0141	19.5
1.0007	2.0	1.0052	7.9	1.0097	13.7	1.0142	19.6
1.0008	2.1	1.0053	8.0	1.0098	13.9	1.0143	19.7
1.0009	2.2	1.0054	8.1	1.0099	14.0	1.0144	19.9
1.0010	2.4	1.0055	8.2	1.0100	14.1	1.0145	20.0
1.0011	2.5	1.0056	8.4	10.101	14.2	1.0146	20.1
1.0012	2.6	1.0057	8.5	10.102	14.4	1.0147	20.3
1.0013	2.8	1.0058	8.6	1.0103	14.5	1.0148	20.4
1.0014	2.9	1.0059	8.8	1.0104	14.6	1.0149	20.5
1.0015	3.0	1.0060	8.9	1.0105	14.8	1.0150	20.6
1.0016	3.2	1.0061	9.0	1.0106	14.9	1.0151	20.8
1.0017	3.3	1.0062	9.2	1.0107	15.0	1.0152	20.9
1.0018	3.4	1.0063	9.3	1.0108	15.2	1.0153	21.0
1.0019	3.5	1.0064	9.4	1.0109	15.3	1.0154	21.2
1.0020	3.7	1.0065	9.6	1.0110	15.4	1.0155	21.3
1.0021	3.8	1.0066	9.7	1.0111	15.6	1.0156	21.4
1.0022	3.9	1.0067	9.8	1.0112	15.7	1.0157	21.6
1.0023	4.1	1.0068	9.9	1.0113	15.8	1.0158	21.7
1.0024	4.2	1.0069	10.1	1.0114	16.0	1.0159	21.8
1.0025	4.3	1.0070	10.2	1.0115	16.1	1.0160	22.0
1.0026	4.5	1.0071	10.3	1.0116	16.2	1.0161	22.1
1.0027	4.6	1.0072	10.5	1.0117	16.3	1.0162	22.2
1.0028	4.7	1.0073	10.6	1.0118	16.5	1.0163	22.4
1.0029	4.8	1.0074	10.7	1.0119	16.6	1.0164	22.5
1.0030	4.0	1.0075	10.8	1.0120	16.7	1.0165	22.6
1.0031	5.1	1.0076	11.0	1.0121	16.9	1.0166	22.7
1.0032	5.2	1.0077	11.1	1.0122	17.0	1.0167	22.9
1.0033	5.4	1.0078	11.2	1.0123	17.1	1.0168	23.0
1.0034	5.5	1.0079	11.4	1.0124	17.3	1.0169	23.1
1.0035	5.6	1.0080	11.5	1.0125	17.4	1.0170	23.3
1.0171	23.4	1.0211	28.6	1.0251	33.8	1.0291	39.0
1.0172	23.5	1.0212	28.8	1.0252	34.0	1.0292	39.2
1.0173	23.7	1.0213	28.9	1.0253	34.1	1.0293	39.3
1.0174	23.8	1.0214	29.0	1.0254	34.2	1.0294	39.4
1.0175	23.9	1.0215	29.1	1.0255	34.4	1.0295	39.6

TABLE B-2 *(continued)*

Corresponding Densitites and Salinities

Density	Salinity	Density	Salinity	Density	Salinity	Density	Salinity
1.0176	24.1	1.0216	29.3	1.0256	34.5	1.0296	39.7
1.0177	24.2	1.0217	29.4	1.0257	34.6	1.0297	39.8
1.0178	24.3	1.0218	29.5	1.0258	34.8	1.0298	39.9
1.0179	24.4	1.0219	29.7	1.0259	34.9	1.0299	40.1
1.0180	24.6	1.0220	29.8	1.0260	35.0	1.0300	40.2
1.0181	24.7	1.0221	29.9	1.0261	35.1	1.0301	40.3
1.0182	24.8	1.0222	30.1	1.0262	35.3	1.0302	40.4
1.0183	25.0	1.0223	30.2	1.0263	35.4	1.0303	40.6
1.0184	25.1	1.0224	30.3	1.0264	35.5	1.0304	40.7
1.0185	25.2	1.0225	30.4	1.0265	35.7	1.0305	40.8
1.0186	25.4	1.0226	30.6	1.0266	35.8	1.0306	41.0
1.0187	25.5	1.0227	30.7	1.0267	35.9	1.0307	41.1
1.0188	25.6	1.0228	30.8	1.0268	36.0	1.0308	41.2
1.0189	25.8	1.0229	31.0	1.0269	36.2	1.0309	41.4
1.0190	25.9	1.0230	31.1	1.0270	36.3	1.0310	41.5
1.0191	26.0	1.0231	31.2	1.0271	36.4	1.0311	41.6
1.0192	26.1	1.0232	31.4	1.0272	36.6	1.0312	41.7
1.0193	26.3	1.0233	31.5	1.0273	36.7	1.0313	41.9
1.0194	26.4	1.0234	31.6	1.0274	36.8	1.0314	42.0
1.0195	26.5	1.0235	31.8	1.0275	37.0	1.0315	42.1
1.0196	26.7	1.0236	31.9	1.0276	37.1	1.0316	42.3
1.0197	26.8	1.0237	32.0	1.0277	37.2	1.0317	42.4
1.0198	26.9	1.0238	32.1	1.0278	37.3	1.0318	42.5
1.0199	27.1	1.0239	32.3	1.0279	37.5	1.0319	42.7
1.0200	27.2	1.0240	32.4	1.0280	37.6	1.0320	42.8
1.0201	27.3	1.0441	32.5	1.0281	37.7		
1.0202	27.5	1.0242	32.7	1.0282	37.9		
1.0203	27.6	1.0243	32.8	1.0283	38.0		
1.0204	27.7	1.0244	32.9	1.0284	38.1		
1.0205	27.8	1.0245	33.1	1.0285	38.2		
1.0206	28.0	1.0246	33.2	1.0286	38.4		
1.0207	28.1	1.0247	33.3	1.0287	38.5		
1.0208	28.2	1.0248	33.5	1.0288	38.6		
1.0209	28.4	1.0249	33.6	1.0289	38.8		
1.0210	28.5	1.0250	33.7	1.0290	38.9		

Note: Density at 15°C. Salinity is expressed in ‰ or g/kg.

Appendix C

Logbook Forms and Useful Tables Applicable in Analytical Work

Each analytical procedure has specified logbook forms. The analyst must complete all parts of the logbook. It should clearly and completely contain all information necessary for validation of the analytical process. The design of the documentation forms included in this Appendix are just examples; each institution and laboratory has its own formats as described and approved in their QA/QC programs.

All documentation related to the raw data and the reported results should be in bound form, easy to identify, and ready for inspection at any time. It should be marked with the date started and ended, identification number started and ended, document title, analytical group, and parameter. Strip charts and print-outs should be stored in file boxes and identified as mentioned above. Selected log forms are shown in Figures C-1 to C-13. Essential factors and information practical in analytical work are presented in Tables C-1 to C-3.

Aqueous Titrimetric Determination of Alkalinity
EPA 310.1

Date: _____
0.1N Std Acid Ref: _____
Std Acid N: _____
Ind Std Reference: _____
Date of Standardization of Acid: _____

Analyst: _____
Equivalent Weight: 50.0mg/meq
Ind Std True Value: _____
0.05N Na$_2$CO$_3$ Ref: _____

H$_2$SO$_4$ Standardization _____ N

Standardization of Acid	Std Na$_2$CO$_3$ Sol'n Vol. (mL)	Std Acid			Std Acid Sol'n Normality (meq/mL)
		Start	End	Total	Avg	
Check #1						
Check #2						
Check #3						

Sample ID	Sample Volume (mL)	Initial pH	Volume of Titrant to pH 8.3 (mL)			Volume of Titrant to pH 4.5 (mL)			Alkalinity Concentration (mg/L)					
			Initial	Final	Total	Initial	Final	Total	T$_{alk}$*	P$_{alk}$*	H$_{alk}$*	C$_{alk}$*	B$_{alk}$*
Blank														
Ind Std										%Rec =				
Dup										%RPD =				

T$_{alk}$ = Total Alkalinity   P$_{alk}$ = Phenolphthalein Alkalinity   B$_{alk}$ = Bicarbonate Alkalinity
H$_{alk}$ = Hydroxide Alkalinity   C$_{alk}$ = Carbonate Alkalinity

Approved By: _____

Figure C-1
Aqueous titrimetric determination of alkalinity. EPA 310.1

Potentiometric Determination of pH
EPA 150.1/9040/9045

4.00 Buffer Reference:	_____	**Date:**	_____
7.00 Buffer Reference:	_____	**Time:**	_____
10.00 Buffer Reference:	_____	**Analyst:**	_____
2.00 Buffer Reference:	_____	**Meter Serial No.:**	_____
13.00 Buffer Reference:	_____		

INITIAL CALIBRATION

Buffer	Reading	Limits	Satisfactory(Y/N)
7.00	7.00	---	---
4.00	4.00	---	---
10.00		9.95-10.05	
2.00		1.90-2.10	
13.00		12.90-13.10	

ANALYSES

Sample ID	Dilution Information (Solids only)		pH Reading
	Weight (g)	Final Volume (mL)	
Dup			

CALIBRATION CHECK

Buffer	Reading	Limits	Satisfactory(Y/N)
7.00		6.95-7.05	
4.00		3.95-4.05	
10.00		9.95-10.05	
2.00		1.90-2.10	
13.00		12.90-13.10	

Page ___ of ___   Data Checked By: _____

Figure C-2
Potentiometric determination of pH. EPA 150.1/9040/9045.

Titrimetric Determination of Hardness as CaCO₃
EPA 130.2

EDTA Buffer Reference:	_____		Time:	_____
Indicator Reference:	_____		Date:	_____
EDTA Titrant Reference:	_____		Analyst:	_____
Ind. Std Reference:	_____		Equivalent Weight:	50.0 mg/meq
			Ind. Std True Value:	_____

EDTA Titrant Standardization

Standardization of Titrant	0.02N CaCO₃ Sol'n Vol. (mL)	EDTA				EDTA Titrant Sol'n Normality (meq/mL)
		Start	End	Total	Avg	
Check #1						
Check #2						
Check #3						

Sample Analyses

Sample ID	Sample Volume (mL)	Dilution	Beg. pH	Volume of the Titrant			Hardness Concentration (mg/L)
				Beg.	End	Total	
Blank							
Independent Std							%Rec =
Dup							
							%RPD =

Page ___ of ___   Data Checked By: _____

Figure C-3
Titrimetric determination of hardness as CaCO₃. EPA 130.2.

Colorimetric Determination of Total and Amenable Cyanide in Waters
EPA 9010A

Wavelength: __578nm__   Cell: __1cm__   Date:_____

Cal. Curve Ref.: _____   Analyst:_____

Sodium Phosphate Ref: _____   Chloramine-T Ref:_____   Pyridine-Barbituric Ref:_____

Ind. Standard Reference: _____   Ind. Std True Value:_____

Spike Reference: _____   Spike Added Conc:_____

Sample ID	pH > 12 (Y/N)	Sulfide Check (Y/N)	Chlorine Check (Y/N)	Sample Vol. (mL)	Dilution Information			Time	Abs	Curve Result (µg)	Total Cyanide Conc (mg/L)	
					Init. Vol. of Dil. (mL)	Final Vol. of Dil. (mL)	Dil. Factor					
Dist.Std Hi =											%Rec	%Rec = 90-110%?
Dist.Std Lo =											%Rec	%Rec = 90-110%?
Blank												
Ind. Std											%Rec	
MS											%Rec	RPD
MSD											%Rec	
Ind. Std											%Rec	

Figure C-4
Colorimetric determination of total and amenable cyanide in waters. EPA 9010A.

CALIBRATION

Standard ID	Conc (mg/L)	Potentiometer Reading (mV)
Calibration Standards	0.10	
	0.20	
	0.40	
	0.60	
	0.80	
	1.00	
	1.20	
	1.60	
	2.00	

"*ln* (Conc) vs Reading" Calibration Curve Verification			
Type of Verification	Value	Acceptance Limits	Acceptable? (Y/N)
Correlation		< -0.995	
Slope Check for Conc's 0.1 & 1.0		50 → 60	
Slope Check for Conc's 0.2 & 2.0		50 → 60	

Analysis Date: ____________

Analyst: ____________

Buffer Reference: ____________

Cal. Std Reference: ____________

Ind. Std Reference: ____________

Ind Std True Value: ____________

Spk Std Reference: ____________

Spk Added Conc: ____________

SAMPLE ANALYSES

Sample ID	Prep Date	pH	Distillation: Init. Vol (mL)	Distillation: Init. Wet Wt (g)	Distillation: Init. Wet Wt (g)	Distillation: Final Vol (mL)	Dilution Info. (mL): Initial Volume	Dilution Info. (mL): Total Volume	Dilution Info. (mL): Dilution Factor	Potentiometer Reading (mV)	Fluoride Concentration (mg/L or mg/kg*)
Blank		---									
Ind. Std		---									%Rec
MS											%Rec
MSD											%Rec
											RPD

QC Calculations

* Indicate units - mg/L for water concentration or mg/kg for solids concentration.

CALIBRATION CHECK

Standard ID	Potentiometer Reading (mV)	Fluoride Concentration (mg/L)	Acceptance Limits (mg/L)	Satisfactory? (Y/N)
1.00mg/L Calibration Std			0.95 → 1.05	

Page __ of __

Data Checked By: ____________

Figure C-5
Potentiometric determination of fluoride. EPA 340.2.

Titrimetric Determination of Total Sulfides in Waters
EPA 9030A

Prep. Date:		Analysis Date:		Analyst:		Ind Std Ref:		Ind Std True Value:	
Prep Method:		Stock Std Ref:		Equivalent Weight:	16.0 mg/meq	Spike Std Ref:		Spike Added Conc:	
Trap Reagent:		Std I₂ Sol'n Ref:		Starch Ind. Ref:		Na₂S₂O₃ Ref:		Na₂S₂O₃ Normality:	0.0250 N

Standard Iodine Solution Standardization

0.0250 N Standard Iodine Solution Check	Std Iodine Sol'n Vol. (mL)	Sodium Thiosulfate Titrant (mL)			Satisfactory (Y/N)	Avg	Std Iodine Sol'n Normality (meq/mL)
		Start	End	Total			
Check #1	5.00						
Check #2	5.00						
Check #3	5.00						

Sample Preparation and Analyses

Sample ID	Sample pH	Sample Vol (mL)	Final Prep Vol (mL)	Aliquot Volume (mL)	Std Iodine Sol'n Added (mL)			Sodium Thiosulfate Titrant (mL)			Sulfide Concentration (mg/kg)
					Init.	Add'l	Total	Start	End	Total	
Blank											
Ind Std											%Rec =
MS											%Rec =
MSD											%Rec = RPD =

Comments: _____

Page ___ of ___

Data Checked By: _____

Figure C-6
Titrimetric determination of total sulfides in waters. EPA 9030A.

Turbidimetric Determination of Sulfate in Solids
EPA 9038M

Date: ______

Curve Std Reference: ______ Analyst: ______

Ind. Std Reference: ______ True Value: ______

Spike Reference: ______ Spike Added Conc: ______

Calibration and Calibration Verification

Sample ID		Nephelometer Reading (NTU)	Curve Result (mg/L)	
Calibration Standards	Blank			
	40ppm SO_4			
	30ppm SO_4			
	20ppm SO_4			
	10ppm SO_4			
	5ppm SO_4			
Sample Blank				
Ind. Std				%Rec =

Sample Analyses

Sample ID	Sample Wt. (g)	Extraction Volume (mL)	Dilution Information			Nephelometer Reading (NTU)			Curve Result (mg/L)	SO_4 Conc (mg/L)	
			Init. Vol.(mL)	Final Vol.(mL)	Dil. Factor	Sample	Blank	Sample - Blank			
Method Blank											
MS											%Rec
MSD											%Rec

MS	mL of Sample =	True Value Background (µg) =	NTUs	%Rec	RPD
	mL of Std =	True Value Spike (µg) =			
MSD	mL of Sample =	True value Background (µg) =	NTUs	%Rec	
	mL of Std =	True Value Spike (µg) =			

Page ___of ___ Data Checked By: ______

Figure C-7
Turbidimetric determination of sulfate in solids. EPA 9038M.

Titrimetric Determination of Sulfite in Waters
EPA 377.1

Analyst:	_____	Date:	_____
Thyodene Reference:	_____	Equivalent Weight:	40.0 mg/meq
Independent Std Ref:	_____	Independent Std True Value:	_____
KI-KIO$_3$ Titrant Ref:	_____	KI-KIO$_3$ Titrant Normality:	_____

Sample Analyses

Sample ID	Sample Volume (mL)	pH	Dilution* Info. (mL)		KI-KIO$_3$ Titrant (mL)				Concentration (mg/L)
			Init. Vol	Total Vol	Start	End	Total	Total - Blank	
Blank									
Independent Std									%Rec =
MS									%Rec =
MSD									%Rec =

*Dilution required if more than 10mL of titrant is required.   RPD =

Page ___ of ___   Data Checked By: _____

Figure C-8
Titrimetric determination of sulfite in waters. EPA 377.1.

Titrimetric Determination of Reactive Sulfide
EPA SW-846 Chapter 7.3/9030A

Date: _____

Starch Indicator Ref.: _____   Analyst: _____

Std Iodine Sol'n Ref.: _____   Equivalent Weight: 16.0mg/meq

Ind. Std Reference: _____   Ind. Std True Value: _____

Spike Std Reference: _____   Spike Added Conc.: _____

Sodium Thiosulfate
Reference: _____   Sodium Thiosulfate
Normality: 0.0250 N

Standard Iodine Solution Standardization

0.0250 N Standard Iodine Solution Check	Std Iodine Sol'n Vol. (mL)	Sodium Thiosulfate Titrant (mL)					Std Iodine Sol'n Normality (meq/mL)
		Start	End	Total	Satisfactory (Y/N)	Avg	
Check #1	5.00						
Check #2	5.00						
Check #3	5.00						

Sample Analyses

Sample ID	Sample pH	Aliquot Volume (mL)	Std Iodine Sol'n Added (mL)			Sodium Thiosulfate Titrant (mL)			Sulfide Concentration (mg/L)
			Init.	Add'l	Total	Start	End	Total	
Blank									
Independent Std									%Rec =
MS									%Rec =
MSD									%Rec =
									RPD =

Page ___ of ___   Data Checked By: _____

Figure C-9
Titrimetric determination of reactive sulfide. EPA 9030A.

Phosphate Prep

Analyst: _____   Date: _____

Preparation Method: _____

Spike Reference: _____

Standard Reference: _____

Sample ID	Sample Amount (indicate units)	Sample Wt (g)		Final Vol (mL)	pH	Comments
		Wet	Dry			

Reviewed By: _____

Page ___ of ___

Figure C-10
Phosphate preparation log.

TKN Prep

Analyst: _____ Date: _____

Preparation Method: _____

Spike Reference: _____

Standard Reference: _____

Sample ID	Sample Amount (indicate units)	Sample Wt (g)		Final Vol (mL)	pH	Comments
		Wet	Dry			

Reviewed By: _____

Page ___ of ___

Figure C-11
Total Kjeldahl nitrogen (TKN) preparation log.

Reactivity, Total Sulfide, and Total Cyanide Prep

Analyst: _____   Date: _____

Preparation Method: _____   Trap Reagent: _____

Spike Reference: _____

Standard Reference: _____

Sample ID	Sample Amount (Indicate units)	Sample Wt (g)		Final Vol (mL)	pH	Comments
		Wet	Dry			

Reviewed By: _____

Page ___ of ___

Figure C-12
Total sulfide and total cyanide preparation log.

PERIODIC TABLE OF THE ELEMENTS

New Notation / Previous IUPAC Form / CAS Version

Key to Chart: Atomic Number (50), Symbol (Sn), 1993 Atomic Weight (118.71); Oxidation States (+2 +4); Electron Configuration (18 18 4)

Group 1 IA	2 IIA	3 IIIA IIIB	4 IVA IVB	5 VA VB	6 VIA VIB	7 VIIA VIIB	8 VIIIA VIII	9 VIIIA VIII	10 VIIIA VIII	11 IB	12 IIB	13 IIIB IIIA	14 IVB IVA	15 VB VA	16 VIB VIA	17 VIIB VIIA	18 VIIA	Shell
1 H +1 -1 1.00794 1																	2 He 0 4.0020602 2	K
3 Li +1 6.941 2-1	4 Be +2 9.012182 2-2											5 B +3 10.811 2-3	6 C +2 +4 -4 12.011 2-4	7 N +1 +2 +3 +4 +5 -1 -2 -3 14.00674 2-5	8 O -2 15.9994 2-6	9 F -1 18.9984032 2-7	10 Ne 0 20.1797 2-8	K-L
11 Na +1 22.989768 2-8-1	12 Mg +2 24.3050 2-8-2											13 Al +3 26.981539 2-8-3	14 Si +2 +4 -4 28.0855 2-8-4	15 P +3 +5 -3 30.973762 2-8-5	16 S +4 +6 -2 32.066 2-8-6	17 Cl +1 +5 +7 -1 35.4527 2-8-7	18 Ar 0 39.948 2-8-8	K-L-M
19 K +1 39.0983 -8-8-1	20 Ca +2 40.078 -8-8-2	21 Se +3 44.955910 -8-9-2	22 Ti +2 +3 +4 47.867 -8-10-2	23 V +2 +3 +4 +5 50.9415 -8-11-2	24 Cr +2 +3 +6 51.9961 -8-13-1	25 Mn +2 +3 +4 +7 54.93805 -8-13-2	26 Fe +2 +3 55.845 -8-13-2	27 Co +2 +3 58.93320 -8-15-2	28 Ni +2 +3 58.6934 -8-16-2	29 Cu +1 +2 63.546 -8-18-1	30 Zn +2 65.39 -8-18-2	31 Ga +3 69.723 -8-18-3	32 Ge +2 +4 72.61 -8-18-4	33 As +3 +5 -3 74.92159 -8-18-5	34 Se +4 +6 -2 78.96 -8-18-6	35 Br +1 +5 -1 79.904 -8-18-7	36 Kr 0 83.80 -8-18-8	-L-M-N
37 Rb +1 85.4678 -18-8-1	38 Sr +2 87.62 -18-8-2	39 Y +3 88.90585 -18-9-2	40 Zr +4 91.224 -18-10-2	41 Nb +3 +5 92.90638 -18-12-1	42 Mo +6 95.94 -18-13-1	43 Tc +4 +6 +7 (98) -18-13-2	44 Ru +3 101.07 -18-15-1	45 Rh +3 102.90550 -18-16-1	46 Pr +2 +3 106.42 -18-18-0	47 Ag +1 107.8682 -18-18-1	48 Cd +2 112.411 -18-18-2	49 In +3 114.818 -18-18-3	50 Sn +2 +4 118.710 -18-18 -4	51 Sb +3 +5 -3 121.760 -18-18-5	52 Te +4 +6 -2 121.760 -18-18-6	53 I +1 +5 +7 -1 126.90447 -18-18-7	54 Xe 0 131.29 -18-18-8	-M-N-O
55 Cs +1 132.90543 -18-8-1	56 Ba +2 137.327 -18-8-2	57* La +3 138.9055 -18-9-2	72 Hf +4 178.49 -32-10-2	73 Ta +5 180.9479 -32-11-2	74 W +6 183.84 -32-12-2	75 Re +4 +6 +7 186.207 -32-13-2	76 Os +3 +4 190.23 -32-14-2	77 Ir +3 +4 192.217 -32-15-2	78 Pt +2 +4 195.08 -32-18-1	79 Au +1 +3 196.96654 -32-18-1	80 Hg +1 +2 200.59 -32-18-2	81 Tl +1 +3 204.3833 -32-18-3	82 Pb +2 +4 207.2 -32-18-4	83 Bi +3 +5 208.98037 -32-18-5	84 Po +2 +4 (209) -32-18-6	85 At (210) -32-18-7	86 Rn 0 (222) -32-18-8	-N-O-P
87 Fr +1 (223) -18-8-1	88 Ra +2 226.025 -18-8-2	89** Ac +3 227.028 -18-9-2	104 Unq +4 (261) -32-10-2	105 Unp (262) -32-11-2	106 Unh (263) -32-12-2	107 Uns (262) -32-13-2	108 Uno (265) -32-14-2	109 Une (266) -32-15-2	110 Uun (269) -32-16-2									O P Q

* Lanthanides	58 Ce +3 +4 140.115 -19-9-2	59 Pr +3 140.90765 -21-8-2	60 Nd +3 144.24 -22-8-2	61 Pm +3 (145) -23-8-2	62 Sm +2 +3 150.36 -24-8-2	63 Eu +2 +3 151.965 -25-8-2	64 Gd +3 157 .25 -25-9-2	65 Tb +3 158.92534 -27-8-2	66 Dy +3 162.50 -28-8-2	67 Ho +3 164.93032 -29-8-2	68 Er +3 167.26 -30-8-2	69 Tm +3 168.93421 -31-8-2	70 Yb +3 +5 173.04 -32-8-2	71 Lu +3 174.967 -32-9-2	N O P
** Actinides	90 Th +4 232.0381 -18-10-2	91 Pa +5 +4 231.03588 -20-9-2	92 U +3 +4 +5 +6 238.0289 -21-9-2	93 Np +3 +4 +5 +6 237.048 -22-9-2	94 Pu +3 +4 +5 +6 (244) -24-8-2	95 Am +3 +4 +5 +6 (243) -25-8-2	96 Cm +3 (247) -25-9-2	97 Bk +3 +4 (247) -27-8-2	98 Cf +3 (251) -28-8-2	99 Es +3 (252) -29-8-2	100 Fm +3 (257) -30-8-2	101 Md +2 +3 (258) -31-8-2	102 No +2 +3 (259) -32-8-2	103 Lr +3 (206) -32-9-2	O P Q

The new IUPAC format numbers the groups from 1 to 18. The previous IUPAC numbering system and the system used by Chemical Abstracts Service (CAS) are also shown. For radioactive elements that do not occur in nature, the mass number of the most stable isotape is given in parentheses.

References
1. G. J. Leigh, Editor, *Nomenclature of Inorganic Chemistry*, Blackwell Scientific Publications, Oxford, 1990.
2. *Chemical and Engineering News*, 63(5), 27, 1985.
3. Atomic Weights of the Elements, 1993, *Pure & Appl. Chem.*, 66, 2423, 1994.

Figure C-13
Periodic Table of the Elements.

TABLE C-1

SI Units and Conversion Factors

		Conversion to Metric Measures			
	Symbol	When You Know	Multiply By	To Find	Symbol
Length	in	inches	2.54	centimers	cm
	ft	feet	30.48	centimeters	cm
	yd	yards	0.9	meters	m
	mi	miles	1.6	kilometers	km
Area	in²	square inches	6.5	square centimeters	cm²
	ft²	square feet	0.09	square meters	m²
	yd²	square yards	0.8	square meters	m²
	mi²	square miles	2.6	square kilometers	km²
		acres	0.4	hectares	ha
Mass (Weight)	oz	ounces	28	grams	g
	lb	pounds	0.45	kilograms	kg
		short tons (2000 lb)	0.9	tonnes	t
Volume	tsp	teaspoons	5	milliliters	mL
	Tbsp	tablespoons	15	milliliters	mL
	fl oz	fluid ounces	30	milliliters	mL
	c	cups	0.24	liters	L
	pt	pints	0.47	liters	L
	qt	quarts	0.95	liters	L
	gal	gallons	3.8	liters	L
	ft³	cubic feet	0.03	cubic meters	m³
	yd³	cubic yards	0.76	cubic meters	m³
Temp.	°F	Fahrenheit temperature	5/9 (after subtracting 32)	Celsium temperature	°C

$$°C = \frac{5}{9}(°F - 32)$$

		Conversions From Metric Measures			
	Symbol	When You Know	Multiply By	To Find	Symbol
Length	mm	millimeters	0.04	inches	in
	cm	centimeters	0.4	inches	in
	m	meters	3.3	feet	ft
	m	meters	1.1	yards	yd
	km	kilometers	0.6	miles	mi
Area	cm²	square centimeters	0.16	square inches	in²
	m²	square meters	1.2	square yards	yd²
	km	square kilometers	0.4	square miles	mi²
	ha	hectares (10,000 m²)	2.5	acres	
Mass (Weight)	g	grams	0.035	ounces	oz
	kg	kilograms	2.2	pounds	lb
	t	tonnes (1000 kg)	1.1	short tons	
Volume	mL	milliliters	0.03	fluid ounces	fl oz
	L	liters	2.1	pints	pt
	L	liters	1.06	quarts	qt
	L	liters	0.26	gallons	gal
	m³	cubic meters	35	cubic feet	ft³
	m³	cubic meters	1.3	cubic yards	yd³
Temp.	°C	Celsius temperature	9/5 (then add 32)	Fahrenheit temperature	°F

$$°F = \frac{9}{5}(°C + 32)$$

TABLE C-2

Molecular and Equivalent Weights

Compound	Formula	Molecular Weight	Equivalent Weight
Acetic acid	CH_3COOH	60.03	60.03
Alum, ammonium	$Al_2(SO_4)_3 \cdot (NH_4)SO_4 \cdot 24H_2O$	906.64	151.11
Alum, crystals	$Al_2(SO_4)_3 \cdot 18H_2O$	666.43	111.07
Alum, potassium	$Al_2(SO_4)_3 \cdot K_2SO_4 \cdot 24H_2O$	948.76	159.79
Aluminum	A	26.97	8.99
Aluminum hydroxide	$Al(OH)_3$	77.99	26.00
Aluminum oxide	Al_2O_3	101.94	16.99
Aluminum sulfate	$Al_2(SO_4)_2$	342.14	57.02
Ammonia	NH_3	17.03	17.03
Ammonium chloride (sal ammoniac)	NH_4Cl	53.50	53.50
Ammonium hydroxide	NH_4OH	35.05	35.05
Ammonium nitrate	NH_4NO_3	80.05	80.05
Ammonium oxalate	$(NH_4)_2C_2O_4$	124.08	62.04
Ammonium oxalate, crystals	$(NH_4)_2C_2O_4 \cdot H_2O$	142.09	71.05
Ammonium sulphate	$(NH_4)_2SO_4$	132.14	66.07
Barium carbonate	$BaCO_3$	197.36	98.68
Barium chloride	$BaCl_2$	208.27	104.14
Barium chloride, crystals	$BaCl_2 \cdot 2H_2O$	244.31	122.15
Barium sulfate (barium white)	$BaSO_4$	233.42	116.71
Boric acid	H_2BO_2	62.02	
Bromine	Br	79.92	79.92
Calcium bicarbonate	$Ca(HCO_2)_2$	162.10	81.05
Calcium carbonate (limestone)	$CaCO_2$	100.08	50.04
Calcium chloride	$CaCl_2$	110.99	55.50
Calcium hydroxide (slaked lime)	$Ca(OH)_2$	74.10	37.05
Calcium oxalate	CaC_2O_4	128.08	64.04
Calcium oxide (quicklime)	CaO	56.08	28.04
Calcium sulfate (gypsum)	$CaSO_4 \cdot 2H_2O$	172.17	86.09
Calcium sulfate (Plaster of Paris)	$CaSO_4$	136.14	68.07
Carbon dioxide	CO_2	44.00	
Chlorine	Cl_2	70.914	35.457
Copper sulfate	$CuSO_4$	159.66	79.83
Copper sulfate (blue vitriol)	$CuSO_4 \cdot 5H_2O$	249.71	124.86
Ferric chloride	$FeCl_3$	162.21	54.07
Ferric chloride, crystals	$FeCl_3 \cdot 6H_2O$	270.30	90.10
Ferric hydroxide	$Fe(OH)_3$	106.86	35.62
Ferric oxide	Fe_2O_3	159.68	26.61
Ferric sulfate	$Fe_2(SO_4)_2$	399.86	66.64
Ferric sulfate, crystals	$Fe_2(SO_4)_3 \cdot 9H_2O$	562.00	93.67
Ferric sulfocyanate	$Fe(CNS)_3$	230.04	76.68
Ferrous ammonium sulfate	$FeSO_4(NH_4)_2 \cdot 6H_2O$	392.13	196.07
Ferrous carbonate	$FeCO_3$	115.84	57.92
Ferrous hydroxide	$Fe(OH)_2$	89.86	44.93
Ferrous sulfate	$FeSO_4$	151.90	75.95
Ferrous sulfate, crystals (copperas)	$FeSO_4 \cdot 7H_2O$	278.01	139.00
Hydrochloric acid (muriatic acid)	HCl	36.47	36.47
Hydrogen	H_2	2.016	1.008
Hydrogen sulfide	H_2S	34.08	17.04
Iodine	I	126.92	126.92
Magnesium ammonium phosphate	$MgNH_4PO_4$	137.38	
Magnesium bicarbonate	$Mg(HCO_8)_5$	146.34	73.17
Magnesium carbonate (magnesite)	$MgCO_3$	95.23	47.62
Magnesium chloride	$MgCl_2$	84.32	47.16
Magnesium hydroxide (milk of magnesia)	$Mg(OH)_2$	58.34	29.17
Magnesium nitrate	$Mg(NO_2)_2$	148.19	74.10
Magnesium oxide	MgO	40.32	20.16
Magnesium pyrophosphate	$Mg_2P_2O_7$	222.68	
Magnesium sulfate	$MgSO_4$	120.36	60.18

TABLE C-2 *(continued)*

Molecular and Equivalent Weights

Compound	Formula	Molecular Weight	Equivalent Weight
Magnesium sulfate, crystals (Epsom salts)	$MgSO_4 \cdot 7H_2O$	246.50	123.24
Manganic oxide	Mn_2O_3	157.86	26.31
Manganous oxide	MnO	70.93	35.47
Manganous sulfate	$MnSO_4$	150.99	75.50
Manganous sulfate, crystals	$MnSO_4 \cdot 4H_2O$	223.05	111.53
Methane	CH_4	16.03	
Molybdic acid	H_3MoO_4	162.02	
Nitric acid	HNO_2	63.02	63.02
Oxalic acid	$H_2C_2O_4$	90.02	45.01
Oxalic acid, crystals	$H_2C_2O_4 \cdot 2H_2O$	126.05	63.02
Oxygen	O_2	32.00	8.00
Phosphoric acid	H_3PO_4	98.04	32.68
Phosphorus pentoxide	P_2O_5	142.05	
Platinic chloride	$PtCl_4$	337.06	84.76
Potassium biniodate	$KIO_3 \cdot HIO_3$	389.95	32.49
Potassium bromate	$KBrO_3$	167.02	167.02
Potassium carbonate	K_2CO_3	138.20	69.10
Potassium chloride	KCl	74.56	74.56
Potassium chloroplatinate	K_2PtCl_6	486.17	
Potassium chromate	K_2CrO_4	194.20	
Potassium cyanide	KCN	65.11	65.11
Potassium dichromate	$K_2Cr_2O_7$	294.20	49.04*
Potassium ferricyanide	$K_3Fe(CN)_6$	329.19	
Potassium hydroxide (caustic potash)	KOH	56.11	56.11
Potassium iodide	KI	166.03	166.03
Potassium permanganate	$KMnO_4$	158.03	31.61*
Potassium nitrate (saltpeter)	KNO_3	101.11	101.11
Potassium oxalate, crystals	$K_2C_2O_4 \cdot H_2O$	184.22	92.11
Potassium oxide	K_2O	94.20	47.10
Potassium sulfate	K_2SO_4	174.26	87.13
Potassium thiocyanate	KCNS	97.17	97.17
Silicon dioxide (silica)	SiO_4	60.06	
Silver chloride	AgCl	143.34	143.34
Silver chromate	Ag_2CrO_4	331.76	
Silver nitrate	$AgNO_3$	169.89	169.89
Siliver nitrite	$AgNO_2$	153.89	76.94
Silver sulfate	Ag_2SO_4	311.82	155.91
Sodium bicarbonate (baking soda)	$NaHCO_3$	84.00	84.00
Sodium carbonate, anhydrous (soda ash)	Na_2CO_3	105.99	53.00
Sodium carbonate (sal soda)	$Na_2CO_3 \cdot 10H_2O$	287.15	143.58
Sodium chloride (common salt)	NaCl	58.45	58.45
Sodium hydroxide (caustic soda)	NaOH	40.00	40.00
Sodium hypochlorite	NaOCl	74.45	
Sodium nitrate (Chile saltpeter)	$NaNO_3$	85.01	85.01
Sodium oxalate	$Na_2C_2O_4$	134.00	67.00
Sodium phosphate, mono-	NaH_2PO_4	120.04	40.01
Sodium phosphate, di-	Na_2HPO_4	142.02	47.34
Sodium phosphate, tri-	Na_2PO_4	164.01	54.67
Sodium sulfate	Na_2SO_4	142.05	71.03
Sodium sulfate (Glauber's salt)	$Na_2SO_4 \cdot 10H_2O$	322.21	161.11
Sodium thiosulfate (hypo)	$Na_2S_2O_3$	158.11	158.11
Sodium thiosulfate, crystals	$Na_2S_2O_3 \cdot 5H_2O$	248.19	248.19
Sulfuric acid (oil of vitriol)	H_2SO_4	98.08	49.04
Sulfurous acid	H_2SO_3	82.08	41.04
Water	H_2O	18.02	9.01

* Oxidation and reduction in acid medium.

TABLE C-3

Gravimetric Factors

Known	Unknown	Factor
$AgNo_2$	Cl	0.2087
Al	Al_2O_2	1.8856
Al_2O_2	Al	0.5303
Al_2O_3	$Al_2(SO_4)_3 \cdot 18H_2O$	6.5375
$Al_2(SO_4)_3 \cdot 18H_2O$	Al_2O_3	0.1530
$Al_2(SO_4) \cdot 18H_2O$	CaO	0.2525
$Al_2(SO_4)_3 \cdot 18H_2O$	$Ca(OH)_2$	0.3335
$Al_2(SO_4)_3 \cdot 18H_2O$	Na_2CO_3	0.4776
$BaSO_4$	SO_4	0.4115
Ca	CaO	1.3390
Ca	$CaCO_2$	2.4967
Ca	$Ca(OH)_2$	1.8487
Ca	$Ca(HCO_3)_2$	4.0443
Ca	$CaCl_2$	2.7693
Ca	$CaSO_4$	3.3967
Ca	$KMnO_4$	1.5771
$CaCl_2$	$CaCO_3$	0.902
$CaCl_2$	Na_2CO_3	0.9550
$CaCO_4$	$Ca(OH)_2$	0.7404
$CaCO_3$	CaO	0.5604
$CaCO_3$	Mg	0.2430
$CaCO_3$	CO_2	0.4395
$CaCO_3$	H_2SO_4	0.9796
$CaCO_3$	Na_2CO_3	1.0591
$CaH(CO_3)_2$	$CaCO_3$	0.617
CaO	Ca	0.7149
CaO	$Ca(OH)_2$	1.3208
CaO	$CaCO_3$	1.7843
CaO	CO_2	0.7846
$Ca(OH)_2$	CaO	0.7571
$Ca(OH)_2$	CO_2	0.5938
$Ca(OH)_2$	$CaCO_3$	1.351
$CaSO_4$	Na_2CO_3	0.7786
$CaSO_4$	$CaCO_3$	0.735
CO_2	CaO	1.2750
CO_2	$Ca(OH)_2$	1.6840
CO_2	$CaCO_3$	2.2750
CO_2	$NaOH$	0.9102
Fe	Fe_2O_3	1.4298
Fe	$FeCl_3$	2.9025
Fe	$FeSO_4 \cdot 7H_2O$	4.9787
Fe	$FeSO_4(NH_4)_2SO_4 \cdot 6H_2O$	7.0224
Fe	$CaCO_3$	1.7941
Fe_2O_3	Fe	0.6994
Fe_2O_3	$FeCl_3$	2.0317
Fe_2O_3	$FeSO_4$	1.9026
$FeSO_4 \cdot 7H_2O$	$FeSO_4$	0.5464
$FeSO_4 \cdot 7H_2O$	Fe	0.2008
$FeSO_4(NH_4)_2SO_4 \cdot 6H_2O$	Fe	0.1424
HCl	Na_2CO_3	1.4534
H_2SO_4	$CaCO_3$	1.0208
H_2SO_4	CaO	0.5720
H_2SO_4	$Ca(OH)_2$	0.7556
H_2SO_4	Na_2CO_3	1.0809
H_2SO_4	$NaOH$	0.8168
H_2SO_4	Mg	0.2480
I	O	0.0630

TABLE C-3 *(continued)*

Gravimetric Factors

Known	Unknown	Factor
I	$Na_2S_2O_3 \cdot 5H_2O$	1.9553
$K_2Cr_2O_7$	$Na_2S_2O_3 \cdot 5H_2O$	2.5307
$KIO_3 \cdot HIO_3$	$Na_2S_2O_3 \cdot 5H_2O$	7.6371
$KMnO_4$	Ca	0.6341
$KMnO_4$	Mn	0.3476
$KMnO_4$	O	0.2531
K_2PtCl_6	K	0.1608
K_2PtCl_6	KCl	0.3067
K_2PtCl_6	K_2O	0.1937
Mg	$CaCO_2$	4.1151
Mg	$Mg(OH)_2$	2.3960
Mg	CaO	2.3059
Mg	$Ca(OH)_2$	3.0467
Mg	$MgSO_4$	4.9490
Mg	$MgCl_2$	3.9105
MgCl	$CaCO_3$	1.051
$Mg(HCO_3)_2$	$CaCO_3$	0.680
Mg(OH)	$CaCO_3$	1.716
$Mg_2P_2O_7$	Mg	0.2184
$MgSO_4$	$CaCO_3$	0.831
N	NH_4Cl	3.8190
N	NH_3	1.2158
N	NO_2	3.2844
N	NO_3	4.4266
Na	NaCl	2.5417
NaCl	Na	0.3934
Na_2CO_3	H_2SO_4	0.9253
Na_2CO_3	$Al_2(SO_4)_3 \cdot 18H_2O$	2.0959
Na_2CO_3	$CaCO_3$	0.9442
NaOH	H_2SO_4	1.2242
NaOH	CO_2	1.0986
$Na_2S_2O_3 \cdot 5H_2O$	I	0.5114
$Na_2S_2O_3 \cdot 5H_2O$	O	0.003223
$Na_2S_2O_3 \cdot 5H_2O$	$K_2Cr_2O_7$	0.3951
$Na_2S_2O_3 \cdot 5H_2O$	$KIO_3 \cdot HIO_3$	0.1309
NH_3	N	0.8225
$(NH_4)_2C_2O_4 \cdot H_2O$	$KMnO_4$	0.4448
NO_2	N	0.3045
NO_3	N	0.2259
O	$Na_2S_2O_2 \cdot 5H_2O$	31.0238
O	I	15.8665
O	$KMnO_4$	3.9507
Pt	K_2O	0.4825

Note: The above factors are derived from the equivalent weights of the various elements, radicals, and compounds. To change from the weight of the known compound to its equivalent of the unknown multiply the weight of the known by the factor.

Appendix D

DOCUMENTATION FORMS FOR QUALITY CONTROL DATA

The responsibilities of the analyst have been discussed in previous chapters. Those responsibilities are to prepare reagents and standards, make the calibrations and calibration checks, run the tests as described in the laboratory SOP, convert raw data to reportable data by using correct calculations, apply all required QC checks according to the approved QA/QC program of the organization to defend the reportable data, and recognize errors and conduct corrective actions.

The summary forms are documentation of quality control data analyzed for a specified sample set. Selected documentation forms are shown in Figures D-1 to D-6.

Project Name: _____ Project # : _____
Lab Number: _____ Instrument ID: _____
Date Analyzed: _____ Matrix: AQ ___ SO ___ NA ___

	Client Station ID	Lab Sample ID
1		
2		
3		
4		
5		
6		
7		
8		
9		
10		
11		
12		
13		
14		
15		
16		
17		
18		
19		
20		
21		
22		
23		

COMMENTS:_____

_____

_____

APPROVED BY:_____

Figure D-1
Method blank summary.

Project Name: _____   Project #: _____

Lab Number: _____   Matrix: AQ _____   SO _____

SAMPLE ID	X_1	X_2	RANGE	AVERAGE*	RPD*	DATE	ANALYST

Formulae:

$$Range = x_1 - x_2$$

$$Average = \frac{(x_1 + x_2)}{2}$$

$$Relative\ Percent\ Difference\ (RPD) = \frac{Range}{Average} \times 100$$

* Do not Calculate if results are ND.
ND - Not detected at detection limit
NA - Not Applicable

APPROVED BY: _____

Figure D-2
Duplicate analyses record.

Project Name: _____   Project #: _____

Lab Number: _____ MS/MSD   Date: _____

MS/MSD Data Apply to the following:

_____ _____ _____ _____
_____ _____ _____ _____
_____ _____ _____ _____
_____ _____ _____ _____
_____ _____ _____ _____

COMPOUND	SPIKE ADDED (mg/L)	SAMPLE CONCENTRATION (mg/L)	MS CONCENTRATION (mg/L)	MS % REC. #	QC LIMITS % REC.

COMPOUND	SPIKE ADDED (mg/L)	MSD CONCENTRATION (mg/L)	MSD% REC. #	% RPD #	QC LIMITS	
					%RPD	%REC.

Column to be used to flag recovery and RPD values with an asterisk.
* Values outside of QC Limits

RPD: _____ out of _____ outside limits

Spike Recovery: _____ out of _____ outside limits

COMMENTS: _____
_____
_____

APPROVED BY:_____

Figure D-3
Water matrix spike/matrix spike duplicate (MS/MSD) recovery.

Project Name: _____ Project #: _____

Lab Number: _____ MS/MSD Date: _____

MS/MSD Data Apply to the following samples:

COMPOUND	SPIKE ADDED (mg/kg)	SAMPLE CONCENTRATION (mg/kg)	MS CONCENTRATION (mg/kg)	MS% REC #	QC LIMITS %REC.

COMPOUND	SPIKE ADDED (mg/kg)	MSD CONCENTRATION (mg/kg)	MSD% REC #	% RPD #	QC LIMITS %RPD %REC.

\# Column to be used to flag recovery and RPD values with an asterisk.
* Values outside of QC Limits

RPD: _____ out of _____ outside limits

Spike Recovery: _____ out of _____ outside limits

COMMENTS:_____

_____

_____

APPROVED BY: _____

Figure D-4
Soil matrix spike/matrix spike duplicate (MS/MSD) recovery.

Date	Analyst	Standard ID	True Value	Value Found	% Rec	Test

Figure D-5
Initial calibration verification standard (ICV).

Limits Table for Nitrate–Nitrite

Analytic	Matrix	Reference Standard %Rec	MS/MSD %Rec	MS/MSD %RPD	PQL
Nitrate–Nitrite	Aqueous	90–102	64–131	21.5	0.05 mg/L
Nitrate–Nitrite	Solids	90–102	50–150	50	10 mg/kg
Nitrite	Aqueous	90–102	85–115	5.8	0.05 mg/L

Figure D-6
Sample limits table for nitrate–nitrite. This table represents the acceptance QC limits for nitrate–nitrite determination for different matrices. These limits are variable according to the used method and laboratories.

REFERENCES

Analytical Chemistry for Technicians, John Kenkel, Lewis Publishers, 1990.

APHA-AWWA-WPCF, *Standard Methods for the Examination of Water and Wastewater*, 17 ed., 1989, and 18th ed., 1994.

Chemical Technicians Ready Reference Handbook, Shugar/Shugar, Bauman/Bauman, McGraw-Hill, 2nd ed., 1981.

Chemistry, The Study of Matter and Its Changes, James E. Brady and John R. Holum, John Wiley & Sons, 1993.

CRC Handbook of Laboratory Safety, A Keith Furr, CRC Press, 4th ed., 1995.

Environmental Ecology, Bill Friedman, Academic Press, 1989.

Environmental Law Handbook, Thomas F. P. Sullivan, Government Institutes, Inc., 1995.

Environmental Sampling and Analysis for Technicians, M. Csuros, Lewis Publishers, 1994.

Environmental Science and Technology Handbook, Ayers et al., Government Institutes, Inc., 1994.

General Chemistry, Darrel D. Ebbing, Houghton Mifflin, 4th ed., 1993.

Handbook for Sampling and Sample Preservation of Water and Wastewater, EPA 600/4-82-029, 1982.

Health Effects of Toxic Substances, M. J. Malakowski and Arlee F. Goldberg, Government Institutes, Inc., 1995.

Introduction to College Chemistry, Drew H. Wolfe, McGraw-Hill, 2nd ed., 1988.

Methods for Chemical Analysis of Water and Wastes, EPA-600/4-79-020, Rev. March 1983.

Quality Assurance in Chemical Measurements, John Keenan Taylor, Lewis Publishers, 1988.

Quantitative Analysis, R. A. Day, Jr. and A. L. Underwood, Prentice-Hall, 6th ed., 1991.

Quantitative Analytical Chemistry, James S. Fritz and George H. Schenk, Allyn and Bacon, 5th ed., 1987.

Test Methods for Evaluating Solid Waste, EPA SW-846, 3rd ed., 1986.

INDEX